American Megafaunal Extinctions at the End of the Pleistocene

Vertebrate Paleobiology and Paleoanthropology Series

Edited by

Eric Delson
Vertebrate Paleontology, American Museum of Natural History,
New York, NY 10024, USA
delson@amnh.org

Ross D. E. MacPhee
Vertebrate Zoology, American Museum of Natural History,
New York, NY 10024, USA
macphee@amnh.org

Focal topics for volumes in the series will include systematic paleontology of all vertebrates (from agnathans to humans), phylogeny reconstruction, functional morphology, paleolithic archaeology, taphonomy, geochronology, historical biogeography, and biostratigraphy. Other fields (e.g., paleoclimatology, paleoecology, ancient DNA, total organismal community structure) may be considered if the volume theme emphasizes paleobiology (or archaeology). Fields such as modeling of physical processes, genetic methodology, nonvertebrates, or neontology are out of our scope.

Volumes in the series may either be monographic treatments (including unpublished but fully revised dissertations) or edited collections, especially those focusing on problem-oriented issues, with multidisciplinary coverage where possible.

For other titles published in this series, please go to
www.springer.com/series/6978

American Megafaunal Extinctions at the End of the Pleistocene

Edited by

Gary Haynes
Anthropology Department, University of Nevada, Reno, Nevada, USA

Springer

Editor
Gary Haynes
Anthropology Department
University of Nevada, Reno
Reno, Nevada
USA

ISBN 978-1-4020-8792-9 ISBN 978-1-4020-8793-6 (eBook)

Library of Congress Control Number: 2008930995

Cover Illustration: The photograph is of Nehimba seep in Hwange National Park, Zimbabwe, where bones can be seen from African elephants that died during drought years of the 1990s. The mammoths are overlain drawings made by an unknown graphic artist in the 1970s, based on cave paintings and numerous reconstructions by other artists. Photo and artwork composition by Gary Haynes.

Printed on acid-free paper

springer.com

List of Contributors

Luis Alberto Borrero
Instituto Multidisciplinario de Historia y Ciencias Humanas, CONICET,
Saavedra 15 – Piso 5,
(1083 ACA) Buenos Aires, Argentina.
Email: dipa.imhicihu@conicet.gov.ar

Alberto L. Cione
División Paleontología de Vertebrados,
Facultad de Ciencias Naturales y Museo,
Universidad Nacional de La Plata, Paseo del Bosque,
1900 La Plata, Argentina.
Email: acione@museo.fcnym.unlp.edu.ar

Stuart Fiedel
Louis Berger Group,
203 East Cary St, Suite 100,
Richmond, VA 23219, USA.
Email: sfiedel@louisberger.com

Daniel C. Fisher
Museum of Paleontology and Department of Geological Sciences,
University of Michigan,
Ann Arbor, MI 48109-1079, USA.
Email: dcfisher@umich.edu

Alex D. Greenwood
Department of Biological Sciences,
Mills Godwin Building, Room 108E,
Old Dominion University,
Norfolk, VA 23529, USA.
Email: agreenwo@odu.edu

Gary Haynes
Anthropology Department,
University of Nevada, Reno,
Reno, NV 89557-0096, USA.
Email: gahaynes@unr.edu

R.D.E. MacPhee
Division of Vertebrate Zoology,
American Museum of Natural History,
Central Park West at 79th St,
New York, NY 10024, USA.
Email: macphee@amnh.org

Leopoldo Soibelzon
División Paleontología de Vertebrados,
Facultad de Ciencias Naturales y Museo,
Universidad Nacional de La Plata, Paseo del Bosque,
1900 La Plata, Argentina.
Email: lsoibelzon@museo.fcnym.unlp.edu.ar

Todd A. Surovell
Department of Anthropology,
University of Wyoming,
1000 East University Avenue,
Laramie, WY 82071, USA.
Email: surovell@uwyo.edu

Eduardo P. Tonni
División Paleontología de Vertebrados,
Facultad de Ciencias Naturales y Museo,
Universidad Nacional de La Plata, Paseo del Bosque,
1900 La Plata, Argentina.
Email: eptonni@museo.fcnym.unlp.edu.ar

Nicole M. Waguespack
Department of Anthropology,
University of Wyoming,
1000 East University Avenue,
Laramie, WY 82071, USA.
Email: nmwagues@uwyo.edu

Contents

1. Introduction to the Volume

Gary Haynes*
Anthropology Department
University of Nevada, Reno
Reno, Nevada 89557-0096, USA
gahaynes@unr.edu

Keywords Theories • Overkill • climate-change • blitzkrieg • sitzkrieg • hyperdisease

Introduction

Imagine elephant-like animals in ancient North America. They are hairier than the elephants of Africa or Asia, and not as easily annoyed by your close approach because they have never seen an animal like you before. They yank up coarse grass with their trunks and thump it against a foreleg to knock off the dirt. Imagine horses, too – thick-bodied with big jaws chewing bite after bite of grass and bark. Imagine one-humped camels listening to the horses bray like zebras. Then imagine something else; imagine these animals and every one like them dead and gone just a short time later.

What Happened?

The ice-age tableau of 13,000 cal bp[1] is not hard to visualize because we have so often seen paintings and reconstructions, but what is much harder to view in our mind's eye is how the end came. Around 30 genera of mammals vanished in North America (Table 1.1) and possibly more than 50 species vanished in South America (see Cione et al., Chapter 7; Cione et al., 2003), all of them apparently rubbed out in geologically quick time. The great vanishing act took place in California and Rhode Island and Texas, in the center of Brazil and along the Pacific coast of Chile, in the inland steppes and pampas and plains of both Americas, in the central river valleys of Alaska, in the cold southern cone of Argentina, in the lowlands and plains, mountains and foothills, everywhere in both continents and at nearly the same time. It is a mystery we cannot solve – a true cold case.

The Theories

Scientists have been arguing for a very long time about what (or who) is to blame for the deaths. The theories tend to be unicausal; some are nearly apocalyptic and others are information-poor. They are vigorously debated by intense and steadfast opponents with well developed mannerisms – the spinning of one side of a case while caricaturing the other side, the rhetorical fudging of facts, the drumming out of a skewed point of view through repetition and eloquence and bombast. Proponents of one tangling theory turn prickly when faced with criticism. Articles are written to declare an end to the debate (Grayson and Meltzer, 2003, for example), yet they fail because the arguing is based on too many unprovable assertions and incomplete evidence, such as the temporal disconnection between extinctions and climate-change or the scarcity of associations of extinct animals with evidence for human killing. Skeptical fellow scientists demand explicit protocols for finding proof, and in reply other scientists may ignore the criticisms or claim to be too experienced to make mistakes of logic and interpretation.

The main unicausal theories seem to fall along the same lines as the causes sought for all earlier extinctions in earth's biotic history. The division is usually between the "exogenous" theories that blame the extinctions on "external stresses imposed on the ecosystem by the environment," and the "endogenous" or "biotic" theories that blame extinctions on "the dynamics of the ecosystem" such as "overzealous predators or the introduction of new competitors into formerly stable systems" (Newman and Palmer, 2003:2). The leading example of an exogenous theory is climate-change at the end of the Pleistocene. In this theory, the largest land mammals in the Americas are acknowledged as dying out within a relatively few millennia – although some such as Grayson (2007) still argue that the extinctions occurred asynchronously over the course of many millennia – because

* Address for correspondence: gahaynes@unr.edu

G. Haynes (ed.), *American Megafaunal Extinctions at the End of the Pleistocene,* 1–20.

TABLE 1.1. Mammalian taxa that became extinct in North America at or near the end of the Pleistocene; generic names are italicized. An asterisk denotes a globally extinct genus (From Koch and Barnosky, 2006 supplemental information: table S1).

Xenarthra		
	Dasypodidae	
		Dasypus (extinct species within genus)
	Glyptodontidae	
		Glyptotherium*
	Megalonychidae	
		Megalonyx*
	Megatheriidae	
		Eremotherium*
		Nothrotheriops*
	Mylodontidae	
		Glossotherium*
	Pampatheriidae	
		Holmesina/Pampatherium*
Rodentia		
	Castoridae	
		Castoroides*
	Hydrocheridae	
		Hydrochaeris (extinct on continent)
		Neochoerus*
Carnivora		
	Canidae	
		Canis (extinct species within genus)
	Felidae	
		Felis
		Homotherium*
		Miracinonyx*
		Panthera (extinct species within genus)
		Smilodon*
	Ursidae	
		Arctodus*
		Tremarctos (extinct on continent)
		Ursus
Proboscidea		
	Gomphotheriidae	
		Cuvieronius*
	Mammutidae	
		Mammut*
	Elephantidae	
		Mammuthus*
Perissodactyla		
	Equidae	
		Equus (extinct on continent)
	Tapiridae	
		Tapirus (extinct on continent)
Artiodactyla		
	Antilocapridae	
		Antilocapra
		Stockoceros*
		Tetrameryx*
	Bovidae	
		Bison (extinct species within genus)
		Bootherium/Symbos*
		Bos (extinct on continent)
		Euceratherium*
		Oreamnos (extinct species within genus)
		Ovibos
		Ovis
		Saiga (extinct on continent)
		Camelidae
		Camelops*

(continued)

TABLE 1.1. (continued)

		Hemiauchenia*
		Paleolama*
	Cervidae	
		Alces (extinct species within genus)
		Bretzia*
		Cervalces*
		Navahocerus*
		Odocoileus
		Rangifer
		Torontoceros*
	Tayassuidae	
		Mylohyus*
		Platygonus*

climate shifts stressed them to the extreme by fragmenting regional populations, thus reducing gene flow within the continents and decreasing genetic diversity or abilities to migrate in search of forage and mates, and locally depressing the quality and perhaps quantity of forage depended on for survival; human hunter-gatherers may have contributed a small amount to the reduction of local subpopulations of large mammals, but did not have a decisive effect.

Other exogenous theories exist besides the climate-change type. Potential explanations range from the effects of an extraterrestrial bolide impacting the earth, to a hypervirulent disease organism brought by dispersing human foragers (or their dogs) that jumped species to the large land mammals.

The main opposing biotic or endogenous theory recognizes that the largest land mammals were continentally stressed by climate shifts at the end of the Pleistocene, although they had survived numerous earlier stresses during glacial-interglacial reversals; but what actually is to blame for killing them out forever near the end of the Pleistocene was the behavior of rapidly dispersing human foragers whose ancestry was in northeast Asia. But a problem with the term "endogenous" applied to the selective extinctions at the end of the Pleistocene is uncertainty about whether humans co-existed with the animals for a long time or only a relatively brief period, which would have made them new additions to the ecosystems. The killing might have been done quickly by active hunting ("blitzkrieg") or very slowly by less direct means such as the burning and alteration of large parts of different habitats while humans also carried out lower-intensity hunting ("sitzkrieg"). If humans had been present in the Americas for millennia, and therefore were already part of the existing ecosystems, they were endogenous factors. But if they rapidly appeared just before the extinctions, they were exogenous.

Thus, the ongoing debate about the causes of the extinction may seem confusing to some observers because it cannot contrast climate-change and human hunting as potential explanations that pit exogenous versus endogenous factors. Nevertheless, these are the two leading contenders to explain the extinctions.

This Volume and its Contributors

This book conspicuously does not include chapters written by the most prominent spokespersons for or against the specific unicausal theories discussed above. One reason is that some were asked to contribute but declined because they thought they had already said everything they could. Another and appropriately weightier reason is that the chief promoters of one theory or another have long been expected to establish rigid positions and most likely would not be able to argue justly anymore – they would pick the easiest points to make and obstinately deflect the criticisms instead of facing them down.

The contributors to this volume are:

Gary Haynes, archeologist and author of a book about the dispersal of people into North America (*The Early Settlement of North America: The Clovis Era*; Cambridge University Press, 2002). One chapter (this rest of this one) concerns the nature of the debate and the theories advanced to explain extinction; a second chapter is a speculative look at the "extinctions risks" of some large mammals that disappeared at the end of the Pleistocene.

Stuart Fiedel, archeologist and author of the one-volume guide *Prehistory of the Americas* (Cambridge University Press, 2nd edition, 1992), a prominent commentator on archeological interpretations of the earliest peopling of North America. His chapter is a survey of the evidence for the timing of the megafaunal extinctions in both North and South America.

Daniel Fisher, paleontologist and author of numerous studies of the paleobiology of mammoths and mastodonts in the northern hemisphere, and well known for developing microscopic methods for examining tusks to determine proboscidean life history traits. His chapter is a survey of his own and colleagues' recent and continuing research and the new information now available about Great Lakes proboscideans.

Todd Surovell and Nicole Waguespack, archeologists known for their work on Paleoindian subsistence behavior, hunter-gatherer theory, and modeling studies. Their chapter reconsiders the pro and con evidence about Clovis Paleoindians as megafaunal hunters.

Alex Greenwood, geneticist with a specialty in ancient DNA studies, known for his work on demography and the population dynamics of large mammals, based on recovered genetic material. His chapter is a summary of the issues and methods of studying ancient DNA and the contributions of those studies to recent interpretations of large-mammal demography.

Alberto Cione, Eduardo Tonni, and Leopoldo Soibelzon, paleontologists known for their studies of megafaunal communities and extinctions in South America. Their chapter is a survey of the mammal communities that changed at the end of the Pleistocene and beginning of the Holocene in South America.

Luis Borrero, archeologist known for both his careful studies of prehistory and his taphonomic and actualistic studies in South America. His chapter surveys the known (or suggested) associations of animal bones with traces of human activities throughout South America.

Ross MacPhee, mammalogist known for his research on extinct animals and also for suggesting a recent alternative theory to explain late Pleistocene extinctions, popularly called hyperdisease. His chapter summarizes new research results on extinct and surviving mammalian taxa from the Caribbean, including the island of Cuba. His chapter is not focused on Pleistocene extinctions, and it is a stretch to describe most of the extinct taxa he discusses as "megafauna," but several factors including his different perspective on extinctions make this contribution relevant.

This Chapter

The literature available about the competing theories is not massive, but most academic and municipal libraries probably contain at least two full-length books either specifically written about the late Pleistocene extinctions or featuring them as notable case studies such as Dave Foreman's *Rewilding North America* (2004), or Peter Ward's *The Call of Distant Mammoths* (1997). Some of the books are meant for popular audiences and some are too technical to attract more than a few hundred readers. Numerous journal articles have appeared in the last quarter-century, rehashing old arguments or presenting new types of evidence to support the old reasoning.

This first chapter is not a complete run-down of all the books and articles that have materialized in the English-language literature, but instead is a personal survey of common recent themes, topics, and noteworthy publications about the theories. Over the last quarter-century, papers, articles, polemical pieces, and opinionated broadsides have appeared in dribs and drabs along with the books with sections that not only try to explain the late Pleistocene extinctions but also treat them as grave lessons for us to absorb about human destructiveness on our endangered planet (e.g., Flannery, 1995, 2001; Ward, 1997; MacPhee, 1999).

There's not enough space here for a comprehensive literature review of the late Pleistocene extinctions, so my chapter is only an entrée into what's been available in the last few years about North America. Here I classify books and papers into three genres or thematic groups – (1) the disputative, (2) the less polemical 'scientific' offerings, and (3) the (ostensibly) neutral reviews. Clearly some of the publications would fit into more than one class, but most references cited here do self-identify themselves as belonging to one of the three genres.

Disputative (Dialectical) Themes

Human Hunting as the Decisive Factor in the Extinctions

Recurring themes in the literature have marked certain writers as partisan to one or another of the possible extinction theories. The place to begin is Paul Martin's blitzkrieg theory of Overkill, as it is frequently nicknamed. The term

"Overkill" may refer to several different variants of human killing (high or low intensity of hunting, quick or gradual impact of hunting on megafaunal populations) as the main cause of extinction, but Martin's blitzkrieg version is the one most often cited – and many opponents seem to believe that if this extreme version of Overkill can be falsified, all other variants by default must also be untrue, which is an error of logic.

In the blitzkrieg version of the Overkill theory, which Martin and colleagues have summarized several times (for example, Martin, 1967, 1973, 1984; Martin and Steadman, 1999) since its original presentation by Martin in the 1960s, human foragers are envisioned as having rapidly entered the Americas by traveling through Beringia in search of new resources, and once they were south of North America's late Pleistocene ice sheets they reproduced rapidly after encountering naïve prey inept at defending against the new predators' efficient hunting abilities and their well engineered weaponry, namely the thin stone spear-hafted bifaces of the Clovis fluted-point tradition and the polished bone/antler points similar to those manufactured for millennia in the Eurasian Upper Paleolithic. Spreading swiftly southwards from their entry point in northwestern North America, the foragers depleted megafaunal prey species whenever they first contacted them, with wasteful and proficient killing practices, leaving behind only a very few camp sites and nearly no kill sites of the many species that had been Overkilled.

A weighty argument advanced against the proposed inevitability of first-contact Overkill is that outside the Americas, in Africa, Europe, and Asia, some megafaunal extinctions occurred only after several thousand years of human co-existence with megafauna, and often – but not always – in sync with post-glacial climate cycles and climate/vegetational changes. Thus it may not seem evident why the first contact between megafauna and human foragers in the Americas should have immediately led to extinction or extirpation locally, since long temporal overlap did occur in the Old World before there were any extinctions. For example, following deglaciation in northern Europe, Scandinavian reindeer recolonized northern ranges about 12,500 BP and were seasonally hunted by people (migratory people, perhaps) even that early, but nevertheless the reindeer survived until 9,200 BP (Aaris-Sørensen et al., 2007).

In response to this line of reasoning, it can be pointed out that some of the species that did manage to survive human hunting for several millennia in Eurasia's Pleistocene-Holocene transition, such as woolly mammoth, horse, and reindeer, were ones that had co-existed for millennia already with humans – and hence the argument can be made that they possessed specific behaviors or biological features that allowed them to withstand human hunting longer than the American species, which never had developed such features.

Note that prey naiveté is an important component of the blitzkrieg theory; non-naïve prey in Eurasia and Africa did not suffer extinctions to the same extent, the story goes, because they co-evolved over many generations with hominin hunters and had developed effective anti-predator behaviors and defences. The New World herbivores, on the other hand, had never seen spear-wielding bipedal predators and did not know how to avoid or escape them. It is still unclear what the specific anti-human responses would have been or how they would have differed from ordinary anti-predator behaviors; there were plenty of four-legged Pleistocene predators in the Americas before humans appeared and the prey animals must have developed various different behaviors to defend themselves, such as habitat selectivity, different degrees of dispersal or aggregation while foraging, and levels of vigilant behavior (see, for example, Creel et al., 2007, on elk defensive tactics when wolves are known to be present). Nevertheless, as I discuss below, this concept of prey naiveté is still plausible and deserves respect as a possible explanatory principle in the "Overkill" theory.

Beyond Paul Martin's writings, support for human-mediated extinctions can be found in a variety of other authors' publications. Overkill seems logical to Burney and Flannery (2005). They point out that temporally stepwise megafaunal collapse correlates closely with first human appearances everywhere in the world. They also point out that no quantitative models of climate changes have been made that track the extinctions either loosely or tightly, as human dispersals do. Climate changes simply do not correlate at all with the extinction chronologies. Burney and Flannery recommend that researchers compare pre-human to post-human ecology in the landmasses affected by late Quaternary extinctions, and precisely determine the time of first human appearances by either directly dating artifacts and sites or, if need be, by seeking signature proxy records, such as the first appearance of introduced animal or plant species, or abrupt changes in levels of preserved *Sporormiella* spores (derived from a fungus that grows where dung deposits are found, hence indicative of megafauna presence or absence – see more information below).

Lyons et al. (2004) performed a quantitative analysis of body-size distribution globally for extinct and extant animal species, and concluded that the appearance of humans must have been a much greater factor in extinctions than changes in climate. Before the extinctions, animal body masses on all continents were similarly distributed, but afterwards clear differences are found in the sizes of extinct surviving species. Size selectivity is obvious, although a definition of the word megafauna would have to differ for each continent since not all affected landmasses had giant mammals. Lyons et al. (2004) also note that extinction threats due to habitat-loss in the historical era are not size selective, although modern species threatened by human hunting are size-selected, supporting the idea that the largest animals in the late Pleistocene/Holocene extinctions were deliberately targeted by humans and not universally disadvantaged due to climate-caused vegetation changes.

The study by Lyons and colleagues concluded that the size selectivity could not have been due to any universally shared ecological traits of the larger animals, because the analysis examined different orders of animals having dissimilar reproductive rates and population densities. Somewhat in disagreement, a study by Johnson (2002) concluded by deciding that body size was not as important as reproductive rate in predicting the extinctions of the late Pleistocene. Larger body mass and slower reproductive rate are of course linked, but even in the much smaller animal taxa the slowest reproducers were hardest hit by the extinctions, and among the survivors, those that have relatively slow reproductive rates are predominantly arboreal or nocturnal or dwell in closed habitats where they would have been harder for human hunters to find. Thus, Johnson does not support a conscious form of human "blitzkrieg" but does allow that even low-level human hunting contributed to the extinctions in the Americas, Australia, Madagascar, Europe, Asia, and Africa.

Another article supporting Overkill (Schuster and Schüle, 2000) proposed that even some of the earliest Plio-Pleistocene extinctions should be attributed to evolving hominins. Some of the authors' logic is a bit offputting – for example, they state that "once the causes [of extinctions] are identified, the mechanisms become explicable," which may seem like cart-before-horse thinking to some; and in a refreshingly candid admission they profess that sometimes a fuzzy or incomplete view of the facts can lead to more satisfying interpretations than over-exact definitions. Their proposals are very similar to those appearing in a later paper by Surovell et al. (2005), namely that large terrestrial mammals anywhere in the world did not react well to new hominin predators wielding spears. In the opinion of Schuster and Schüle (2000), the earliest hominins must have hunted (rather than passively scavenged), and therefore the largest mammals were depleted within the progressively expanding range of the genus *Homo*, while outside that range there were still big-game reserves of sorts where the animals could survive.

Surovell et al. (2005) explicitly proposed that big-game animal ranges were defined by hominin presence and absence. Whenever climate prevented hominin geographic expansions or reversed them, big game animals were restored to ranges (and the opposite happened at other times). The authors analyzed a sample of 41 fossil proboscidean kill or scavenge sites, which typically contained the remains of few animals spatially associated with artifacts, the bones often partly articulated. The results of the analysis demonstrated a linear latitudinal trend of concurrent hominin range expansion and proboscidean range retraction or extinction. This interpretation has been challenged by Ugan and Byers (2007).

Other examples of literature written by non-archeologists and non-paleontologists support the idea that humans did have either some or a major influence on the distribution and survival of large terrestrial mammals. For example, Charles Kay, wildlife ecologist and a specialist in the politics of conservation, has examined Native American actions such as hunting and habitat burning, and found that aboriginal land management practices clearly shaped American habitats and animal distributions in prehistory (Kay 1994, 2002, among numerous other papers).

More support for Overkill came in a book about American biogeography written by an Australian ecologist T. Flannery (2001). He attributed the end-Pleistocene megafaunal extinctions in the Americas to pioneering stone-age humans experiencing ecological release from competing predators, parasites, and diseases. The pioneers brought with them an urge to exploit the perceived boundless resources encountered in an empty continent.

Simulations and Models in Support of Overkill

Simulation models have been run to test the possibility of Overkill. Early models such as in Mosimann and Martin (1975) and Whittington and Dyke (1984) depended upon estimates of North America's carrying capacity to predict the moments when human killing led to irreversible population collapses of megafauna; but these models are not realistic because carrying capacity must have greatly varied in space and time through the continent. The models are best appreciated as demonstrations of Overkill's plausibility (under certain conditions; for example, Mithen, 1993) rather than insightful predictors of the actual events and processes.

More recently, increasingly sophisticated computer/numerical models that input climate changes, shifts in the distribution or productivity of vegetational communities, and changing animal distributions and densities have been devised with various different emphases, and they too conclude that adding humans to the mix is extra destructive for megafauna. Alroy (1999, 2001), for example, factored many variables into his model, including changes in human reproductive rates in parallel with changing hunting success rates, to conclude that over the course of hundreds of years many thousand human hunters could have wiped out most of the extinct species. This version of Overkill may not qualify to be called 'blitzkrieg,' since it required a millennium of killing to reach its end point, but in geological time the interval is indeed relatively brief. Alroy (1999) also pointed out how extremely different the late Pleistocene extinctions were in comparison to the many episodes of faunal turnover and species disappearances recorded over the course of earth history. The late Pleistocene extinctions were "extraordinarily selective" for prey body size and therefore unique and plainly "unnatural" (Alroy, 1999:132, 133).

Whitney-Smith (1995) devised a model she called 'second-order predation,' in which human competition and active killing of predators in the Americas led to population declines in the predators, in turn allowing megafaunal herbivore populations to expand to the point where vegetation suffered, resulting in die-offs and die-outs among the animal species that reproduced at the lowest rates. In the end, after extinctions, vegetational communities were altered by combined factors of overgrazing/browsing and climate change.

Another quantified model based on economic principles was devised by Bulte et al. (2006); in it, humans, being omnivores, were able to take advantage of their increased encounters with small mammals by eating more of them as they pursued higher-ranked megafauna. This is in opposition to the predicted behavior of plant collectors or groups with dependence on cultivation, because any supplemental foraging by these latter types of human groups would have led to fewer encounters with both larger and smaller mammals. Therefore, in the authors' logic, it seems more likely that the Americas were first colonized by people who might have preferred to specialize in big-game animals, but were also satisfied to take small game; when hunting pressure diminished the supply of large animals the response was to take even more small game. The theory as presented has no concern with the complications that might have arisen over varying local conditions in the continent, or changes in geographic dispersal rates, ongoing and inconstant climate changes, and so forth, although it can be said in the model's defense that no qualitative models of the evolution of human foraging during the colonization of North America (such as Meltzer, 2004) has ever looked at all the possible localized variability either (although some models do try to factor in numerous variables; for example, see Diniz-Filho, 2004, for a South American example). The Bulte et al. (2006) model predicts that the archeological record of first entry will yield evidence of what appears to be a generalized diet, as many anti-Overkill authors predict based on ethnographic analogy and analysis of the available data (Cannon and Meltzer, 2004).

Koch and Barnosky (2006, supplemental table S4) list 12 simulation models and their results; seven models support the possibility of Overkill.

Variant Theories in Which Humans are to Blame, But Hunting is not the Main Destructive Process

Human killing is given a background role in a variety of theory which attributes extinctions to overall ecosystemic changes created by the new presence of dispersing humans. This type of theory, in which extinctions are still considered to be human-caused, is neither 'blitzkrieg' nor 'sitzkrieg,' the latter being a variety of theory in which temporally drawn-out human-killing is cumulatively decisive. Instead, it sees the extinctions as the direct result of (1) competition from or predation by life-forms introduced by humans (examples would be dogs and rats), or (2) habitat destruction or alteration by people clearing land for one reason or another, such as woodland-clearing to prepare fields for planting or pastures for livestock, or (3) diseases introduced by humans and their commensals. The extinctions are therefore partly direct and partly indirect results of the human presence. Steadman (Steadman et al. 2002; Steadman and Martin 2003; Steadman 2007) clearly saw the appearance of first humans in Pacific Islands as the factor behind the dramatic disappearance of so many native bird species in a short period. Not only did killing contribute to the extinctions, but also habitat alteration by early farmers who cleared lands, and the depredations of rats and dogs which seriously depleted the native wildlife populations. An example of a postulated broad human impact (in Australia) is in Miller et al. (1999; 2005).

A large set of publications favors a 'mixed' model of first-contact killing and climate stress to explain the extinctions, such as my own (Haynes, 2002a, b, 2005), but I ultimately pin the blame for extinctions on human predation. Other examples are Koch and Barnosky (2006) and Barnosky et al. (2004), which are described below in the section discussing the "review" genre of literature

The Anti-Overkill Voices

The reception given pro-Overkill theories and arguments has sometimes been extremely negative. Native American author (and Christian theologian) Vine Deloria (1969:112) strongly insisted that indigenous people must be defended against damage to their image by "right-wing fanatics" and scholars who would seize upon the Overkill theory as proof that Indians lacked moral fiber and ethical concern for the earth. His defensive reaction is understandable because media discussions of Overkill are so often cast in terms of ancient hunters' guilt or moral culpability, as though the moral stature or land rights of Australian Aborigines or Native Americans would be undermined if their ancestors were shown to have been less than perfect stewards of their respective continents' wildlife.

Deloria punches wild and hard in his anti-Overkill denouncements – he ridicules the scientific practices and principles of geology, radiometric dating, taphonomy, and archeology – with an aim of shutting down further dialogue in the argument. He might as well ridicule electricity, modern medicine, and genetics, since in effect physics and chemistry are not to be trusted, if radiometric dating is nonsense. To account for the existence of extinct animals, he suggests that some megafauna must have died in catastrophes en masse, killed by volcanoes, floods, atmospheric changes, and earthquakes, while others may have survived until recently to be killed by benevolent natural forces.

Some archeologists who ostensibly have more trust in the empirical record also argue against human influence on the extinctions. One thread in the anti-Overkill argument is that nearly "invisible" earlier colonists (that is, pre-Clovis peoples) lived beside megafauna for thousands of years, and hence it is argued that the long temporal overlap of people and megafauna (without extinctions) proves that blitzkrieg and sitzkrieg did not happen.

Numerous examples of these hard-to-find pre-Clovis-age possibilities appear regularly in the literature. Some are unmistakably shaky, but some seem plausible. Alex Krieger (1964) and others such as Louis Leakey (Leakey et al., 1968),

Richard MacNeish (1976, 1982), and William Irving (1985) envisioned numerous phases of the human presence in North America, dating back to the beginning of MIS 2 (the last glacial stage) and well beyond in some papers, each phase having different technology, subsistence behavior, and geographic distribution. More recently, papers such as Arroyo-Cabrales et al. (2006) interpret instances of mammoth sites as possible pre-Clovis evidence, including some very early material, such as bones found around a "hearth" dated > 31,000 BP at El Cedral, Mexico Holen (2006) interprets pre-Clovis (ca. 18,000 BP) broken mammoth bones as items produced by humans making tools in the central United States. Sites such as these, if they actually were culturally produced and are not accidents of nature or created by the means of their discovery (such as distortion in the ground under the weight of heavy equipment), imply that the pre-Clovis populations that left them were not scattered and inconsequential, but were widely distributed, relatively large in size, and so far have been very poorly recognized because of their unusual nature, especially their scarcity of diagnostic lithics.

Recent archeological research effort has notably centered on finding alternative routes of pre-Clovis human entry into North America. This has come about mainly because the possible presence of cultures dating to the LGM can not be explained by a human dispersal through the Canadian inter-ice-sheet corridor. According to some but not all geologists and paleoecologists, the corridor was completely closed until just after about 14,000 BP and was impassably bleak, wet, and unproductive for another 2 kyr afterwards, only becoming a feasible route of dispersal around 11,000 BP, some time after the beginning of the Clovis era (Duk-Rodkin and Hughes 1991; Mandryk, 1992, 1996; Mandryk et al. 2001; but for other points of view, see MacDonald and McLeod, 1996; Wilson and Burns, 1999; and the deglaciation database and animated maps prepared by Dyke et al., 2003). The corridor opened rapidly between 12,500 and 11,500 BP, according to the most recent work of Canadian geologists such as Dyke (Dyke et al., 2003; Dyke 2004). Southern lobes of the Laurentide ice sheet were in especially rapid retreat just after 12,000 BP (13,950 cal bp) (Lepper et al., 2007), when Pleistocene Lake Agassiz formed.

A possible dispersal route down the northwest coast of North America seems to offer a little more time than the interior corridor for pre-Clovis human entry, but not by much, perhaps only a thousand years (Clague et al., 2004; Fedje et al., 2004). It appears certain that Native American people are descended from an ancestral population that originated in interior northern Asia (Jobling et al., 2004; Powell, 2005; Schurr, 2004; Turner and Scott, 2007; Tamm et al., 2007), so where else could the Asia-Americas connection be located except along the coast? Stanford and Bradley (2002; Bradley and Stanford, 2004) think a better connection is along the North Atlantic ocean-ice zone, based mainly upon similarities in Clovis and Solutrean stone-working techniques, but their opinion hasn't persuaded all American archeologists because of chronological problems, implausibility of North Atlantic voyages in skin boats, and particularly because of the genetic and dental data pointing unambiguously to an Asian rather than European origin for Native Americans (Straus et al, 2005). Thus the Pacific coastal route is by default the new main focus in the search for pre-Clovis dispersal events.

If indeed there were pre-Clovis human dispersals, either down the coast during the latter Late Glacial or through the interior corridor before the Last Glacial Maximum when the land route was open and productive, the populations who filtered into the lower 48 states of North America were virtually invisible, as Meltzer (1997), among others, has proposed, based on the scarcity of archeological discoveries, the widely scattered nature of the few potential sites that are marginally supportable, and the fact that stratified, buried Clovis sites nearly always have no earlier human occupations below the fluted-point level. However, a couple of possible exceptions to this latter point are known, namely Meadowcroft Rockshelter, Pennsylvania (Adovasio and Page, 2002, specifically chapter 7) where no diagnostic Clovis or even Early Archaic artifacts are reported; Topper, South Carolina (Goodyear et al., 2005); Gault, Texas (Collins, 2002), where supposed pre-Clovis materials are found under fluted point levels; and Cactus Hill, Virginia (McAvoy and McAvoy 1997). Even these may not contain true pre-Clovis artifacts. The lower level at Cactus Hill may simply be an early Clovis or formative [proto-]Clovis horizon or down-drifted later artifacts, and in the case of Topper the lithic materials in the lowermost levels are almost certainly, even in the opinion of most pre-Clovis proponents, noncultural debris and small pieces displaced from the overlying Clovis levels.

A problem with the idea that an invisible population preceded Clovis in North America is the belief can never be falsified, which means it is not a scientific hypothesis. Paul Martin's Overkill theory is also accused of being unscientific because it predicts a lack (or more correctly an invisibility due to nonpreservation) of megafaunal killsites, which clearly cannot be falsified, either. As physicist Wolfgang Pauli remarked after reading a paper lacking both falsifiable and predictive statements, if you can't demonstrate whether an idea is wrong or right, it's not even wrong (Peierls, 1960). I think the prevailing view of most professional archeologists is that the enormous literature about a deeper-time pre-Clovis American population is repetitive and frequently based on little but the perceived authority of specific believers rather than on clinching arguments.

Whatever the extent of the earlier dating of a human presence in the Americas, the argument that any amount of temporal overlap of humans and megafauna – whether extended in time as in Africa and Eurasia or relatively brief as in the Americas – automatically falsifies Overkill as not scientifically acceptable. Certainly the blitzkrieg model of Overkill does require first-contact human hunting to be abruptly initiated and rapidly effective, but there's no logical reason why a more drawn-out kind of human killing could not have contributed decisively to megafaunal extinctions. A reliance on temporal overlap of humans and megafauna as an argument against Overkill makes for a too conveniently shiftable set of goalposts on the debate's playing field. Who is to decide whether an overlap of 1,000 years is an adequately robust demonstration of proof against Overkill, or whether 100 years is sufficient. Critics may first demand that kill sites be discovered in order to demonstrate any human-megafauna contemporaneity, then upon such discovery may shift gears to state that a period of coexistence long enough for such sites to be created precludes decisive human impact.

An arbitrary selection of time-limits-as-proof-against-Overkill must be accompanied by careful and realistic modeling of changing human population numbers and hunting offtake and the effects on megafaunal populations, if it is to be considered as more than mere opinion. The modeling must incorporate reasonable estimates of faunal numbers, distributions, and reproductive rates in different environmental regions of the Americas – which has not been done.

Another tack in the anti-Overkill approach is to argue that big-game specialization is not possible or rational (see Surovell and Waguespack, Chapter 5, for more discussion). Grayson (2001) for example reviewed the impacts people may have had on animal populations in prehistory and professed he saw no convincing reason to think human hunting had measurable effects. Grayson and Meltzer (2002) also found the Overkill explanation for extinctions to be wanting in persuasiveness or supporting evidence. Cannon and Meltzer (2004) argued that small fauna outnumbered megafauna in Paleoindian sites, although much of their sample of faunal remains was poorly documented and may not have been behaviorally associated with fluted-point people. Byers and Ugan (2005) quantified the costs involved in hunting and fully processing proboscidean carcasses and concluded that prehistoric hunters would not have wanted to go to all that trouble, and instead would have rationally favored pursuit of the smaller animals.

These sorts of anti-specialization scenarios and models actually can be accommodated by the paleoeconomic model of Bulte et al. (2006) and the other pro-Overkill models, in that hunters who consciously want to kill megafauna will nevertheless still be willing to take time and procure other (smaller) animals during the pursuit of highest-ranked big game. One can also envision a simple division of labor wherein women and children would have taken smaller game while men hunted megafauna. An applicable study by Grant et al. (2005) also supports the feasibility of passive/fortuitous predator "specialization" in big game, in spite of the extreme costs and the supposed irrationality of targeting such dangerous and relatively scarce animals. In Grant et al.'s analysis of Serengeti (nonhuman) predators, prey was selected on the basis of accessibility rather than abundance; perhaps the Americas' largest mammals were much more accessible to Clovis foragers because they were less mobile in the late Pleistocene as they stuck closely to shrinking refugia and optimal habitats or water sources (G. Haynes, Chapter 3; Haynes 2004).

Another recent set of research studies by Bliege Bird et al. (2002) and Bird and Bliege Bird (2000) also supports the idea that a preference for big-game is not necessarily impossible or too irrational to appear in human behavior. In their work the human practice of what is called costly signaling sometimes allows seemingly irrational preferences in foraging (such as choosing megafauna over smaller, more abundant animals), because the aim of a big-game hunter would be to let fellow social-group members know the hunter is highly skilled, intelligent, willing to take great risks on behalf of the group, and is therefore an unusually valuable member of the band with consequent sexual and reproductive advantages.

Some anti-Overkill authors have contributed commentary on much more than the extinctions; Meltzer (2004) has written abundantly about the entire range of processes involved in the transition from late Pleistocene to early Holocene in America. Some of his commentary has to do with how human settlement in new ranges would make Overkill an impossibility. For example, in his 'landscape-learning" models (Meltzer, 2003, 2004), he imagines that in the first exploratory phases of the Clovis dispersal, individuals carried technologically complex, durable and maintainable tools in standardized forms, preferentially hunted the highest ranked animals, and relied on exotic (nonlocal) raw materials that required long-distance return visits to the sources. Eventually the residentially mobile founding groups began using lower-quality (and local) raw materials, broadened their diet breadth, and made less standardized and less complex tools, while logistical mobility increased. If any human overhunting occurred in this model, it would have had to be very early in the pioneering phase, before local human subpopulations actually "settled in". Yet the number of kill sites seems too meager to Meltzer (2004) who cannot imagine Clovis foragers continuing to pursue highest-ranked but increasingly scarce large prey after beginning to learn about other local resources.

Commentary from an outspoken anti-Overkill voice, Stephen Wroe (2006; Wroe et al., 2004) generally concerns the Australian setting of extinctions, but his reasoning is applicable to the direction the debate has taken in the Americas, too. In a press interview (Amos, 2005), he has stated as if from personal experience that "pointy sticks" (wooden spears without specialized lithic points) would not be effective on large animals, and humans could not have possibly impacted megafaunal populations because of their technological backwardness in late Pleistocene Australia. Such statements sidestep the discovery of wooden spears at the much older European sites of Schöningen and Lehringen (both in Germany, both dating to the middle Pleistocene), in association with fossil bones of horses and elephant and indicating successful hunting was possible without the use of atlatl, bow and arrow, or stone-tipped weapons. In both cases, pre-modern hominins used close-quarter thrusting and throwing spears against big, dangerous animals. Wroe (2006) wrote a guest editorial for the journal *Quaternary*

Australasia in which he says the debaters – clearly meaning in this case the pro-Overkill voices – overstate their cases and need to wind in their necks about the effects of human hunting on large mammal populations, but the article seems to me to be a self-referencing defense of the anti-Overkill position he's known for, although it is presented as a plea for multicausal explanations. He mentions a "growing consensus" that single-causes are impossible (but provides only one citation to support this straw vote), and claims the dating of extinctions to 46 ka in Australia is unacceptable in spite of rigorous standards applied by Roberts et al.(2001) where this age was published. He also claims that most American megafauna would be demonstrated to have gone extinct before Clovis if the same rigorous standards were to be applied in their radiometric dating, but this is misleading. One of the major standards applied in the Australian study was the presence of articulated bones as opposed to unarticulated, supposedly a clue to the undisturbed/un-redeposited nature of the best bones for dating, which for many large taxa are often more numerous in Clovis times. Again, his argument comes back around to the unprovable propositions that humans were at much too low density to kill out species and genera, lacked killing technology, and were unfamiliar with Australia's interior anyway because they stuck to the coasts for millennia. He also claims that the LGM in Australia was unique and more severe than the earlier glacial periods that had been survived by megafauna, based on one citation that is described as "mounting evidence." These arguments are noticeably similar to the ones used in North America to attack Overkill (see Grayson and Meltzer, 2002, for example). However, more and more evidence shows that Australia's past climate shifts did not seriously affect the fauna in the Pleistocene (see Roberts et al., 2001; Gillespie, 2002; Prideaux et al., 2007a, b)

Climate-Change as Explanation for Extinctions

A number of authors have chosen to avoid making anti-Overkill comments when writing in favor of climate changes as the main explanation for the extinctions. A prime example is Kelly and Todd's (1988) much-cited paper, in which megafauna are thought to have reacted to Late Glacial climate stresses by becoming scarcer locally and thus inspiring human foragers to be extremely brief occupiers of localities and regions.

A number of publications favoring climate-change over human hunting have provided theoretical models behind extinction, but none have quantified the mechanisms or mapped out the progression of events in a realistic manner. Some different models include Guthrie's (1984) ideas about late Pleistocene mosaics of diverse nutrients developing into zones of less diverse plant communities whose anti-herbivory defenses were too potent for megafauna to survive on; Graham and Lundelius's (1984) ideas about co-evolutionary disequilibrium, or the falling apart of the disharmonious faunas of the late Pleistocene in the face of climate-caused vegetational changes; Owen-Smith's (1987) ideas (almost an Overkill variant) about how certain keystone species (those that helped to engineer habitats, such as mammoths) died out due either to human hunting or climate changes, which subsequently led to dramatic transformations in local habitats and the dying out of other species in response; and ideas that refer to self-organized instability (Solé et al., 2002) which is a sort of sensitivity that ecosystems may have to so-called thresholds of change, not all of which appear large or even significant but which trigger massive collapses or shifts within the systems. This latter kind of modeling is not yet very helpful in mapping out the extinction events, but it does seem to be based on an apparently real kind of instability found in many contemporary ecosystems. However, Grayson (2007) seems to be saying recently that the ecosystem or community concept is passé, and we should be looking for unique extinction causes on a species by species basis.

Guthrie (2003, 2006 for example) has been especially persistent and painstaking in his search for empirical facts that might reveal the processes leading up to extinctions in Beringia. He has documented a rapid body size reduction of Alaska *Equus* in the millennia before the genus disappeared in the north. In his view, "horses are almost obligatory grazers" (Guthrie, 2003: 170) so the disappearance of Beringia's well-drained steppe was a critical blow to their survival. A major pollen shift 13,000–12,500 BP occurred just as horses were disappearing, and a tad before humans arrived, according to the radiocarbon dates Guthrie has amassed. However, later papers by Solow et al., 2006, and Buck and Bard, 2007, convincingly show there was a likely human/horse/mammoth temporal overlap (see Fiedel, Chapter 2, for more discussion). In Guthrie's mind, the questions of Beringian extinction circulate around the issue of whether caecalids (such as horses and mammoths) were replaced by ruminants everywhere as forage conditions changed so profoundly. It is a neat and supportable scenario, and it can explain the Beringian extinctions well, but the model does not explain the lower 48 states' extinctions so neatly or even very well, if the diet studies of Feranec, Hoppe, and others are examined (discussed below). Another possible weakness in the case is the emphasis on size reduction of the Beringian horses as a clearcut predictor of extinction. In fact, animal body mass can evolve relatively quickly in response to environmental factors. For example, red deer on the Channel Island of Jersey between France and Britain were reduced in size by a much greater proportion than the Alaskan horses, at about 121 ka, but survived the severe ecological stresses affecting them (Lister, 1989) without going extinct. On the other hand, the body mass of every now-extinct terrestrial mammal clearly did not reduce in size before the species died out, which (by this logic) would weaken the case for climate-caused factors being behind the extinctions. For example, the giant deer *Megaloceros* in the Ural mountains of Eurasia disappeared after the transition to Holocene environmental conditions and never decreased in size before its extinction occurred (Stuart et al., 2004). It is also worth noting that many species that survived the extinction interval are notably smaller in the Holocene than their Pleistocene ancestors (such as *Bison bison* versus *Bison antiquus*), and it seems clear that size reduction is only an evolutionary tactic in the face of changing conditions and does not mean extinction must follow.

The Graham and Lundelius (1984) model depends as much on ecological reasoning as on empirically based logic. Graham, like Guthrie, has also been persistent in his search for climate-change links to the extinctions (Graham, 2001, 2003; Graham et al., 2002; Graham and Lundelius, 1984). In his view, ecosystems were 'progressively stressed...with each interglacial' and eventually collapsed at the final transition. Graham has not published many of these studies yet, but has presented them at conferences often accompanied by press releases: In 2001, he presented at the Boston meeting of the Geological Society of America a GIS-based model of changes in species distribution over time in the Pleistocene, linked to body size; in the model, he reasoned that large animals needed larger ranges to support viable populations, and the loss or fragmentation of ranges was the fatal blow to species. However, I point out that mammoths of reduced size survived well into the Holocene on tiny islands in the north, effectively demolishing this argument.

No unique climate changes occurred or were required at the end of the Pleistocene to account for the North American extinctions, according to Graham, but instead a threshold effect occurred after the species' ranges had slowly reduced in size as climate changes occurred throughout the late Pleistocene. Each reversal of climate taxed the large animals by fragmenting ranges and altering forage availability, among other effects, although the many species managed to stabilize population sizes or recover during the various interstadials and interglacials. But at the rather abrupt onset of the cold Younger Dryas, an ecological trigger that tipped the balance and finished the process of reducing populations to extinction, the many species disappeared from all the fragmented ranges they had occupied for so long.

In 2003, Graham presented another possible wrinkle in his ideas about extinctions – this time he argued that the megafauna had survived an earlier transition from glacial to interglacial MIS 5e because this stage had warmer winters than the later MIS 1 (the most recent interglacial, which megafauna did not survive into), and therefore its differences may explain the fact that megafauna avoided extinction. He also suggested, based on a much poorer fossil record than we have for MIS 2 and 1, that the MIS 5e faunas were distributed differently, with greater chances for interchanges among biomes, and hence had an advantage over MIS 1 fauna. The next step in this direction of the argument would be to quantify any possible differences in the faunal proxy records to determine how much more or less survivable such possible distinctions would have been.

Some authorities have proposed that the Pleistocene-Holocene transition occurred in a uniquely rapid manner, a mere 40-year period of dramatic changes in climate and vegetation. Again, it would be useful to quantify the earlier shifts and flip-flops in late Pleistocene climate to compare how much more or less rapid and cumulative they were, and whether the last shift from MIS 2 to 1 would have created the very patchy environments where human foragers could notice changes in animal biomass and distribution during their lifetimes (see Haynes, Chapter 3; Haynes, 2002b). Graham (2006a, b) has argued that proboscideans' greater niche partitioning in the latest Pleistocene and the species' separate geographic distributions were unique in time and made all the taxa much more sensitive to climate-induced changes in vegetational communities, thus uniquely increasing their extinction vulnerabilities.

Kiltie (1984) proposed a more specific climate-related cause for the extinctions of the large animals in the Americas. He suggested that increased atmospheric heat during the last deglacial interval and the transition to the most recent Interglacial period had the deleterious effect of hindering gestation, thus slashing reproductive rates in the megafaunal populations – leading to species extinction. The suggestion may make sense, although the Last Interglacial was warmer than the Holocene or the Bølling-Allerød warming stage. The idea also fails to explain extinctions in the tropics or the staggered extinctions in the Caribbean (see MacPhee, Chapter 9). Kiltie's model does not quantify how this most recent glacial-interglacial transition differs from the earlier cycles of warming that the same species had survived.

It may seem these sorts of climate-based scenarios which reason extinction occurred due to a long-term progression of events and trends may be supportable with data about shrinking ranges, as Graham (2001) attempted, but some species' ranges actually seemed to increase just before extinction; Agenbroad (2005) demonstrated that North American mammoths 10,000–15,000 BP had expanded their range to its largest size for a long time. Of course, a large geographic range (or "extent of occurrence") does not necessarily translate into a large population. If habitats were indeed fragmented in the Late Glacial, the many widely scattered subpopulations spread out over a large part of the continent need not have added up to all that many animals (see Haynes, Chapter 3).

A variation of these arguments for climate-caused extinctions could be found in the concept of Natural Turnover of ecosystems/taxa (summarized in Vrba, 1993), a general theory which does not attempt to trace exact processes of change but instead models the overall trends in species appearance and disappearance over very long time intervals. However, the quantified normal extinction rates in natural turnovers are usually much lower, cited in Schuster and Schüle (2000) as 1 extinction per 40,000 years, when compared to the 1 extinction per 30 years at end of Pleistocene; and certainly these observed earlier cycles of turnover are not so obviously size-selective as during the late Pleistocene in the Americas. As well, natural turnovers are seen as affecting all global sectors, from terrestrial to marine – yet there were no marine extinctions at the end of the Pleistocene. Vrba's (1993) Faunal Turnover Pulse hypothesis may be applied to the situation in the Americas but it does not explain the population fluctuations, range changes, or feeding ecology of species during the final millennia of the Pleistocene.

An expression of this vague general theory appeared in Cupper and Duncan (2006) to explain late Pleistocene extinctions in SE Australia. The authors reason that it is not human hunting or any specific climate changes but instead it is ecosystem 'senescence' that must explain the Australian extinctions – meaning presumably that the animal species had reached their age limits under the faunal-turnover theory. It should be pointed out that the bonesite they studied may not be anything other than one in a long list of such fossil deposits, which had been accumulating at various times for many thousands of years in the region, for one reason or another; as in the Americas, the lack of evidence for human hunting is almost universal in Australia's fossil deposits, but so is the lack of evidence for major climate-caused processes that would have killed so many species all at once or gradually.

Faunal-turnover arguments to explain the extinctions do not hold up to scrutiny. Alroy's (1999) analysis of faunal turnovers and extinction events in the last 55–60 Ma showed that the American end-Pleistocene extinctions are unique in the series of all the other extinctions; hence the climate-change argument about faunal turnovers cannot be seen as "normal" or expectable at the Pleistocene-Holocene transition.

Other Exogenous Theories

A bolide theory and other possible unicausal themes are discussed in the next section.

The Ostensibly Less Polemical (or More "Scientific") Themes

These themes also may be associated with certain authors who favor single causes, but they do not immediately appear to be openly partisan or biased:

Chronology

One major issue in the extinction debate has been the chronology of the disappearances. At this time, both precision and accuracy in dating are too loose to allow unassailable correlations to be made or to strongly enough implicate specific causes such as first human regional appearances or particular climate events. Russ Graham's and Tom Stafford's efforts to discern a two-step extinction process in North America seem particularly futile, given the effects of both limited precision and calibration. However, Fiedel (Chapter 2) makes a case for the extinctions occurring in parallel with first human appearances (see also Fiedel 1999, 2006).

Ancient Genetic Material as Clues to Megafaunal Ecology

Preserved genetic material in the megafaunal fossils might hold clues to population health in the Pleistocene. Greenwood (Chapter 6) briefly summarizes the methods and results of such studies. A trend seen in all the research has been the loss of genetic diversity in the late Pleistocene, possibly following range fragmentation. Beringian frozen-ground preservation of aDNA is exceptional, and studies have been done of northern bison (Shapiro et al., 2004; Drummond et al., 2005) that show a population boom and bust pattern, with rapid growth 75–25 ka, doubling every 10,000 years, but followed by a rapid decline after 45 ka when tree cover likely increased replacing open steppe-tundra. Genetic diversity in Beringian bison (*B. priscus*) had been high until 37 ka when it began declining greatly, before humans appeared. Therefore, it has been argued (Shapiro et al., 2004) that environmental changes before the Last Glacial Maximum (about 21,000 cal bp) were to blame for the eventual loss of genetic diversity – although loss of genetic diversity does not always lead to extinction, as seen clearly in the case of *Homo sapiens*, a species that is much less diverse than *Pan troglodytes*, for example – and perhaps by implication also to blame for the extinctions of other taxa, too – although questionably artifactual fossil materials from Old Crow in the Yukon (Morlan, 1980; Irving, 1985) are also mentioned as a potential signal of early human presence at the time the bison population began crashing. The Old Crow materials are not universally acknowledged to be true artifacts and may be naturally broken megafaunal bones. A later study by Drummond et al. (2005) proposed that the bison population bottleneck was most severe at about 10 ka, definitely during a time of early human presence in North America.

Genetic bottlenecks, like body mass shrinkage, may not always be preludes to extinction, so possible climate-caused bottlenecking cannot be automatically identified as the reason for all subsequent extinctions. For example, the cataclysmic Mt. Toba eruption in SE Asia arguably created a possibly severe human genetic bottleneck at 75 ka (Ambrose, 1998; but see Petraglia et al., 2007, for a suggestion that the eruption had no effect on humans in Middle Stone Age India), yet humans survived and populated the globe afterwards. Geneticists agree that there was a severe bottleneck for humans at the time of the modern *Homo sapiens* global dispersal around 50,000–60,000 BP. MacPhee and Greenwood (2007) (also see Greenwood, Chapter 6) studied the loss of genetic diversity in *Ovibos* tundra musk-oxen and emphasized that the biological loss was associated with the end of the Pleistocene in the face of almost no archeological evidence for hunting the taxon. Yet in spite of the major reduction in diversity and the better evidence for later hunting by humans the species has survived to the present day. O'Brien (2005) has shown that Old World cheetahs suffered a late Pleistocene bottleneck, and have very reduced diversity, but are still around.

Genetic studies also have been done of fossil brown bears (*Ursus*) in northwest North America; Barnes et al. (2002) showed that major geographic changes in population distribution occurred before the LGM (35,000–21,000 cal bp), with little change afterwards. Such a pattern is not explained by climatic/environmental events, and some other reason must be sought. The bears apparently were unhunted by humans, whose presence cannot be shown until much later. The authors note that brown bears were absent in the region 35,000–21,000 cal bp, but because they are very tolerant of different environments and only minor climate events are known to have occurred at the time, their absence must have been due to a factor entirely within the ecosystem. The suggestion is that brown bears could not compete against the indigenous and hypercarnivorous short-faced bear (*Arctodus*), and when that taxon became locally extinct or extirpated in Beringia after 21,000 cal bp (they survived until 13,000 cal bp south of the ice sheets) perhaps brown bears expanded into the range. This does not explain the extinction of short-faced bear, but it does make a logical case that climate was not always a factor in the distribution and survival of large mammalian species.

Isotope Ecology and Megafauna Diets

Recent studies have often focused on the feeding ecology of extinct taxa, as reconstructed through analysis of isotopes in the fossil bones and teeth. A noticeable patterning is that many different species now extinct survived on quite different diets in ranges with different forage types in the local plant communities. Feranec (2003) showed that diets of *Hemiauchenia* were variable individually and did not necessarily directly reflect local habitat dominance; this and another study (Feranec 2004) also showed that dissimilar plant communities could support the same animal associates. Therefore changes in flora at the time of late Pleistocene climate shifts cannot simply be used to explain extinctions. The extinct animals were flexible feeders and before the argument can be made that changes in plant communities killed the animals, it must be demonstrated explicitly that the changes were far more severe than the species could survive.

Coltrain et al. (2004) have shown that Rancho La Brea's fossil carnivores competed for similar herbivore prey, and that the herbivores mostly fed on C3 plants in open environments around the time of the LGM when ruminant and non-ruminant diets converged, although later the diets differed a bit more. The diets never predominantly included C4 grasses. An implication of this study is that many different kinds of diet – grazing, browsing, mixed feeding – co-existed in one locality through time, yet all the different species suffered the same fate at the end of the Pleistocene when plant communities were altered by climate change. *Bison* migrated in and out of the area, *Camelops* with its hypsodont teeth survived by browsing rather than grazing, *Equus* was a year-round resident with 50% browse in its diet plus grazing, and ground sloth was either a grazer or a mixed-feeder. One must wonder about the nature of vegetational changes at the end of the Pleistocene when these taxa became extinct – how could a shift to proportionately more woody vegetation locally, or alternatively to much more grass cover, indiscriminately have killed all the different grazers and browsers and mixed feeders?

Isotope studies are providing important new clues to late Pleistocene faunal ecologies, but the field is still developing its methods and standards. Hoppe et al. (2004) showed that the relationship between modern feral horse tooth enamel carbonate isotopes and the environment is not straightforward and clear – isotope values from multiple samples (> 9 animals) must be derived to gain a fair picture of average diets. The importance of the study is that fossil isotope analysis may underestimate variability and the amount of C4 grasses in the diet by 10% or so. The authors also demonstrate that ^{18}O in bones varies for reasons not directly reflecting local rainfall, which many analysts assume to be the main factor in its variability. Another important outcome of the study is the clear demonstration that modern feral horses can and do change their diet (adding C3 plants to a mostly C4 diet), and the genus is not always an obligate grazer. Any shift of late Pleistocene vegetation such as from open grassland to mosaic or wooded habitats should not have led inevitably to the extinction of horse in the Americas, unless it can be demonstrated quantifiably that the developing plant communities did not provide adequate nutrition for *Equus*.

Matheus et al. (2003a) carried out a study that revealed important relationships within the eastern Beringian faunal communities of the late Pleistocene. The authors found that O and N values from fossil bone collagen indicated lions, scimitar cats, short-faced bears, wolves, and wolverines ate mostly bison, horse, and mammoth, and less so caribou and muskox. Scimitar cats ate the most mammoth; lions ate mostly bison; short-faced bears scavenged diverse animal prey carcasses, and wolves also had diverse prey choices, with lots of mammoths (probably scavenged, in the authors' opinion). Brown bears did not eat salmon, as they habitually do today, and were omnivorous (as they are today). Clearly, the loss of herbivorous megafauna would have seriously affected the carnivores, leading them to extinction as well perhaps. The study does not indicate the ultimate cause of extinctions, but does help us understand community dynamics before the extinctions occurred.

Another study by Matheus et al. (2003b) reconstructed eastern Beringian herbivore diets from isotopes (using C and N). Bison had a narrow diet (mostly grasses and sedges); mammoths with their unusually high ^{15}N values had a diet that cannot be reconstructed; horses ate shrubby browse, *Artemisia*, and grasses; muskox ate varying amounts of lichen, graminoids, forbs, and willow. Changes in the diets of these taxa over time could not be clearly reconstructed due to climatically driven shifts in plant isotope ratios at the same times as the diets also would have been shifting. These sorts of studies are valuable contributions to our understanding of ecosystem changes in the late Pleistocene of the north, and may help clarify the nature of stresses undergone by the animals, even if they do not clarify the events of the process of extinction.

Fox-Dobbs et al. (2006) examined fossil avian carnivores from California to understand why condors survived the extinctions in the North American West but not in the rest of the continent. The authors conclude that the California condor did not become extinct because it had been able to change its diet after the megafaunal extinctions. California's avian carnivores (teratorns, condors, eagles, vulture) had to switch to feeding on marine mammals after the terrestrial megafauna disappeared, although half of the avian taxa went extinct anyway, indicating the cascading effect of removing so many herbivores. The authors do not think humans hunted the avian carnivores, which died out with the disappearance of so much of their food source.

Other Proxy Records of Megafaunal Population Health

Other types of new studies have provided proxy records of megafaunal population levels, herd structure, and possible migration patterns. For example, Matheus et al. (2003c), using radiocarbon dating frequencies from eastern Beringia, proposed that (1) horse numbers increased in the early LGM, (2) caribou populations may have risen after the LGM – around 17,000–15,000 BP – when other taxa were declining, and (3) muskox were not very common in the later Pleistocene.

Fossil animal trackways may yield clues to megafaunal population size and health. At St. Mary reservoir in southern Alberta, McNeil et al. (2005) analyzed mammoth tracks found in a 1.5 m thick wind-deposited sediment dated around 11,300–11,000 BP. The area had once been a grassy plain, according to the interpretations of root etching on bones found in the sediments. The mammoth track sample was about 500 "tracks and trackways" (McNeil et al., 2005: 1255), of which only 51 could be measured. The footprint sizes were categorized into four "age classes" based on size, with 33% of the measured footprints appearing to belong to animals under 12 years old. This percentage struck the authors as too low a proportion for a healthy population (McNeil et al., 2005). The size distribution indicated to the authors that a declining population left the tracks. Perhaps one could balk at accepting this conclusion from such a tiny sample, but while the study is certainly not decisive it is valuably suggestive, in spite of the potential stochasticity/sampling limitations that the authors briefly discuss in the article. The locality, which also yielded possible Clovis-killed horses (Kooyman et al., 2001; Kooyman et al. 2006), may indeed contain evidence of declining megafaunal numbers in a habitat that should have supported healthy populations, at the same time as human foragers were present.

Migration Patterns

Isotopes and trace elements may reveal not only megafaunal diets but also migration patterns. Hoppe (2004) studied the mammoth bones from three Clovis-associated sites (Dent, CO, Miami, TX, and the Clovis type site Blackwater Locality Number 1, NM), to compare with one possible saber-tooth den in a cave (Friesenhahn Cave, TX) and another noncultural site where mammoths had died en masse during a natural disaster such as a flood (Waco, TX); she found high variability in C, O, and Strontium in the tooth enamel bioapatite from the Clovis mammoths, indicating they were unrelated individuals who might have died at different times. She proposed that the Clovis site mammoths had come from subpopulations living in separate home ranges that did not mix and did not make long migratory movements. The mammoths of the Southern High Plains did not contact the mammoths of the central High Plains.

A Fungus that Lives in Megafauna Dung

Another fairly recent development in reconstructing late Pleistocene ecologies is the discovery that a fungal spore which is often preserved in old sediments has the potential for indicating the presence (and relative density) of megafauna when no more direct evidence can be found, such as fossil bones. This spore, from the *Sporormiella* fungus, has been found where large-mammal dung is deposited in recent sites such as waterholes frequented by cattle (Burney et al., 2003; Davis 1987; Davis and Shafer 2006) and it is likely that the spore's presence in ancient sediments also indicates that large-mammal dung was locally abundant. A study in southeastern New York (Robinson et al., 2005) revealed an abrupt decline in spore density, presumably indicative of a reduced megafaunal population about 1,000–1,500 years before fluted-point-making peoples appeared in the northeastern United States. The authors connect the fungal spore and inferred megafaunal decline to the initial impact of cryptic human hunters, although it also corresponds in time to Bølling climatic warming. After the fungal spore decrease, charcoal particles increase in the same pollen cores, interpreted by Robinson et al. (2005) as human-induced burning of vegetation which had become more abundant after the herbivore collapse. However, final extinction of proboscideans in this region only occurred later, about 11,000 BP, as shown by direct dates on mastodont bones.

A Bolide Impact

A dissatisfaction with the main theories of extinction – human hunting, climate changes, hyperdisease or other ("grassy knoll") theories (see Haynes, 2002b: 235) – may lead researchers to seek other potential causative factors. One cause for the extinctions has been proposed by Firestone and Topping (2001), Firestone and West (2005), Firestone et al. (2007a, b), namely extraterrestrial impact. The impact point of the hypothesized comet or meteor has not been identified.

One of the authors of the studies, R. Firestone, had earlier gained some notoriety for co-authoring a speculative paper that proposed our use and understanding of radiocarbon dating in the late Pleistocene must be wrong for reasons that involve extraterrestrial processes (Firestone and Topping, 2001).

Firestone and colleagues argue that several lines of evidence point to the extinctions resulting from the impact of a 10-km wide comet hitting North America around 12,900 cal bp, or alternatively a supernova's ejecta impacting earth at that time after its much earlier explosion. The evidence consists of (1) abundant rounded and magnetic particles in mammoth-bearing sediments of Clovis-era sites; (2) embedded micrometeorites in cherts from seven Clovis sites; and (3) elemental proportions in the magnetic particles that are similar from all sites (suggesting a single source) but different from earth's crustal or solar abundances. An impact point from a comet or meteor has not been identified but the meeting with earth may have happened in northeastern North America. Iron-rich grains also had impacted mammoth tusks from Alaska and Siberia about 34,000 BP, at the same time that a 175% peak in radiocarbon occurred in earth sediments, perhaps caused by a supernova that had exploded 7,000 years earlier, and apparently setting the precedent for a later supernova/comet whose impacts were felt at the Clovis era. Note that studies such as at 49 ka Barringer Meteorite Crater in east-central Arizona show local effects of such impacts must have been serious, but at a distance beyond a few hundred kilometers from the impact point any extirpations of plants and animals or other perturbations in the fossil record have not been detected.

Some arguments immediately come to mind in opposition to the bolide theory (S. Fiedel, 2007, personal communication):

1. In spite of what the comet theorists claim, Paleoindians thrived after the time of the supposed comet impact: Clovis culture was transformed into Folsom, Dalton, and Eastern US variants, and all of these are much more numerous than Clovis, suggesting a human population increase, not collapse.
2. Megafauna including giant sloths were wiped out as far south as Florida by 12,900 years ago, but medium-sized sloths in the Caribbean islands (including Cuba) survived thousands of years longer, which is hard to explain.
3. South American megafauna survived until at least 12,500 years ago, and actually probably longer—why didn't they suffer a synchronous collapse with the North American animals?
4. Bison and grizzly survived the so-called impact event in North America, and elk and moose entered North America at around the time of the so-called impact – how could these species be immune to such an important event?
5. Mammoths survived on Wrangel Island in the Russian Arctic and probably in pockets in Siberia well past the date of the supposed comet; why wouldn't an impact in the eastern Arctic have wiped them out, too?
6. The Younger Dryas cold period (supposedly started by the comet impact) ended at 11,590 years ago even more abruptly than it began; why doesn't this require another hypothetical impact? If it doesn't, then why does the onset call for an extraterrestrial trigger?

A bolide impact as close as Hudson Bay cannot cleanly account for later extinctions elsewhere in the world where the events are similarly abrupt and track human arrivals but at different times. The bolide hypothesis does have some explanatory value, but is not perfect; for example, the South American extinctions lagged in time by a millennium or so, which the hypothesis does not explain, and the early Holocene extinctions in northern Eurasia are also not so neatly explained. Ironically, the idea of earth's fauna being wiped out by an extraterrestrial object brings to mind the flip-side concept that all life on earth actually originated from bacteria and viruses carried by comets or meteorites, a theory championed by Sir Fred Hoyle (who coined the term "Big Bang") (Hoyle and Wickramasinghe 1978).

The Commentary Papers

A section of the recent literature appears by design to be neutral, usually indicated by the tactic of favoring multicausal explanations – even when written by authors identified for the most part with one of the major competing theories.

An essay by Brook and Bowman (2004) is Australia-centered but the work is relevant because their review of dating and underlying assumptions would lead to a conclusion that extinctions are not proven to be due to Overkill in the Americas as in Australia. The authors criticize several aspects of the well-known and oft-cited Alroy (1999, 2001) simulations because of certain assumptions, such as human hunting was done only for subsistence and not for social display reasons, or that human population density was dependent on prey density. The idea of prey naiveté is also considered weak, since it has not been proven whether or not all prey animals are effectively hardwired to be vigilant or to remain naïve when encountering a strange new predator; also criticized is the interpretation that humans and megafauna never co-existed for a long interval before extinctions (in Australia). The authors propose a probability of overlap, based on simulations of "radiometrically instantaneous intervals" (determined by seeing if mean dated ages of archeological sites and extinct fauna are separated by at least three times the sum of standard deviations [Webb, 1998], which they're not). Since the 2004 article, Brook and Bowman (2005; Brook et al., 2007) have apparently become even stronger advocates of Overkill.

Koch and Barnosky (2006) observe correctly that the results of the literature's various simulations have varied quite a bit because of different inputs. They distinguish the main climate-change explanations as being of three types (as I've described above): the keystone species concept of Owen-Smith, which is weak because the most extreme keystone species – mammoths and mastodonts – did not die first but appear to be the last of the megafauna to die out (Graham et al., 2002); the coevolutionary disequilibrium idea of Graham and Lundelius (1984), which is also weak because the last glacial-interglacial transition was not so much more severe than the others the megafauna had lived through; and the mosaic-nutrient theory of Guthrie (1984), which also does not adequately explain why the animals had survived so many earlier changes in forage ecology. Koch and Barnosky (2006: 73–74) see the human involvement case as much stronger now than in earlier years, but their ultimate judgment is that climate-change exacerbated the effects of human hunting and therefore it must be considered the decisive factor. I would note that Barnosky et al. (2004) prominently give equal weight to the opposite view – that because earlier climate stresses were survivable, the ultimate decision must be that human hunting exacerbated climate stresses, and hence (philosophically) it is ultimately to blame. Koch and Barnosky (2006) also mention possible "threshold effects" where ecosystem states become unstable even before changes are fully expressed. Such hard-to-measure "susceptibility to collapse" is speculated but I wonder if it is quantifiably known anywhere in the world at the ecosystem level. The authors mention the recent results of isotopic studies of megafaunal diets, tests of chronology, which are sometimes clear and sometimes not, and archeological tests. The infamous lack of kill sites does not disprove Overkill, in their view, which is a fair and scientifically valid assessment of the nature of evidence; even a small hunting input would have been serious stress to late Pleistocene megafaunal subpopulations, and kill sites need not be abundant. In their table 3 they show that only two taxa – *Mammut* and *Platygonus* – do not seem to be potentially associated often enough with human hunting to explain their extinction.

Koch and Barnosky (2006) cite the work of D. Fisher (Chapter 4, for example) – whose studies distinguish between resource stress and the effects of human hunting on American mastodonts (*Mammut americanum*) – to support the acceptance of a human factor in the extinctions, which other researchers refuse to believe. Koch and Barnosky (2006) predict that future work will deal with translating fossil abundance into estimates of real animal abundance (see Haynes, Chapter 3) and will involve far more dating.

Some of the data and assumptions in Koch and Barnosky (2006) can be corrected since the paper was published – such as the mistaken interpretation that some megafauna (horse and mammoth) do not overlap in time with humans in Beringia; we now know they did (Solow et al., 2006; C. Holmes, 2006, personal communication; Buck and Bard, 2007). It is also a theme of dispute that prey naiveté is not a supportable concept capable of explaining how humans might have hunted them to extinction; in fact, megafaunal defense mechanisms do vary among members of the same species, seen for example in Africa – such as in the learned responses of some elephant groups exercising extra vigilance against two-legged human predators and a heightened wariness about allowing the youngest animals to feed too far apart from adults. Humans with spears (or rifles) can create serious wounds in big animals from an unseen distance or while hiding, while the other predators with four legs must latch on to the prey animal to hurt it or bring it down; hence, wariness against lions might require a fairly large separation from them for elephants to feel safe, while with humans the distance of spatial separation might have been allowed to be much smaller. The relatively wide spatial dispersal of a feeding mammoth herd might have been advantageous for humans to approach very closely but the animals never would have allowed other (more familiar) carnivores to do so. Clues to the implications of such differential discrimination by megafauna can be seen nowadays with horseback tours where rhinos and other animals do not spook when humans on horse approach closely but do run to escape when they perceive humans on foot approaching from a much greater distance.

Barnosky et al. (2004) wrote a shorter paper for *Science* which summarized most of the arguments and conclusions later provided by Koch and Barnosky (2006). The widely read article states that in North America climate changes plus human effects led to the extinctions, but in the southern hemisphere (Australia and South America) the causes are still very unclear.

Other recent papers that agree about the additive effects of human hunting and climate change include Burney and Flannery (2005) (and see the response by Wroe et al., 2006) which points out the worldwide association of first humans and the extinctions of the biggest members of each animal family or order on different landmasses. The authors evaluate different possible mechanisms such as hunting, climate changes, habitat changes, and fire, before concluding that they all might have had an effect.

My last example of a recent review paper that is ostensibly neutral about the causes of extinction actually says little about them but propagates some important errors in its (mis) understanding of the archeology that contributes to the irresolution about extinctions. Klein and Schiffner (2003) are social historians who reviewed the debate about the origins of the first American people, but they make the mistake of assuming that a pre-Clovis "consensus" has been reached among archeologists and that there is a pre-LGM dating of the first immigrants from Asia. It is no wonder that the public is confused, professional archeologists are baffled, and interested nonspecialists cannot make heads or tails of where the debate is heading, when the literature of prejudice and bias is read as if it were the literature of proof.

Conclusion

A much larger literature exists than what I have been able to survey here, and each contribution to the debate may offer a new perspective but also sometimes little more than a new opinion based on personal preferences. The debate is fascinating, seemingly endless, and frustrating, all at once. I do not conclude this chapter by deciding which explanation for the extinctions is correct. No one can do that, and no one can decide which explanation is incorrect, either.

Note

1. The chonometric ages in this book are reported in three ways: the notation ka (kiloannum) means "thousands of years before now." The notation BP refers to uncalibrated (radiocarbon) years ago, and is used when the age estimates are derived from radiocarbon dates using 1950 CE (Common Era, equivalent to AD dates) as the "present." The notation cal bp refers to calibrated radiocarbon years, using 1950 CE as the "present." Another notation, kyr, refers to the duration of an event or process, and not to its age.

References

Aaris-Sørensen K, Mühldorff R, Petersen EB (2007) The Scandinavian reindeer (*Rangifer tarandus* L.) after the Last Glacial Maximum: Time, seasonality and human exploitation. J Arch Sci 34: 914–923

Adovasio JM, Page J (2002) The First Americans: In pursuit of archaeology's greatest mystery. Random House, New York

Agenbroad LD (2005) North American proboscideans: Mammoths: The state of knowledge, 2003. Quat Intl 126–128:73–92

Alroy J (1999) Putting North America's end-Pleistocene megafaunal extinction in context: Large scale analyses of spatial patterns, extinction rates, and size distributions. In: MacPhee RDE (ed) Extinctions in near time: Causes, contexts, and consequences. Plenum, New York, pp 105–143

Alroy J (2001) A multi-species Overkill simulation of the end-Pleistocene megafaunal mass extinction. Science 292:1893–1896

Ambrose SH (1998) Late Pleistocene human population bottlenecks, volcanic winter, and the differentiation of modern humans. J Human Evol 34:623–651

Amos J (2005) Climate 'key to beasts' demise.' BBC News online, http://news.bbc.co.uk/2/hi/science/nature/4594793.stm, accessed 31 May, 2004

Arroyo-Cabrales J, Polaco OJ, Johnson E (2006) A preliminary view of the coexistence of mammoth and early peoples in Mexico. Quat Intl 142–143:79–86

Barnes I, Matheus P, Shapiro B, Jensen D, Cooper A (2002) Dynamics of Pleistocene population extinctions in Beringian brown bears. Science 295:2267–2270

Barnosky AD, Koch PL, Feranec RS, Wing SL, Shabel AB (2004) Assessing the causes of late Pleistocene extinctions on continents. Science 306:70–75

Bird DW, Bliege Bird R (2000) The ethnoarchaeology of juvenile foragers: Shellfishing strategies among the Meriam children. J Anthropol Archaeol 19:461–476

Bliege Bird R, Bird DW, Smith EA, Kushnick GC (2002) Risk and reciprocity in Meriam food sharing. Evol Hum Behav 23:297–321

Bradley B, Stanford D (2004) The North Atlantic ice-edge corridor: A possible Paleolithic route to the New World. World Archaeol 36:459–478

Brook BW, Bowman DMJS (2004) The uncertain blitzkrieg of Pleistocene megafauna. J Biogeograp 31:517–523

Brook BW, Bowman DMJS (2005) One equation fits Overkill: Why allometry underpins both prehistoric and modern body size-biased extinctions. Popul Ecol 47:137–141

Brook BW, Bowman DMJS, Burney DA, Flannery TF, Gagan MK, Gillespie R, Johnson CN, Kershaw P, Magee JW, Martin PS, Miller GH, Peiser B, Roberts RG (2007) Would the Australian megafauna have become extinct if humans had never colonized the continent? Comments on "A review of the evidence for a human role in the extinction of Australian megafauna and an alternative explanation" by S Wroe and J Field. Quat Sci Rev 26:560–564

Buck CE, Bard E (2007) A calendar chronology for Pleistocene mammoth and horse extinction in North America based on Bayesian radiocarbon calibration. Quat Sci Rev 26:2031–2035

Bulte E, Horan RD, Shogren JF (2006) Megafauna extinction: A paleoeconomic theory of human Overkill in the Pleistocene. J Econ Behav Org 59:297–323

Burney DA, Flannery TF (2005) Fifty millennia of catastrophic extinctions after human contact. Trends Ecol Evol 20(7):395–401

Burney DA, Robinson GS, Burney LP (2003) *Sporormiella* and the late Holocene extinctions in Madagascar. Proc Natl Acad Sci 100(19):10800–10805

Byers DA, Ugan A (2005) Should we expect large game specialization in the late Pleistocene? An optimal foraging perspective on Early Paleoindian prey choice. J Archaeol Sci 32:1624–1640

Cannon MD, Meltzer DJ (2004) Paleoindian foraging: Examining the faunal evidence for large mammal specialization and regional variability in prey choice. Quat Sci Rev 23:1955–1987

Cione AL, Tonni EP, Soibelzon LH (2003) The broken zig-zag: Late Cenozoic large mammal and turtle extinction in South America. Revista del Museo Argentino de Ciencias Naturales "Bernardino Rivadavia" 5:1–19

Clague JJ, Mathewes RW, Ager T (2004) Environments of northwestern North America before the Last Glacial Maximum. In: Madsen DB (ed) Entering America: Northeast Asia and Beringia before the Last Glacial Maximum. University of Utah Press, Salt Lake City, pp 63–96

Collins M (2002) The Gault Site, Texas, and Clovis research. Athena Rev 3(2):24–36

Coltrain JB, Harris JM, Cerling TE, Ehleringer JR, Dearing M-D, Ward J, Allen J (2004) Rancho La Brea stable isotope biogeochemistry and its implications for the palaeoecology of late Pleistocene coastal southern California. Palaeogeogr, Palaeocl, Palaeoecol 205:199–219

Creel S, Christianson D, Liley S, Winnie JA Jr (2007) Predation risk affects reproductive physiology and demography of elk. Science 315(5814):960

Cupper ML, Duncan J (2006) Late Glacial megafaunal death assemblage and early human occupation at Lake Menindee, southeastern Australia. Quat Res 66:332–341

Davis OK (1987) Spores of the dung fungus *Sporormiella*: Increased abundance in historic sediments and before Pleistocene megafaunal extinction. Quart Res 28:290–294

Davis OK, Shafer DS (2006) *Sporormiella* fungal spores, a palynological means of detecting herbivore density. Palaeogeogr, Palaeocl, Palaeoecol 237(1):40–50

Deloria V (1969) Custer died for your sins: An Indian manifesto. Macmillan, New York

Diniz-Filho JA (2004) Macroecological analyses support an Overkill scenario for late Pleistocene extinctions. Brazil J Biol 64(3A):407–414

Drummond AJ, Rambaut A, Shapiro B, Pybus OG (2005) Bayesian coalescent inference of past population dynamics from molecular sequences. Mol Biol Evol 22(5):1185–1192
Duk-Rodkin A, Hughes OL (1991) Age relationships of Laurentide and montane glaciations, Mackenzie Mountains, Northwest Territories. Géographie Physique et Quaternaire 45:79–90
Dyke AS, Moore A, Robertson L (2003) Deglaciation of North America. Geological Survey of Canada Open File 1574
Dyke A (2004) An outline of North American deglaciation with emphasis on central and northern Canada. In: Ehlers J, Gibbard PL (eds) Quaternary glaciations – extent and chronology, Part II, vol. 2b. Elsevier, Amsterdam, pp 373–424
Fedje DW, Mackie Q, Dixon EJ, Heaton TH (2004) Late Wisconsin environments and archaeological visibility on the northern Northwest Coast. In: Madsen DB (ed) Entering America: Northeast Asia and Beringia before the Last Glacial Maximum. University of Utah Press, Salt Lake City, pp 97–138
Feranec RS (2003) Stable isotopes, hypsodonty, and the paleodiet of *Hemiauchenia* (mammalian: Camelidae): A morphological specialization creating ecological generalization. Paleobiology 29(2):230–242
Feranec RS (2004) Geographic variation in the diet of hypsodont herbivores from the Rancholabrean of Florida. Palaeogeogr, Palaeocl, Palaeoecol 207(3–4):359–369
Fiedel SJ (1999) Older than we thought: Implications of corrected dates for Paleoindians. Am Ant 64(1):95–116
Fiedel SJ (2006) Points in time: Establishing a precise hemispheric chronology for Paleoindian migrations. In: Morrow JE, Gnecco C (eds) Paleoindian archaeology, a hemispheric perspective. University Press of Florida, Gainesville, pp 21–43
Firestone RB, Topping W (2001) Terrestrial evidence of a nuclear catastrophe in Paleoindian times. Mammoth Trumpet 16(2):9–16
Firestone RB, West A (2005) Evidence for the extinction of mammoths by an extraterrestrial impact event. Paper presented at Second International World of Elephants Congress, Sept. 22–25, Hot Springs, SD
Firestone RB, West A, Kennett JP, Becker L, Bunch TE, Revay ZS, Schultz PH, Belgya T, Kennett DJ, Erlandson JM, Dickenson OJ, Goodyear AC, Harris RS, Howard GA, Kloosterman JB, Lechler P, Mayewski PA, Montgomery J, Poreda R, Darrah T, Que Hee SS, Smith AR, Stich A, Topping W, Wittke JH, Wolbach WS (2007a) Evidence for an extraterrestrial impact 12,900 years ago that contributed to the megafaunal extinctions and the Younger Dryas cooling. Proc Natl Acad Sci USA 104:16016–16021
Firestone RB, West A, Revay Z, Belgya T, Smith A, Que Hee SS (2007b) Evidence for a massive extraterrestrial airburst over North America 12.9 ka Ago. Eos Trans. AGU, 88(23), Jt. Assem. Suppl., Abstract PP41A-01
Flannery TF (1995) The future eaters: An ecological history of the Australasian lands and people. G. Braziller, New York
Flannery TF (2001) The eternal frontier: An ecological history of North America and its people. Grove Press, New York
Foreman D (2004) Rewilding North America: A vision for conservation in the 21st Century. Island Press, Washington
Fox-Dobbs K, Stidham TA, Bowen GJ, Emslie SD, Koch PL (2006) Dietary controls on extinction versus survival among avian megafauna in the late Pleistocene. Geology 34(8):685–688
Gillespie R (2002) Dating the first Australians. Radiocarbon 44(2):455–472
Goodyear AC, Steffy K, Pickering AD, Sweeney KB (2005) Clovis in Allendale: the Topper and Big Pine Tree sites. Paper presented at the Clovis in the Southeast Conference, Columbia, SC, October 26–29
Graham RW (2001) Triggers, thresholds, Russian roulette, and Pleistocene extinctions. Paper number 122–0, presented at the Annual Meeting of the Geological Society of America, Boston, MA, November 1–10
Graham RW (2003) Environmental models for late Pleistocene megafaunal extinctions and MIS 5E. Paper presented at the XVI INQUA Congress, 23–30 July, 2003, Reno, NV. Abstract published in XVI INQUA Congress Programs with Abstracts, p 89
Graham RW (2006a) Late Pleistocene environments: Shifting resources and ecological crisis. Paper presented at the Annual Meeting of the American Association for the Advancement of Science, St. Louis, MO, February 16–20
Graham RW (2006b) Niche partitioning in Late Pleistocene proboscideans in North America north of Mexico. American Quaternary Association Program and Abstracts of the 19th Biennial Meeting, Aug. 17–20, Bozeman, MT, pp 89–90
Graham RW, Lundelius EL Jr (1984) Coevolutionary disequilibrium and Pleistocene extinctions. In: Martin PS, Klein RG (eds) Quaternary extinctions: A prehistoric revolution. University of Arizona Press, Tucson, pp 211–222
Graham R, Stafford T, Lundelius E, Semken H, Southon J (2002) C-14 Chronostratigraphy and litho-stratigraphy of Late Pleistocene megafauna extinctions in the New World. Paper presented at 67th annual meeting of Society for American Archaeology, Denver, CO, Mar. 20–24
Grant J, Hopcraft C, Sinclair ARE, Packer C (2005) Planning for success: Serengeti lions seek prey accessibility rather than abundance. J Animal Ecol 74:559–566
Grayson DK (2001) The archaeological record of human impacts on animal populations. J World Prehist 15:1–67
Grayson DK (2007) Deciphering North American Pleistocene extinctions. J Anthropol Res 63(2):185–214
Grayson DK, Meltzer DJ (2002) Clovis hunting and large mammal extinction: A critical review of the evidence. J World Prehist 16:313–359
Grayson DK, Meltzer, DJ (2003) A requiem for North American Overkill. J Archaeol Sci 30:585–593
Guthrie RD (1984) Mosaics, alleochemics, and nutrients: An ecological theory of late Pleistocene megafaunal extinctions. In: Martin PS, Klein RG (eds) Quaternary extinctions: A prehistoric revolution. University of Arizona Press, Tucson, pp 259–298
Guthrie RD (2003) Rapid body size decline in Alaskan Pleistocene horses before extinction. Nature 426:169–171
Guthrie RD (2006) New carbon dates link climatic change with human colonization and Pleistocene extinctions. Nature 441:207–209
Haynes G (2002a) The catastrophic extinction of North American mammoths and mastodonts. World Archaeol 33:391–416
Haynes G (2002b) The early settlement of North America: The Clovis era. Cambridge University Press, Cambridge
Haynes G (2004) Clovis megafauna-hunting: Opportunistic, not always optimal, and a better explanation for extinctions. Paper presented at the 69th Annual Meeting of the Society for American Archaeology, Mar. 31–Apr. 4, Montreal, Quebec
Haynes G (2005) Mammoth landscapes: Good country for hunter-gatherers. In: Storer J (ed) Proceedings of the 3rd International

Mammoth Conference, May 23–26, 2003, Dawson City, Yukon. Quat Intl 142–143:20–29

Holen SR (2006) Taphonomy of two Last Glacial Maximum mammoth sites in the central Great Plains of North America: A preliminary report on La Sena and Lovewell. Quat Intl 142–143:30–43

Hoppe KA (2004) Late Pleistocene mammoth herd structure, migration patterns, and Clovis hunting strategies inferred from isotopic analyses of multiple death assemblages. Paleobiology 30(1):129–145

Hoppe KA, Amundson R, Vavra M, McClaran MP, Anderson DL (2004) Isotopic analysis of tooth enamel carbonate from modern North American feral horses: Implications for paleoenvironmental reconstructions. Palaeogeogr, Palaeocl, Palaeoecol 203:299–311

Hoyle F, Wickramasinghe NC (1978) Lifecloud – The origin of life in the universe. J. M. Dent & Sons, London

Irving WN (1985) Context and chronology of Early Man in the Americas. Annual Rev Anthropol 14:529–555

Jobling MA, Hurles ME, Tyler-Smith C (2004) Human evolutionary genetics: Origins, peoples & diseases. Garland Science, New York

Johnson CN (2002) Determinants of loss of mammal species during the Late Quaternary 'megafauna' extinctions: Life history and ecology, but not body size. Proc Roy Soc London B 269:2221–2227

Kay C (1994) Aboriginal Overkill: The role of Native Americans in structuring western ecosystems. Human Nature 5(4):359–398

Kay C (2002) False gods, ecological myths, and biological reality. In: Kay C, Simmons R (eds) Wilderness and political ecology: Aboriginal influences and the original state of nature. University of Utah Press, Salt Lake City, pp 238–261

Kelly RL, Todd LC (1988) Coming into the country: Early Paleoindian hunting and mobility. Am Ant 53(2):231–244

Kiltie RA (1984) Seasonality, gestation time, and large mammal extinctions. In: Martin PS, Klein RG (eds) Quaternary extinctions: A prehistoric revolution. University of Arizona Press, Tucson, pp 299–314

Klein HS, Schiffner DC (2003) The current debate about the origins of the Paleoindians of America. J Social Hist 37: 483–492

Koch PL, Barnosky AD (2006) Late Quaternary extinctions: State of the debate. Annual Rev Ecol, Evol Syst 37:215–260

Kooyman B, Newman ME, Cluney C, Lobb M, Tolman S, McNeil P, Hills LV (2001) Identification of horse exploitation by Clovis hunters based on protein analysis. Am Ant 66:686–691

Kooyman B, Hills LV, McNeil P, Tolman S (2006) Late Pleistocene horse hunting at the Wally's Beach site (DhPg-8), Canada. Am Ant 71(1):101–122

Krieger AD (1964) Early Man in the New World. In: Jennings JD, Norbeck E (eds) Prehistoric man in the New World. University of Chicago Press, Chicago, pp 23–81

Leakey LSB, Simpson RD, Clements T (1968) Archaeological excavations in the Calico Mountains, California: Preliminary Report. Science 160(3831):1022–1023

Lepper K, Fisher TG, Hajdas I, Lowell TV (2007) Ages for the Big Stone Moraine and the oldest beaches of glacial Lake Agassiz: Implications for deglaciation chronology. Geology 35(7):667–670

Lister AM (1989) Rapid dwarfing of red deer on Jersey in the Last Interglacial. Nature 342:539–542

Lyons SK, Smith FA, Brown JH (2004) Of mice, mastodons, and men: Human-mediated extinctions on four continents. Evolut Ecol Res 6:339–358

MacDonald GM, McLeod TK (1996) The Holocene closing of the 'ice-free' corridor: A biogeographical perspective. Quat Intl 32:87–95

MacNeish RS (1976) Early Man in the New World. Am Scientist 63:316–327

MacNeish RS (1982) A late commentary on an early subject. In: Ericson JE, Taylor RE, Berger R (eds) Peopling of the New World, pp 311–315. Ballena Press. Anthropological Papers 23

MacPhee RDE (ed) (1999). Extinctions in near time: Causes, contexts, and consequences. Kluwer/Plenum, New York

MacPhee RDE, Greenwood AD (2007) Continuity and change in the extinction dynamics of late Quaternary muskox (*Ovibos*): Genetic and radiometric evidence. Bulletin of the Carnegie Museum of Natural History 39:203–212

Mandryk CAS (1992) Paleoecology as Contextual archaeology: Human viability of the Late Quaternary ice-free corridor, Alberta, Canada. Ph.D. Dissertation, University of Alberta

Mandryk CAS (1996) Late Wisconsinan deglaciation of Alberta: Processes and paleogeography. Quat Intl 32:79–85

Mandryk CAS, Josenhans H, Fedje DW, Mathewes RW (2001) Late Quaternary paleoenvironments of northwestern North America: Implications for inland versus coastal migration routes. Quat Sci Rev 20:301–314

Martin PS (1967) Pleistocene Overkill. Nat Hist 76:32–38

Martin PS (1973) The discovery of America. Science 179(4077): 969–974

Martin PS (1984) Prehistoric Overkill: The global model. In: Martin PS, Klein RG (eds) Quaternary extinctions: A prehistoric revolution. University of Arizona Press, Tucson, pp 354–403

Martin PS, Steadman DW (1999) Prehistoric extinctions on islands and continents, In: MacPhee RDE (ed) Extinctions in near time. Kluwer/Plenum, New York, pp 17–55

Matheus P, Guthrie RD, Kunz ML (2003a) Predator-prey links in Pleistocene East Beringia: Evidence from stable isotopes. Third International Mammoth Conference, 2003: Program and Abstracts: 80. Occasional Papers in Earth Sciences No. 5, Palaeontology Program, Government of the Yukon

Matheus P, Guthrie RD, Kunz ML (2003b) Isotope ecology of Late Quaternary megafauna in eastern Beringia. Third International Mammoth Conference, 2003: Program and Abstracts: 80–82. Occasional Papers in Earth Sciences No. 5, Palaeontology Program, Government of the Yukon

Matheus P, Kunz ML, Guthrie RD (2003c) Using frequency distributions of radiocarbon dates to detect relative changes in Pleistocene mammal populations – a test case from northern Alaska. Third International Mammoth Conference, 2003: Program and Abstracts: 83–84. Occasional Papers in Earth Sciences No. 5, Palaeontology Program, Government of the Yukon

McAvoy JM, McAvoy LD (eds) (1997) Archaeological investigations of site 44SX202, Cactus Hill, Sussex County, Virginia. Virginia Department of Historic Resources, Research Report Series Number 8. Richmond

McNeil P, Hills LV, Kooyman B, Tolman SM (2005) Mammoth tracks indicate a declining late Pleistocene population in southwestern Alberta, Canada, Quat Sci Rev 24:1253–1259

Meltzer DJ (1997) Monte Verde and the Pleistocene peopling of the Americas. Science 276:754–755

Meltzer DJ (2003) Lessons in landscape learning. In: Rockman M, Steele J (eds) Colonization of unfamiliar landscapes: The archaeology of adaptation. Routledge, London, pp 222–241

Meltzer DJ (2004) Modeling the initial colonization of the Americas: Issues of scale, demography, and landscape learning. In: Barton CM, Clark GA, Yesner DR, Pearson GA (eds) The settlement of the American continents: A multidisciplinary approach to human biogeography. University of Arizona Press, Tucson, pp 123–137

Miller GH, Fogel ML, Magee JW, Gagan MK, Clarke SJ, Johnson BJ (2005) Ecosystem collapse in Pleistocene Australia and a human role in megafaunal extinction. Science 309:287–290

Miller GH, Magee JW, Johnson BJ, Fogel ML, Spooner NA, McCulloch MT, Ayliffe LK (1999) Pleistocene extinction of *Genyornis newtoni*: Human impact on Australian megafauna. Science 283:205–208

Mithen S (1993) Simulating mammoth hunting and extinction: Implications for the Late Pleistocene of the Central Russian Plain. In: Peterkin GL, Bricker HM, Mellars P (eds) Hunting and animal exploitation in the later Palaeolithic and Mesolithic of Eurasia. Anthropol Pap Am Anthropol Assoc 4:163–178

Morlan RE (1980) Taphonomy and archaeology in the Upper Pleistocene of the Northern Yukon Territory: A glimpse of the peopling of the New World. National Museum of Man [Canada] Mercury Series, Archaeological Survey of Canada No. 94

Mosimann JE, Martin PS (1975) Simulating Overkill by Paleoindians. Am Scientist 63:304–313

Newman MEJ, Palmer RG (2003) Modeling extinction. Oxford University Press, New York

O'Brien SJ (2005) Tears of the cheetah, and other tales from the genetic frontier. St. Martin's Press, New York

Owen-Smith N (1987) Pleistocene extinctions: The pivotal role of megaherbivores. Paleobiology 13:351–362

Peierls R (1960) Wolfgang Ernst Pauli, 1900–1958. Biograp Mem Fellows Roy Soc 5:174–192

Petraglia M, Korisettar R, Boivin N, Clarkson C, Ditchfield P, Jones S, Koshy J, Lahr MM, Oppenheimer C, Pyle D, Roberts R, Schwenninger J-L, Arnold L, White K (2007) Middle Paleolithic assemblages from the Indian subcontinent before and after the Toba super-eruption. Science 317:114–116

Powell JF (2005) The first Americans: Race, evolution, and the origin of Native Americans. Cambridge University Press, Cambridge

Prideaux GJ, Roberts RG, Megirian D, Westaway KE, Hellstrom JC, Olley JM (2007a) Mammalian responses to Pleistocene climate change in southeastern Australia. Geology 35(1):33–36

Prideaux GJ, Long JA, Ayliffe LK, Hellstrom JC, Pillans B, Boles WE, Hutchinson MN, Roberts RG, Cupper ML, Arnold LJ, Devine PD, Warburton NM (2007b) An arid-adapted middle Pleistocene vertebrate fauna from south-central Australia. Nature 445:422–425

Roberts RG, Flannery TF, Ayliffe LK, Yoshida H, Olley JM, Prideaux GJ, Laslett GM, Baynes A, Smith MA, Jones R, Smith BL (2001) New ages for the last Australian megafauna: Continent-wide extinction about 46,000 years ago. Science 292:1888–1892

Robinson G, Burney LP, Burney DA (2005) Landscape paleoecology and megafaunal extinction in southeastern New York. Ecol Monograp 75(23):295–315

Schurr TG (2004) The peopling of the New World: Perspectives from molecular anthropology. Ann Rev Anthropol 33:551–583

Schuster S, Schüle W (2000) Anthropogenic causes, mechanisms and effects of Upper Pliocene and Quaternary extinctions of large vertebrates. Oxford J Archaeol 19(3):223–239

Shapiro B, Drummond AJ, Rambaut A, Wilson MC, Matheus PE, Sher AV, Pybus OG, Gilbert MTP, Barnes I, Binladen J, Willerslev E, Hansen AJ, Baryshnikov GF, Burns JA, Davydov S, Driver JC, Froese DG, Harington CR, Keddie G, Kosintsev P, Kunz ML, Martin LD, Stephenson, RO, Storer J, Tedford R, Zimov S, Cooper A (2004). Rise and fall of the Beringian steppe bison. Science 306:1561–1565

Solé RV, Alonso D, McKane A (2002) Self-organized instability in complex ecosystems. Philos Trans Roy Soc S B 357:667–681

Solow AR, Roberts DL, Robbirt KM (2006) On the Pleistocene extinctions of Alaskan mammoths and horses. Proc Natl Acad Sci USA 103(19):7351–7353

Stanford D, Bradley B (2002) Ocean trails and prairie paths? Thoughts about Clovis origins. In: Jablonski NG (ed) The First Americans: The Pleistocene colonization of the New World. Memoirs No. 27. California Academy of Sciences, San Francisco, pp 255–271

Steadman DW (2007) Extinction and biogeography of tropical Pacific birds. University of Chicago Press, Chicago

Steadman DW, Martin PS (2003) The Late Quaternary extinction and future resurrection of birds on Pacific Islands. Earth-Sci Rev 61:133–147

Steadman DW, Pregill GK, Burley DV (2002) Rapid prehistoric extinction of birds and iguanas in Polynesia. Proc Natl Acad Sci USA 99:3673–3677

Straus L, Meltzer DJ, Goebel T (2005) Ice age Atlantis? Exploring the Solutrean-Clovis connection. World Archaeol 37(4):507–532

Stuart AJ, Kosintsev PA, Higham TFG, Lister AM (2004) Pleistocene to Holocene extinction dynamics in giant deer and woolly mammoth. Nature 431:684–689

Surovell TA, Waguespack NM, Brantingham PJ (2005) Global archaeological evidence for proboscidean Overkill. Proc Natl Acad Sci 102:6231–6236

Tamm E, Kivisild T, Reidla M, Metspalu M, Smith DG, Mulligan CJ, Bravi CM, Rickards O, Martinez-Labarga C, Khusnutdinova EK, Fedorova SA, Golubenko MV, Stepanov VA, Gubina MA, Zhadanov SI, Ossipova LP, Damba L, Voevoda MI, Dipierri JE, Villems R, Malhi R (2007) Beringian standstill and spread of Native American founders. PLoS ONE 2(9):e829. doi:10.1371/journal.pone.0000829

Turner CG II, Scott GR (2007) The dentition of American Indians: Evolutionary results and demographic implications following colonization from Siberia. In: Henke W, Tattersall I (eds) Handbook of paleoanthropology, vol 3, Phylogeny of the Hominines. Springer-Verlag, Heidelberg, pp 1901–1941

Ugan A, Byers D (2007) Geographic and temporal trends in proboscidean and human radiocarbon histories during the Late Pleistocene. Quat Sci Rev 26:3058–3080

Vrba ES (1993) The pulse that produced us. Natural Hist 5/93:47–51

Ward PD (1997) The call of distant mammoths: Why the Ice Age mammals disappeared. Copernicus, New York

Webb RE (1998) Problems with radiometric 'time': Dating the initial human colonization of Sahul. Radiocarbon 40(2):749–758

Whitney-Smith E (1995) Pleistocene extinctions: The case of the aboricidal megaherbivores. Accessed 7 Dec. 2007, Electronic Publication http://www.well.com/user/elin/mstry.htm

Whittington SL, Dyke B (1984) Simulating Overkill: Experiment with the Mossiman and Martin model. In: Martin PS, Klein RG (eds) Quaternary extinctions: A prehistoric revolution. University of Arizona Press, Tucson, pp 451–465

Wilson MC, Burns JA (1999) Searching for the earliest Canadians: Wide corridors, narrow doorways, small windows. In: Bonnichsen R, Turnmire KL (eds) Ice Age people of North America: Environments, origins and adaptations. Oregon State University and Center for the Study of the First Americans, Corvallis, pp 213–248

Wroe S (2006) On little lizards and the big extinction blame game. Quat Australasia 23(1):8–12

Wroe S, Field J, Fullagar R, Jermin LS (2004) Megafaunal extinction in the late Quaternary and the global Overkill hypothesis. Alcheringa 28:291–331

Wroe S, Field J, Grayson DK (2006) Megafaunal extinction: Climate, humans and assumptions. Trends Ecol Evol 21(2):61–62

2. Sudden Deaths: The Chronology of Terminal Pleistocene Megafaunal Extinction

Stuart Fiedel*
Louis Berger Group
203 East Cary St, Suite 100
Richmond, VA 23219, USA
sfiedel@louisberger.com

Keywords Extinction process • North and South America • human impacts on megafauna

Terminal Pleistocene Extinction: Abrupt Event or Slow Process?

If we ever hope to ascertain the cause(s) of the extinction of North American megafauna at the end of the Pleistocene, a necessary first step is to establish the chronology of this occurrence. Was it an abrupt event, in which about 30 or more genera disappeared simultaneously within no more than several hundred years, or instead a long, drawn-out, gradual process, with each species dying out independently and asynchronously, over the course of millennia?

Those who advocate a vague climatic/environmental cause favor the latter gradual scenario; they recognize that if the extinctions were shown instead to be abrupt and synchronous, it would compel them to "attribute to the extinction 'event' .... speed and taxonomic breadth .... Once that is done, explanations of the extinctions must be structured to account for these assumed properties, whether those explanations focus on people, climate...or disease" (Grayson and Meltzer, 2002:347).

Grayson (1993:68) observed that by 1969 radiocarbon dating had established that mammoth, mastodont, camel, horse, tapir, and Shasta ground sloth became extinct between 12,000 and 10,000 BP.[1] However, he questioned the assumption that other genera had died out at the same time. "Most telling is the fact that that, although it was easy to place six of the genera between 12,000 and 10,000 years ago, it has been extremely difficult to place any more in that time span. In the more than 20 years that have passed since 1969 – when the last of the six was placed between 12,000 and 10,000 years ago – we have been able to date only three more – the sabertooth cat, the giant short-faced bear, and the stout-legged llama – to that 2,000-year slot. As a result, it is beginning to look more and more as if scientists dealing with the extinctions were too hasty in assuming that all 35 genera of mammals became extinct within a narrow 2,000-year time span. At the very least, if the extinctions did occur that quickly, we cannot show it, and it appears equally likely that the extinctions were spread over a much longer period of time, perhaps on the order of several thousand years longer."

Grayson was wrong on two counts. First, the 2,000-year extinction "slot" appears to have been even shorter than he allowed – less than 400 years, in fact. Second, only four years after these remarks were published, evidence was presented of the persistence of another eight genera until ca. 11,000 BP (these include the sloth *Megalonyx*, the peccaries *Mylohyus* [long-nosed] and *Platygonus* [flat-headed], the giant beaver *Castoroides,* the elk-moose *Cervalces*, the shrub ox *Euceratherium*, and Harlan's musk ox, *Bootherium*). As more dates accumulate (Table 2.1), it becomes increasingly evident that megafaunal extinction was an abrupt, very rapid event encompassing at least 17 genera and probably all 35 in North America as well as at least 37 genera in South America (Cione et al., 2003); that it was precisely coeval with rapid human colonization of the Americas (by Clovis people and their descendants); and it is best explained as a result of human hunting and ancillary activities.

Unfortunately, most of the crucial data for fine-tuning of the extinction chronology remain unpublished. Stafford and colleagues have obtained radiocarbon dates directly from dozens of bones of extinct megafauna, and presented these dates in graphic form in several oral presentations over the past decade (e.g., Graham, 1998; Graham et al., 1997, 2002),

* Address for correspondence: sfiedel@louisberger.com

G. Haynes (ed.), *American Megafaunal Extinctions at the End of the Pleistocene*, 21–37.

TABLE 2.1. Youngest published credible dates BP for extinct megafauna.

Species	Date BP	Material dated	Site	Reference
Mammuthus columbi	10,930 ± 100 (TO-8514)	Bone	Bindloss Gravel Pit, Alberta	Hills and Harington, 2003
Mammuthus columbi	10,960 ± 30 (UCIAMS-11341)	Bone (Clovis artifacts associated)	Domebo, Oklahoma	Waters and Stafford, 2007
Mammuthus columbi	10,710 ± 130 (UCIAMS-11344), 11,110 ± 40 (UCIAMS-11345)	Bone (Clovis artifacts associated)	Lange-Ferguson, South Dakota	Waters and Stafford, 2007
Mammuthus columbi	10,790 ± 30 (UCIAMS-11342), 10,950 ± 30 (UCIAMS-11343), 10,864 ± 141 (SMU-254)	Bone (Clovis artifacts associated)	Colby, Wyoming	Waters and Stafford, 2007
Mammuthus columbi	11,065 ± 35 (UCIAMS-11339), 10,940 ± 30 (UCIAMS-11340)	Bone (Clovis artifacts associated)	Dent, Colorado	Waters and Stafford, 2007
Mammuthus exilis	11,030 ± 50 (CAMS-71697)	Bone	Santa Rosa Island, California	Agenbroad et al., 2005
Equus conversidens	10,870 ± 45 (CAMS-82411)	Bone	Pashley Gravel Pit, Alberta	Hills and Harington, 2003
Equus conversidens	11,330 ± 70 (TO-7696)	Bone	Wally's Beach, Alberta	McNeil et al,. 2005
Equus	11,210 ± 50 11,350 ± 40 (UCR-3783/ CAMS- 59288)	Bone	Fishbone Cave, Nevada	Grayson, 2006
Bootherium bombifrons	10,980 ± 80 (TO-7691)	Bone	Wally's Beach, Alberta	McNeil et al, 2005
Camelops	11,390 ± 60	Bone	Sunshine, Long Valley, Nevada	Huckleberry et al., 2001
Camelops	11,190 ± 50	Bone	Casper, Wyoming	Frison, 2000
Cervalces scotti	11,405 ± 50 (CAMS-82932) and11,260 ± 70 (CAMS-09476)	Bone	Lang Farm, Illinois	Schubert et al., 2004
Cervalces scotti	11,500 ± 130	Bone	Licking County, Ohio	Dyer et al., 1986
Cervalces scotti	10,950 ± 150 (I-4016)	Bone	Dewey-Parr site, Orange Co., NY	Funk et al., 1970
Megalonyx jeffersonii	11,430 ± 60 (CAMS-33974), 11,485 ± 40 (CAMS-82933)	Bone	Lang Farm, Illinois	Schubert et al,. 2004
Platygonus compressus (flat-headed peccary)	11,130 ± 60 and 11,060 ± 60	Bone	Sheriden Cave, Ohio	Redmond and Tankersley, 2005
Platygonus compressus	11,340 ± 50	Bone	Franklin Peccary, Idaho	McDonald, 2002
Mylohyus nasutus (long-nosed peccary)	11,860 ± 40	Bone	Sheriden Cave, Ohio	Redmond and Tankersley, 2005
Castoroides ohioensis (giant beaver)	10,850 ± 60	Bone	Sheriden Cave, Ohio	Redmond and Tankersley, 2005
Arctodus simus (short-faced bear)	11,610 ± 90, 11,570 ± 50, 11,570 ± 70, 11,480 ± 60	Bone	Sheriden Cave, Ohio	Redmond and Tankersley, 2005
Arctodus simus	10,870 ± 75	Bone	Huntington Mammoth Site, Utah	Madsen, 2000
Mammut americanum (mastodont)	10,850 ± 140,10,790 ± 70 (CAMS-27143), 10,705 ± 80 (AA-6968), 10,630 ± 80 (CAMS-17407), 10,515 ± 120 (Beta-24412)*	Bone (*poor condition) #conifer twig	Hiscock, NY	Laub, 2006, personal communication; CARD (Canadian Archaeological Radiocarbon Database)
Mammut americanum	10,970 ± 40	Bone	Otisville, NY	Robinson et al., 2005
Mammut americanum	11,000 ± 80 (NZA-12584)	Bone	Temple Hill, NY	Robinson et al., 2005
Mammut americanum	10,860 ± 70 (Pitt-830), 11,390 ± 80 (NSRL-283/ AA-6980), 11470 ± 90 (Pitt-841) *, 11,720 ± 110 (Beta-35045)*	Bone, *twigs	Burning Tree, Ohio	CARD
Mammut americanum	10,395 ± 100 (Beta 1388) (*terminus ante quem*)	Wood	Pleasant Lake, Michigan	CARD
Mammut americanum	10,780 ± 60 (CAMS-24944), 10,910 ± 60 (CAMS-24428), 11,140 ± 60 (CAMS-24945); 9480 ± 100*	Bone, *rejected	Deerfield, Wisconsin	CARD
Mammut americanum	10,920 ± 190 (Beta-15265), 11,320 ± 140 (Beta-15266)	Bone (tusk and cranial fragments)	Grandville, Michigan	CARD
Nothrotheriops shastensis	11,360 ± 260 (A-1202), 11,005 ± 100, 11,080 ± 90	Dung	Gypsum Cave, Nevada	Hofreiter et al., 2000

Nothrotheriops shastensis	11,020 ± 200 (A-1068), 11,000 ± 140 (A-1066), 10,780 ± 200 (A-1067), 10,500 ± 180, 10,400 ± 275 (I-442), 10,035 ± 250 (L-473 A)*	Dung, *contaminated?, 10,940 ± 60 on *Ephedra* twig from same sample	Rampart Cave, Arizona	Martin, 2005
Nothrotheriops shastensis	11,140 ± 160 (A-1212); 11,060 ± 240 (A-2627); 10,650 ± 220 (A-2626)	Dung	Muav Caves, Arizona	Mead and Agenbroad, 1992
Nothrotheriops shastensis	11,330 ± 370 (A-1878)	Dung	Shelter Cave, New Mexico	Martin, 2005; Steadman et al., 2005
Nothrotheriops shastensis	11,080 ± 200 (Y-1163B) 9,840 ± 160* (Y-1163A)	Dung, *body tissue, possibly contaminated	Aden Crater, New Mexico	Martin, 2005
Nothrotheriops shastensis	10,750 ± 140 (A-1583), 10,780 ± 140 (A-1534), 11,060 ± 180 (A-1584)	Dung	Upper Sloth Caves, Texas	Steadman et al., 2005
Nothrotheriops shastensis	11,140 ± 320 (A-1589)	Dung	Williams Cave, Texas	Steadman et al., 2005
Nothrotheriops shastensis	10,950 ± 70	Dung	Crescendo Cave, Grand Canyon, AZ	Martin, 2005
Oreamnos harringtoni (extinct mountain goat)	10,870 ± 200 (A-1155), 11,490 ± 180 (1852)	Dung, horn sheath	Stanton's Cave, Grand Canyon, AZ	Mead et al., 1986
Euceratherium (shrub-ox)	11,950 ± 50 (UCR-3780/CAMS 59155)	Bone	Falcon Hill, Nevada	Dansie and Jerrems, 2004
Euceratherium collinum	11,630 ± 150	Dung	Bechan Cave, Utah	Mead and Agenbroad, 1992
Smilodon fatalis	11,130 ± 275	Bone carbonate, dubious	La Brea, California	Marcus and Berger, 1984
Canis dirus	9,850 ± 550, 10,710 ± 320	Bone carbonate, dubious	La Brea, California	Marcus and Berger, 1984
SOUTH AMERICA				
Nothrotherium maquinense	12,200 ± 120 (NZA-6984)	Dung	Gruta de Brejoes, Brazil	Steadman et al., 2005
Catonyx cuvieri	9,580 ± 200	Associated charcoal	Lapa Vermelha IV	Neves and Pilo, 2003
Catonyx cuvieri	9,990 ± 40 (Beta-165398)	Bone	Cuvieri Site	Neves and Pilo, 2003
Eremotherium laurillardi	11,340 ± 50	Bone	Itaituba fossil quarry, on Tapajos River, Brazil	Rossetti et al., 2004
Smilodon populator	9,260 ± 150 (Beta-174722)	Bone	Escrivania 5, Brazil	Neves and Pilo, 2003
Glossotherium robustum	10,500 ± 90 (AA-9049)	Bone	Arroyo Seco 2, Argentina	Borrero et al., 1998
Mylodon darwinii	10,200 ± 400 (SA-49) 10,575 ± 400 (GX-6248) 10,400 ± 330 (A-1391)*	Dung, *hide	Cueva del Milodon, Chile	Steadman et al., 2005
Sloth (species?)	10,200 ± 300 (A-1636), 10,285 ± 240 (A-9494), 10,610 ± 210 (A-1351), 10,900 ± 185 (A-9493), 10,950 ± 60 (GrN-5558)	Dung	Gruta del Indio, Argentina	Long et al., 1998; Steadman et al., 2005
Sloth (species?)	10,570 ± 130 (AA-65707)	Ossicles	El Trebol, Argentina	Hajduk et al., 2006
Megatherium americanum (?) also present, *Equus neogeus, Toxodon, Hemiauchenia, Glossotherium, Glyptodon*	10,190 ± 120 (AA-19291) and 10,440 ± 100 (AA-39363).	Burned bones associated with fishtail point	Paso Otero 5, Argentina	Martinez, 2001
Hippidion saldiasi	10,925 ± 65 (OxA8528)	Bone	Piedra Museo, Argentina	Alberdi et al., 2001
glyptodont, equid (?)	ca. 9100	Charcoal, stratigraphic association	Pay Paso 1, Uruguay	Suarez, 2003
Hippidion saldiasi	10,710 ± 100 (NUTA-1811), 10.860 ± 160 (NUTA-2331)	Bone	Cueva del Medio	Massone, 2003
Onohippidion saldiasi	10,780 ± 60 (OxA-8635), 10,310 ± 160 (OxA-9504)	Bone	Lago Sofia 1	Massone, 2003
Hippidion (?) equid	10,685 ± 70 (OxA-9247)	Bone	Tres Arroyos 1 Va, Argentina	Borrero, 2003
Lama cf. *owenii*	10,430 ± 100 (NUTA-1734), 10,960 ± 150 (NUTA-2330)	Bone	Cueva del Medio	Massone, 2003

but still nowhere on paper. What one can gather is that at least 17 genera went extinct abruptly and simultaneously at about 11,000 BP. Stafford and colleagues actually envision a two-step process, with die-offs of most species about 11,300 BP, followed by the proboscideans at 10,800 BP, but such a division of earlier and later die-offs is untenable, as will be shown later in this chapter.

There are some instructive similarities between the arguments concerning terminal Pleistocene extinctions and those provoked by the Alvarezes' bolide impact theory of the Cretaceous/Tertiary (K/T) dinosaur extinctions. Indeed, Firestone and West and colleagues (Firestone and West, 2005; Firestone et al., 2007a, b) have recently speculated that a bolide impact, perhaps in the vicinity of Hudson's Bay, was also the cause of the terminal Pleistocene extinctions – although they must admit that the supposed impact cannot account for the slightly later extinctions in South America. In each case, there is a clear stratigraphic boundary – iridium-rich layers in the case of the K/T impact, and in the late Pleistocene case, the black mats in the Southwest and Southern Plains (Haynes, 1991). The black mats probably mark the onset of the regional equivalent of the Younger Dryas (Fiedel, 1999a). These features are attributable to algal growth following increased spring discharge (Quade et al., 1998), and thus imply a higher water table, increased precipitation, and reduced evaporation resulting in relatively wet and cold conditions. Clovis artifacts and mammoth bones occur just beneath the black mats, but never above them; only Folsom points and bison are found above the mats, which date to about 10,700 BP.

Skeptics of the K/T impact theory have argued that gradual environmental changes were the actual cause of asynchronous dinosaur extinctions, and point to the last occurrences of some species, millions of years before the impact. An answer to this criticism was provided by P. W. Signor and J. H. Lipps (1982): species that are rare in the fossil record may appear to drop out long before their actual time of demise in an abrupt trans-species mass extinction. This further implies that, purely on statistical grounds, the latest known dated specimen of a relatively rare species is most unlikely to actually be the last individual of that species. Turning to the late Pleistocene case, we should therefore not be surprised if the latest dated specimen of, say, the apparently always rare Aztlan rabbit, is ca. 18,000 BP (Grayson and Meltzer, 2003). It would be more damaging to the assumption of synchronous extinction if a demonstrably common, large-bodied mammalian genus – e.g., *Equus*, *Mammuthus*, *Mammut* (Grayson and Meltzer, 2003: Figure 1) – had disappeared before 11,000 BP, but in fact, those common genera all have late terminal dates, as anticipated.

It seems obvious that the best source of dates to track the extinction event should be the bones of the animals themselves. Unfortunately, bones of such age frequently have not survived at all, or if they have, their collagen has been lost or degraded and contaminated. The contaminants, mainly humates, are usually younger than the bone. Therefore, radiocarbon dates on collagen are generally suspected to be misleadingly young. Vance Haynes, for one, has always taken the oldest of multiple assays on a given bone sample to be the most accurate (Haynes, 1999, personal communication). Two different approaches have been taken recently to surmount this technical problem. Tom Stafford has run dates on individual amino acids that can confidently be assumed to derive from the collagen itself (Stafford, 1994). He has also attempted to exclude contaminants by using an XAD resin filter. However, when dates differ significantly among the various amino acids, it is not always clear which should be preferred. Stafford has reported, with evident confidence in the results, dates on XAD-purified total collagen (e.g., Schubert et al., 2004). At the Oxford AMS laboratory, collagen samples are now processed by ultrafiltration (Ramsey et al., 2004). The filters retain molecules with a weight of more than 30,000 Da; this includes undegraded collagen chains, with a weight of about 100–110,000 Da (Higham et al., 2006). Dates for ultrafiltered samples of Terminal Pleistocene age typically come out at least hundreds of years earlier than they did in previous assays of the same bones (e.g., *Megaloceros*: a specimen previously dated to 9,430 ± 65 BP yielded new dates of 10,585 ± 65 BP and 10,257 ± 75 BP; another, previously 9,225 ± 85 BP, was dated to 11,495 ± 65 BP and 11,159 ± 74 BP after ultrapurification [Higham, 2004]). There has been no direct inter-lab comparison of Stafford's individual amino acid technique vs. the Oxford method. A good test case would be the elk (*Alces alces*) from Miesenheim IV, a known-age animal buried by Laacher See tephra in Germany about 200 years before the onset of the Younger Dryas. The Laacher See eruption dates to 11,063 ± 12 BP (Kromer et al., 2004); previous dates for the elk bones, run at Oxford in 1992, long before installation of the new ultrapurification process, were 11,190 ± 90 BP (OxA-3584), 11,310 ± 95 BP (OxA-3585), and 11,190 ± 100 BP (OxA-3586).

Even if we could assume that every bone date was accurate, we would still be confronted by an unavoidable statistical and logical problem. Given that even a dwindling yet barely viable mammal population on its way to extinction must have numbered in the hundreds or thousands, and that taphonomic factors would have ensured the survival of skeletal remains of only a few of these individuals, the likelihood is very slim that any one skeletal find represents the last surviving individual of the species (unless the mass death resulted from an instantaneous catastrophe of global scale, such as a bolide impact, that left no survivors). Thus, any ostensibly terminal date must be treated as a *terminus post quem*. Using the same logic, Roberts and Solow (2003) contend that the dodo probably went extinct 28 years after the last reliable sighting of a living bird. Extending this approach to the Pleistocene fossil record, Solow et al. (2006) recently argued that the hiatus of several hundred years that appears to separate the dates of horse and mammoth extinction in eastern Beringia (ca. 12,400 vs. 11,500 BP) is not long enough to exclude the possibility that the actual times of extinction overlap with each other and with the arrival of human hunters (around 12,000 BP) (see also Buck and Bard, 2007). Hunting could therefore be the cause of both species' extinctions (*contra* Guthrie, 2004). The recent discovery of a horse tooth at Swan Point (Holmes, 2006, personal communication) has apparently established that humans and horses did, in fact, coexist briefly in Alaska ca. 12,000 BP.

As Solow et al. (2006) acknowledge, the best and perhaps the only way to circumvent this statistical problem is to supplement the analysis with unambiguous stratigraphic data. For dinosaur extinction, we have the K/T iridium-rich layer at 65 Ma; dinosaur fossils are always below it, and apparently never above it. For the Terminal Pleistocene extinctions, the analogous phenomenon is the "black mat" of the Southwest (Haynes, 1998, n.d.; Quade et al., 1998). Unfortunately, though widespread in the west, examples of black mats do not represent an event of global scope; indeed, they are not even trans-continental. Furthermore, some black mats were formed both before and after the main occurrence at ca. 10,700 BP (at the start of the Younger Dryas) so strict contemporaneity of deposits cannot be assumed. Although some mats appear to have been formed by algae in standing water, some black layers that occur in corresponding stratigraphic position may be the products of burning vegetation instead. Another problem with the use of these stratigraphic markers arises from their limited geographic scope. If mammoths' bones are not found above black mats in southern Arizona, how can we be sure that the animals did not simply move to a refugium in northern Mexico for a while?

A similar objection can be raised to the interpretation of sloth dung deposits in Southwestern caves (e.g., Rampart Cave in the Grand Canyon). At these sites, dung deposits accumulated over thousands of years of occupation by Shasta ground sloths (or mammoth, at Bechan Cave), only to cease abruptly at about 11,000 BP (Martin, 2005). Although it is very tempting, and quite possibly correct, to infer that this date marks the extinction event for this species, a minimalist alternative explanation is that the local population simply moved away to another den for reasons unknown. However, when the same observation of abrupt termination is made repeatedly at sites throughout a region, as is the case with sloth dung, the inference of extinction becomes increasingly credible.

The relative abundance of spores of the *Sporormiella* fungus in Late Pleistocene pond sediments provides a corollary to the argument from dung deposits for an abrupt extinction event. This fungus lives on modern cattle dung, and at Bechan Cave in Utah, it was found in well-preserved mammoth dung (Davis, 1987; Davis and Shafer, 2006). In the west, in several lakes, the frequency of fungal spores falls off dramatically at 10,800 BP. The logical inference is that the population of mammoths and other megafauna that formerly defecated at lake shores as they fed and drank suddenly disappeared from the region. A skeptic might counter with the hypothesis that the *local* animal population simply moved off to another spot not yet found or sampled, due to changes in rainfall and/or vegetation. We will never have a comparable record from every body of standing water that existed across western North America at the Pleistocene-Holocene transition; nevertheless, the fact that Davis has documented the same pattern in California as in Colorado indicates that the phenomenon of abrupt disappearance of megamammal feces and the fungus that thrived on it was not merely local.

The peak abundance of *Sporormiella* in the western cores occurs just before extinction. A comparable frequency is not found in the earlier Pleistocene sediments. The high spore densities at that moment (11,000 BP) are comparable to those seen in ponds located beside 19th–20th-century livestock corrals (Davis, 1987). Two explanations may be considered for this extraordinary frequency: (1) There is no indication of a gradual decline of megaherbivores during the Late Glacial. Rather, their population and range seem to have been increasing at that time, as Agenbroad (1984, 2005) and Ferring (2001) have suggested; (2) The extraordinary concentration of fungal spores (and by inference, megaherbivore dung) may be an index of an unprecedented aggregation of animals around standing water, in response to the hypothesized "Clovis drought" of the late Allerød. Megafauna clumped at these oases would have been easy targets for human hunters (Haynes, 2002). With tongue firmly planted in my cheek, I could propose a third possibility: if climate and vegetation changes were forcing megaherbivores to consume plants to which their digestive tracts were poorly adapted (as Guthrie suggests for Beringian mammals), and/or they were suffering from a lethal "virgin soil" epidemic of some flu-like virus (MacPhee and Marx, 1997), maybe the fungus peak denotes a wave of fatal diarrhea. Braun and Pfeiffer (2002) attribute a mass die-off of large mammals at a Pleistocene site in Germany to a seasonal bloom of cyanobacteria (blue-green algae) in a lake. Given the identification of black mats as probably algal products (Haynes, 2007), one must wonder if a substantial number of mammoths in the Southwest were being poisoned, around 11,000 BP, by ingesting pond scum.

In southeastern New York, a similar decline of *Sporormiella* has been documented by Guy Robinson and David and Lida Burney (2005). They see the same temporal pattern here that they have reported previously from Madagascar (Burney et al., 2003): a fungus (feces) decline, followed by charcoal increase. They interpret this sequence as indicating an initial blitzkrieg-like hunting episode that causes a steep decline (but not total extinction) of megaherbivores, followed by a build-up of combustible vegetation, no longer being consumed by the megafauna, which soon burns (producing the charcoal particles); followed by extinction at ca. 10,800 BP. It should be noted that Davis also observed a rise in charcoal particles associated with the *Sporormiella* decline in the western lakes. The problem raised by the northeastern sequence is the date of ca. 12,500 BP attributed to the fungal spore decline. This is some 2,000 calendar years before the actual megamammal extinction in this region (14,700 vs. 12,700 cal bp, e.g., 10,970 ± 40 BP for the Otisville, New York, mastodont). To fit a model of human-induced extinction, Robinson and the Burneys must invoke the presence of cryptic human hunters, who left behind no campsites or even unambiguous kill and butchering sites for 2,000 years. In defense of this dubious proposition, one could point to the very few putative pre-Clovis sites in the east: Cactus Hill (ca. 15,000 BP) (McAvoy and McAvoy, 1997; McAvoy et al., n.d.), Meadowcroft Rockshelter (ca. 16,000–12,000 BP) (Adovasio et al., 1999), and, particularly, the ostensible mammoth kill-sites in Wisconsin (Hebior, Schaeffer, Mud Lake) dated to ca. 13,000–12,500 BP and possibly associated with stone tools of the Chesrow lithic complex (Overstreet and Stafford, 1997; Joyce, 2006). I am personally unconvinced of the validity of these supposed pre-Clovis sites (e.g., Fiedel, 2000, 2002). If the dates from New York are correct, the fungal decline may only indicate a local-scale shift of mastodont ranges in response to climate and vegetation changes caused by the onset of Bølling warming (12,500 BP, or 14,700 cal bp).

The Arrival of Humans as a Possible Factor in the Extinctions

The Northern Route of Entry

A human presence at or before 12,500 BP (14,700 cal bp) is problematic because the terrestrial route from Beringia through the ice-free corridor between the Laurentide and Cordilleran ice sheets probably was blocked by coalescent ice between about 27,000 and 14,000 or even 13,500 cal bp (Dyke et al., 2002a, b). The supposed human occupation of Monte Verde at about 12,400 is now often cited as conclusive proof that humans either arrived prior to 27,000 cal bp, or voyaged by boat along the Pacific Coast around 15,000 cal bp. This is a dubious argument, because Monte Verde is a very peculiar site that lacks the expected hallmarks of a human presence – particularly, lithic debitage. In the massive two-volume site report (Dillehay, 1989, 1997, 2002), not a single map or photograph provides unambiguous provenience for the literal handful of indisputable chipped stone artifacts said to have been found amid a jumble of wood, plant remains, gomphothere bones, and stream gravels, all preserved in peat. My skepticism about Monte Verde (Fiedel, 1999b) remains a minority opinion; nevertheless, I confidently predict that no similar "settlement" will ever be found in the Southern Cone. As site after site in that region yields basal dates of ca. 11,000–10,700 BP for Fell 1 assemblages (Steele et al., 2001; Jackson et al., 2007), a *cordon sanitaire* is being drawn ever more tightly around this anomaly.

Despite a recent speculative volume (Madsen, 2004), no credible case can be made for a human presence in the Americas before coalescence of the ice sheets about 27,000 cal bp. Leaving Monte Verde in limbo, is there any way to account for the Wisconsin mammoth sites, or the hypothesized cryptic hunters in New York at ca. 12,500 BP?

The recent resuscitation by Stanford and Bradley (2002; Bradley and Stanford, 2004) of the notion that Solutreans paddled across the North Atlantic to eastern North America warrants mention here, if not serious consideration (see Straus et al., 2005 for an effective critique). There are undeniable similarities in the techniques of biface thinning used by Solutrean and Clovis toolmakers (although the Solutreans did *not* make fluted points). The last vestiges of the Solutrean industry in France and Spain (16,500 BP, ca. 19,000 cal bp) are separated by 6,000 years from the earliest traces of Clovis culture. Given clear genetic evidence of the ancestral roots of Amerind populations in southern Siberia, a European origin for Clovis would require postulation of a subsequent replacement of the original American population, all the way to Tierra del Fuego, by later Asian immigrants of whom there is no archeological trace. So, it remains most parsimonious to look for Paleoindian ancestors in Beringia and, before that, in Siberia. At present, it seems that there were no humans in eastern Beringia prior to 12,300 BP. The same date seems to be about the earliest that movement southward, past the ice sheets, might have been possible.

Openings in the Ice

Recent research has shown that several islands along the coast of Alaska and British Columbia were ice-free faunal refugia during the Late Glacial, although they seem to have been over-ridden by the Cordilleran ice sheet at the time of its maximum expansion between ca. 19,000 and 12,700 BP (Clague et al., 2004; Heaton and Grady, 2003). Much has been made of the presence of bears on the islands off southeast Alaska; where these big omnivores lived, it is argued, so could people. Perhaps so; but there is a gap in dated bear bones from the Alexander Archipelago between 27,000 and 12,300 BP (Heaton and Grady, 2003). The oldest date for a human presence in that area is 10,300±50 BP, for a tool probably made from a bear rib. Human skeletal elements from the same context (in On Your Knees Cave, on Prince of Wales Island) were dated to about 9,800 BP, but recognition of the largely marine diet of this individual requires reservoir correction to about 9,200 BP, consistent with three charcoal dates for this occupation zone (Dixon, 1999).

Farther south, the Queen Charlotte Islands were ice-free by about 14, 000 BP (17,000 cal bp), but again, the earliest evidence of human presence is dated no earlier than about 10,200 BP (a basalt flake recovered from a deeply submerged surface [Fedje and Josenhans, 2000]) or perhaps as early as 10,800 BP (biface fragments associated with bear bones; Fedje, 2006, personal communication). Dates of first human occupation from the coasts of Oregon and Washington are no earlier. A possible bone or antler point embedded in a mastodont bone from Manis, Washington, may date to about 12,000 BP (M. Waters, 2007, personal communication), but the artifactual status of this object requires further verification. In California, there are a few clusters of Clovis points. A human leg bone from Santa Rosa Island off Santa Barbara, known as Arlington Springs Woman (but very recently identified as a male), has been dated to about 11,000 BP. Interestingly, the latest date for a pygmy mammoth (*M. exilis*) from the island is 11,030 ± 50 (CAMS 71697) (Agenbroad et al., 2005).

Most archeologists (*contra* Cinq-Mars and Morlan, 1999) now regard the broken bones from the Old Crow basin in the Yukon as naturally created and thus not indicative of human presence 40,000 BP. Bones dated earlier than 20,000 BP at Blue Fish Caves are also dubious (Dixon, 1999). Leaving these aside, the earliest dates for human occupation of Eastern Beringia are 12,360 ± 60 and 12,110 ± 50 BP, from Swan Point, Alaska. Several additional dates from this site cluster around 11,700 BP (Crass and Holmes, 2004). A few similar dates have been reported from the Broken Mammoth and Mead sites, also in the Tanana river valley of Alaska. The arrival of humans seems to have been, if not the primary cause, at least a significant factor in an abrupt ecological turnover in eastern Beringia between 12,600 and 11,500 BP.

Extinctions in Eastern Beringia

While the dates amassed by Stafford and colleagues remain unavailable in print, R. Dale Guthrie (2003, 2004, 2006) has been steadily publishing in *Nature* the results of his ambitious effort to date dozens of Alaskan mammals of the late Pleistocene. Guthrie has long favored an explanation of extinctions *in that region* that entails primarily vegetation changes and the differential capabilities of large mammals' digestive tracts to cope with those changes. It is not surprising that he interprets the new radiocarbon dates in this framework. Guthrie's dates indicate that Alaskan horses decreased in size through the Late Glacial, and possibly went extinct ca. 12,500 BP (12,482 ± 80 BP is the latest date). Mammoths seem to have died out later (Guthrie, 2006) (the last date is 11,500 + 160 BP). Ostensible mammoth tracks are dated to ca. 11,600 BP, and ivory used by humans at the Broken Mammoth site was dated to 11,540 ± 140 BP – one of the very few indications of human-mammoth co-existence in eastern Beringia. Recently, Guthrie (2006) has presented a set of radiocarbon dates that appear to document a surge of bison population after 13,500 BP, followed by an explosive expansion of elk (wapiti) at ca. 12,600 BP, then the appearance of moose at about 12,300 BP. Prior to the 12,600 BP expansion, the latest previous date for elk is 18,210 ± 225 BP. It seems that the 12,600 BP dates demarcate an intrusion of *Cervus elaphus* from Siberia after 7,000 years (ca. 21,500–14,700 cal bp) of absence.

Before examining these data and Guthrie's inferences more closely, it must be emphasized that Eastern Beringia (Alaska and the Yukon) represents a special, isolated situation that may not be very informative about extinction processes occurring south of the ice sheets. Most importantly, Eastern Beringia at the LGM and Late Glacial was a cul de sac, blocked off by coalescent ice sheets along its eastern edge from about 27,000 to 13,000 cal bp. On the other hand, it was accessible to penetration by fauna from areas to the west, although an ecological barrier (perhaps a zone of mesic woodlands across central Beringia) seems to have prevented the eastward movement of some Eurasian species, such as the woolly rhinoceros (Guthrie, 2001). In some earlier periods, evidently, American species (including horse and camel) were able to traverse the land bridge to colonize Eurasia. The possibility must be emphasized that Terminal Pleistocene environmental stress or opportunities might as easily have induced movement of horses *westward* as the documented migration of elk *eastward*. The radiocarbon evidence from the high Arctic Taimyr Peninsula suggests a common pattern of ebb and flow of big mammals into and out of the far north in response to climate change (MacPhee et al., 2002). The result is a radiocarbon record with long hiatuses and sudden reappearances. Notably, horse and muskox disappeared from the Taimyr Peninsula around 12,000 BP only to reappear there millennia later, just after 4,000 BP. This pattern is consistent with the widely accepted model, borne out by numerous genetic studies, that assumes the contraction of animal species into discrete southern refugia during glacial cycles, followed by expansion under interglacial conditions. As sea level rose after 19,000 cal bp, Beringia would have suffered continual loss of land mass, culminating in the severing of the land bridge by the Bering Strait around 10,500 BP (12,500 cal bp) (Elias et al., 1996). At this far northern latitude, changes of a few degrees of temperature or a few millimeters of precipitation, coupled with the ongoing loss of terrestrial habitats, could have had much more dramatic effects than in less extreme and marginal environments to the south.

Guthrie (2003) has previously suggested that a marked diminution of foot bones over the course of nine millennia (about 20,000–12,500 BP, or 24,000–14,500 cal bp) since the glacial maximum indicated dwindling body size of Alaskan horses as they suffered a steady and "rapid" decline toward ultimate extinction. This is a dubious argument on both theoretical and empirical grounds. Decreasing body size – even dwarfing – can be an effective adaptation to a variety of environmental pressures, and not a sign of the morbidity of a population. For one example, the millions of bison roaming the Great Plains in the late Holocene were, as individuals, much smaller than the extinct *Bison antiquus* that they had either descended from or replaced. Objectively, a close scrutiny of Guthrie's data shows that (1) foot bones dating to about 20,000 BP range between 195 and 220 mm; (2) the smallest bones (190 and 191 mm) date to 16,480 and 15,370 BP, respectively; (3) the latest bone in the series (12,480 ± 80) measures 200 mm – within the range of the 20,000 BP bones. In any case, Guthrie's latest paper (2006) shows no significant decrease in the numbers of dated horse bones after 18,000 BP. In fact, the number of dated specimens seems to be increasing at ca. 12,500 BP, just prior to complete cessation.

Until recently, it seemed that humans had arrived in Alaska after the local extinction of mammoths. Ivory deposited at early human occupation sites, such as Broken Mammoth, typically yielded dates several thousand years older than the cultural context. This suggested that all of the ivory had been scavenged from long-dead animals. However, a date of 11,540 ± 140 BP has been obtained on ivory collagen from Broken Mammoth, and another date of 12,060 ± 60 BP for ivory from Swan Point, indicating contemporaneity of living mammoths with the earliest human settlers. A recent Bayesian analysis of the Alaskan radiocarbon record suggests that the date of horse extinction is ca. 14,200 cal bp, "broadly contemporary with the arrival of humans in the area," while extinction of woolly mammoth occurred later, ca. 13,300 cal bp (Buck and Bard, 2007).

At first glance, Guthrie's dates would lead one to conclude that mammoths had gone abruptly extinct in eastern Beringia at 11,500 BP. In fact, while this seems to have been the case on the Alaskan mainland, astonishing dates from two islands demonstrate the survival of isolated populations well into the Holocene (Vartanyan et al., 1995; Arslanov et al., 1998). On Wrangel Island, off the north coast of Siberia, the last date for a small mammoth is 3,685 ± 60 BP. On St. Paul Island, one of the Pribilof Islands located in the Bering Sea, 500 km off the Alaska coast, dates of 7,908 ± 100, 8,015 ± 85, and 8,010 ± 40 BP were obtained for split samples of one mammoth bone (Guthrie, 2004). An even later age of ca. 5,700 BP also has been reported for another sample (Crossen et al., 2005). Several issues are raised by these dates: (1) These are small animals that survived the Pleistocene-Holocene transition, while full-size relatives on the mainland succumbed; so much for the assumption that decreasing body size is the prelude to extinction! (2) These islands were, in fact, attached to the northern and southern edges, respectively, of Beringia prior to inundation of the land bridge about 10,500 BP (12,500 cal bp). Therefore, the founders of these populations, unlike their descendants, had not been protected from human predators (or mainland climate and vegetation changes) by isolation. So, what accounts for their survival? As Guthrie (2004) notes, coastally-oriented human hunters would surely have encountered the ancestors of the Pribilof mammoths and would have decimated them. So, their survival can be cited as evidence for an interior settlement focus of the ancestral Paleoindians, consistent with the ice-free corridor hypothesis as opposed to the alternative coastal route migration theory. (3) There is a puzzling hiatus in the Wrangel Island date sequence from about 12,000 to 7,700 BP (MacPhee, 2003). On the assumption that all the dates are accurate – and there is no obvious reason to reject them – it would appear that the island mammoth population was extirpated at the same time as the mainland extinction, but returned (by swimming? or walking on ice?) 5,000 years later, from some unidentified mainland refugium. (4) While the Wrangel population may have been finally hunted to extinction by humans, the St. Paul mammoths seem to have died out due to natural causes (such as loss of habitat due to sea level rise), as the Pribilofs were uninhabited prior to the placement of Aleut settlers there by the Russians in the 19th century. (5) Survival of mammoths on these grass-covered islands might support the theory that it was the dung and trampling of mammoths and other large grazers that maintained the "mammoth steppe" on the mainland, and that their demise – due largely to hunting – allowed the appearance of dense forests in the Holocene. Zimov (2005) has advocated this counter-intuitive theory – megafauna extinction induced vegetation changes, not *vice versa* – for western Beringia. He hopes to create a Siberian "Pleistocene Park," re-stocked with horse and bison, to test this model. (6) The Wrangel Island dates are so recent that one would suspect contamination or laboratory error, were it not for the multiple broadly congruent dates by different laboratories (Arslanov et al., 1998). In the 1950s and 1960s, numerous Holocene-age dates from ca. 9,500 to as late as 5,000 BP were obtained for mastodont tusks and associated wood fragments from North American sites (see Dreimanis, 1968). A review of the dates compiled by R. Morlan in the online Canadian Radiocarbon Database reveals an astonishing number of post-10,000 BP proboscidean dates. However, these dates have been simply dismissed as inaccurate (e.g., by Martin,1967); "They moulder in the graveyard of unverified measurements" (Martin and Stuart, 1995:7). Rejection of all these late dates was probably correct, but we now have to ask, what criteria should be used to assess the validity of anomalous dates, particularly those lacking unambiguous stratigraphic contexts? If re-dating of museum specimens of proboscideans using the most up-to-date protocols were to produce a handful of 9,000 BP or later dates, without stratigraphic support, would survival into the Holocene have been demonstrated? R. Laub (2006, personal communication) submitted twigs from the Hiscock site in western New York, on the assumption that they represented mastodont digesta. Five dates came back in the range of ca. 9,500–9,000 BP. Laub is properly skeptical about the implications; no credible mastodont bone-derived date from Hiscock is later than 10,600 BP. Further, the vagaries of the dating process are illustrated by cases where radiocarbon dates from unimpeachable contexts turn out to be egregiously wrong. One glaring example is the date of 5,215 ± 90 BP (Beta-43663) for a tool made of proboscidean bone from the East Wenatchee Clovis cache (expected age ca. 11,100 BP) (Gramly, 2004). The sample may have been contaminated by organics sprayed on the orchard that surrounds the site.

The Ice-Free Corridor and Chronology of Clovis Expansion

In the 1960s, Vance Haynes (1964, 1966) adduced evidence of the rapid expansion of Clovis mammoth-hunters, and linked their arrival to the opening of the corridor between the Laurentide and Cordilleran ice sheets. Paul Martin (1967) soon suggested that these hunters had perpetrated a "blitzkrieg" that caused megafaunal extinction, and he modeled the hunters' progress as an ever-expanding deadly wavefront that radiated southward from the mouth of the corridor (Mosimann and Martin, 1975).

As of now, the earliest Clovis-associated radiocarbon dates are two from Aubrey, Texas, that average about 11,550 BP

(Ferring, 2001). But these dates were run on charcoal samples that were not securely associated with the human occupation (i.e., not from recognizable hearths) and they may be "old" wood that had inherent age at the time of burning. All other credible Clovis dates are in the range of 11,200–10,750 BP (about 13,200–12,800 cal bp) (Fiedel, 2006; Waters and Stafford, 2007). There is no clear pattern of earlier western vs. later eastern dates for Clovis *sensu stricto;* classic sites such as Lehner and Murray Springs in Arizona date to ca. 10,900 BP, but so does Shawnee-Minisink in eastern Pennsylvania (Dent, 1999). Shawnee-Minisink is now the most precisely dated Clovis site: 10,935 ± 15 BP, by averaging of several dates on burnt seeds (Waters and Stafford, 2007).

The corridor seems now to have opened some time prior to 11,000 BP (*contra* Arnold, 2002). Recent tentative identification of elk (wapiti, *Cervus elaphus*) antler "foreshafts" (their presumed function) in the Anzick infant burial in Montana suggests that elk and other large mammals could have traversed the corridor at the same time that humans did. Two dates for the antler tools are both 11,040 BP (±40 and ±60) (Morrow and Fiedel, 2006). Even if large mammals were deterred from migration for several centuries by wet and cold conditions on the newly deglaciated landscape, human migrants could have survived on waterfowl, which they had already been in the habit of taking in central Alaska (Fiedel, 2007).

If Martin's blitzkrieg model is correct as originally formulated (Martin, 1967; Mosimann and Martin, 1975), we would anticipate a pattern of decreasing age of final megafauna with increasing distance south and east of the Paleoindians' presumed entry point in southern Alberta. However, Beck (1996), using the dates then available and unaware of calibration and plateau effects, did not observe this expected pattern. Four dates have been reported recently on bones from the Wally's Beach site at the St. Mary Reservoir, not far from the Alberta/Montana border (McNeil et al., 2005). In stratigraphic order within a 1.5-m column, these are: 11,350 ± 80 (TO-8972), on caribou; 11,330±70 (TO-7696) on *Equus conversidens* (extinct "Mexican" horse); 11,130 ± 190 (TO-7693) on *Bison antiquus*; and 10,980 ± 80 (TO-7691) on *Bootherium bombifrons* (extinct musk ox). Remarkably, the reservoir silts and sands retain preserved tracks of late Pleistocene mammals, mainly in the zone dated between 11,350 and 10,980 BP, although rare and indistinct tracks extend above the latter date horizon. Based upon the relative number of footprints attributable to juvenile vs. adult woolly mammoths, McNeil et al. (2005) conclude that the mammoth population was "stressed and declining," whether due to climate change or selective human hunting of juveniles. An apparently tight association of debitage and a chopper with skeletal elements of seven equids *(E. conversidens*), and looser association with fluted points (Kooyman et al., 2006), provides the first convincing evidence of human hunting of horse in North America (some horse teeth were found at Murray Springs). Two additional Clovis-era dates recently reported from southern Alberta are 10,930 ± 100 BP (TO-8514) for a Columbian mammoth (*M. columbi*) molar from the Bindloss Gravel Pit, and 10,870 ± 45 BP (CAMS-82411) for an astragalus of horse (*E. conversidens)* from the Pashley Gravel Pit (Hills and Harington, 2003). Taken collectively, these dates from Alberta appear to demonstrate that the megafauna living at the mouth of the ice-free corridor did not become locally extinct measurably earlier than their congeners farther south.

Stratified Sites South of the Corridor

To the south and southeast of the corridor's mouth, in the United States, indubitable kill or butchery sites of the Clovis era (11,200–10,800 BP, or 13,200–12,800 cal bp) are rare (Grayson and Meltzer, 2002, 2003). At these sites, stone tools are most often associated with mammoth or mastodont (e.g., Kimmswick). Next most common is *Bison antiquus* (e.g., Murray Springs, Jake Bluff [Bement and Carter, 2003]). Although the scarcity of sites has been marshaled as evidence against a significant human role in megafauna extinction, their number is actually impressive (Surovell and Waguespack, 2004) in view of both the narrow time-frame involved (only 400 years) and the likely proportion of human-induced to natural deaths. Even if humans caused a significant 5% increase in the natural death rate (which included carnivore kills), 95% of the animals available for possible skeletal preservation would have shown no sign of human actions.

The most precisely dated western Clovis sites are Lehner and Murray Springs, both mammoth killsites in the San Pedro Valley of southern Arizona. The radiocarbon date for Lehner, averaged from 12 charcoal assays, is 10,930 ± 40 BP. The date for Murray Springs, averaged from eight assays, is 10,900 ± 50 BP (Haynes, 1993:221). The most precise of the Lehner dates are 10,950 ± 90 (SMU-290), 10,940 ± 100 (A-378), 10,950 ± 110 (SMU-194), and 10,710 ± 90 BP (SMU-340). The most precise date from Murray Springs is 10,840 ± 70 BP (SMU-41). Most of the samples were pieces of charcoal found in eroded deposits. The only two dates from definite Clovis cultural contexts were 10,760 ± 100 and 11,150 ± 450 BP. One of the mammoth skeletons at Murray Springs lay directly in contact with the overlying black mat. Judging from the lack of weathering of the bones, it seems that no more than a few months can have elapsed between the animal's death and the deposition of the mat (Haynes, 2007:46). I have previously hypothesized (Fiedel, 1999a) that the abrupt rise of the regional water table in the Southwest, which led to black mat formation, was synchronous with the onset of the Younger Dryas; this was expressed as both increased precipitation and colder temperatures (which reduced evaporation). If this inference is correct, the mammoth and the overlying black mat should date to about 12,900 cal bp. This is consonant with current calibration of the radiocarbon dates.

Sheriden Cave in Ohio (Redmond and Tankersley, 2005) seems to have been either a natural trap or predator den for centuries before humans arrived in the area. In addition to charcoal-derived dates, assays on bone or tooth collagen from the cave's lower levels include three dates for *Cervalces scotti* (stag moose) (12,840 ± 100, 12,590 ± 450, and 12,520 ± 170 BP), one for *Mylohyus nasutus* (long-nosed peccary) (11,860 ± 40 BP) and four for *Arctodus simus* (short-faced bear) (11,610 ± 90, 11,570 ± 50, 11,570 ± 70, and 11,480 ± 60 BP). An extensive but discontinuous charcoal layer, dated to 10,840 ± 80 and 10,960 ± 60 BP, may mark the arrival of fire-starting humans. Whether or not the burning was anthropogenic, above this lens, two beveled-based bone points, a reworked Gainey-like fluted point, and two possibly human-altered bones (a cut snapping turtle vertebra and a perforated ilium of a peccary) were found within a ca. 2-m^2 area. Based upon overlying dates (unfortunately including a few anomalies due to evident water disturbance), the cultural material seems to date to ca. 10,800–10,900 BP. The dated samples from the overlying strata include two bones of the extinct flat-headed peccary (*Platygonus compressus*) (11,130 ± 60 and 11,060 ± 60 BP), a bone of extinct giant beaver (*Castoroides ohioensis*) (10,850 ± 60 BP), and a bone of caribou (*Rangifer tarandus*) dated to 10,440 ± 40 BP. The latter is obviously not an extinct species, but it's been quite a while since caribou lived in Ohio.

Several caves in the Southwest contain stratified, dated sequences that are merely suggestive of a Terminal Pleistocene catastrophe, because no evidence of human habitation or butchery is present. In these sites, steady deposition of dung through millennia by Shasta ground sloth (*Nothrotheriops shastensis*) stops abruptly at 11,000–10,800 BP. Relatively precise terminal dates on dung include: Gypsum Cave, NV, 11,005 ± 100, 11,080 ± 90 BP (Hofreiter et al. 2000); Rampart Cave, AZ, 10,940 ± 60, 11,000 ± 140 BP; Muav Caves, AZ, 11,140 ± 160, 11,060 ± 240, 10,650 ± 220 BP; Aden Crater, NM, 11,080 ± 200 BP; Upper Sloth Caves, TX, 10,750 ± 140, 10,780 ± 140, and 11,060 ± 180 BP (Long and Martin, 1974; Martin, 2005).

To summarize the evidence presented to this point, the most important conclusions are:

1. Megafauna bones and tracks are dated at the mouth of the corridor to ca. 11,300–10,800 BP (Wally's Beach). Terminal dates for megafauna farther south fall within the same ca. 400-calendar-year window (13,300–12,900 cal bp). Extinction appears to have been effectively simultaneous across North America, and it was precisely coeval with the arrival and expansion of Clovis hunters. However, the lack of a clear southward-radiating pattern is more consistent with a "leapfrog" movement of Paleoindians, rather than the wave as formerly envisioned.
2. Multiple dates indicate that other megafauna species (horse, musk-ox, sloth, flat-headed peccary, giant beaver) lasted as long as the proboscideans, with dates as late as ca. 10,800 BP. There is no evidence of a two-step extinction process with later survival of proboscideans, nor is it plausible to suggest that megafauna with last known dates of ca. 11,400 or 11,300 BP did not survive into the period of Clovis expansion. Radiocarbon ages of 11,300 and 11,050 (Clovis-age, e.g., Anzick) BP are separated by only about 80 years of real time (13,300 to 13,220 cal bp) (Kromer et al. 2004).

Is Climate Change a Credible Cause of Megafauna Extinction?

Before the 1990s, advocates of climate change as the cause of megafauna extinction envisioned a stark dichotomy of two successive climate extremes (cold Pleistocene vs. warm Holocene), with a gradual transition at about 10,000 BP. A dominant model, expounded by Graham (e.g., Graham et al., 1997), pointed to the change to less equable climate in the early Holocene – warmer summers, colder winters – as a cause of vegetation change and faunal extinction. However, our understanding of late Pleistocene climate change has been profoundly altered in the last decade by new data from the Greenland ice core projects (GISP2, GRIP, and NGRIP). Abrupt and sharp warming and cooling episodes – glacials and interglacials and shorter stadials and interstadials – occurred repeatedly throughout the late Pleistocene. We can now ask, if Holocene warming was so disastrous for megafauna, why wasn't there a wave of extinction around 125 ka in the last interglacial? The climate was even warmer than present then, with higher sea levels. Or again, if warming was so deleterious, why don't the extinctions cluster at the onset of the Bølling warming at 14,700 cal bp (12,500 BP)? Instead, they cluster at the time of the Younger Dryas onset at 12,900 cal bp (10,900–10,600 BP). In most parts of the Northern Hemisphere, this marked a return to very cold conditions, to which animals such as mammoth should have been well adapted (but in some places, due to regional atmospheric circulation patterns, the YD seems to have been rather warm and wet; for example, in central Florida [Grimm et al., 2006] and in the central Appalachians [Kneller and Peteet, 1999]). The Younger Dryas lasted for 1,300 years. This was a time of very cold winters but relatively warm summers. With the ice-free corridor open, perhaps this was the first time that the "Arctic express" cold fronts from Siberia hit the Plains (Yu and Wright, 2001). In any case, judging from Greenland ice, the Younger Dryas was no colder than previous stadials, none of which had caused widespread extinction of large North American mammals. It is also time to recognize that Holocene climate cannot have been a causative factor in the extinctions. The Younger Dryas ended at 11,600 cal bp (10,000 BP), with a rapid and sharp warming that marked the beginning of the Holocene. By that time, the megafauna had already been gone for about 1,000 years.

Extinctions in South America

It is not frequently noted that more mammalian genera (at least 37, perhaps as many as 58) died out in South America than in North America at the end of the Pleistocene (see Cione et al., Chapter 7, and Borrero, Chapter 8). If climate played an important role in the South American extinction, its effects should be evident considerably earlier there. The climate of northern South America seems to have been in-phase with North America (e.g., the El Abra stadial of

Colombia was synchronous with the Younger Dryas, and the grey scale record of sediments in the Cariaco Basin off Venezuela is precisely in sequence with Greenland ice cores [Hughen et al., 2000]), but the southern tip of the continent was in phase with Antarctica (Sugden et al., 2005), where gradual warming began about 18,000 cal bp. A minor cold reversal in the south was synchronous with the Bølling/Allerød warming in the north (Morgan et al., 2002); it ended as the Younger Dryas began. But the South American megafauna did not die out until about 12,500–12,000 cal bp (500 years after Paleoindian hunters arrived). They may have lingered until ca. 10,500 cal bp or even later in parts of Brazil (where the absence of fishtail points may indicate humans arrived after ca. 12,500 cal bp), Uruguay, and Argentina. Late dates from Brazil include: 9,580 ± 200 BP on charcoal associated with the sloth, *Catonyx cuvieri*, at Lapa Vermelha IV; another date on bone for *C. cuvieri* of 9,990 ± 40 BP; and a collagen date of 9,260 ± 150 BP for the only South American sabertooth species, *Smilodon populator* (Neves and Pilo, 2003). At Pay Paso 1 in Uruguay, a date of ca. 9,100 BP is reported for charcoal from a late fishtail assemblage associated with bones of glyptodont and possibly equid (Suarez, 2003). Three anomalous late dates for sloths at Gruta del Indio, Argentina are 8,990 ± 90 on dung (LP-925), 9,560 ± 90 on bone (GrN-5772) and 9,650 ± 800 on dung (A-1282) (García, 2003; Long et al., 1998). Steadman et al. (2005) cautiously state that "Until the ages of these samples are verified independently, we question their validity, although leaving open the possibility that sloth extinction in South America occurred as much as a millennium later than in North America." At Campo Laborde in Argentina, *Megatherium americanum* was dated by AMS at 8,080 ± 200 (AA-55118) and 7,750 ± 250 (AA-55117) BP; a similar date of 7,320 ± 50 for this species from Arroyo Seco 2 was rejected by the excavator (Politis et al., 2004). Four dates of ca. 6,500–7,500 BP for glyptodont (*Doedicurus clavicaudatas*) were obtained from La Moderna (Politis et al., 2004). An even later date for survival of glyptodonts in Argentina (4,300 BP) has been the subject of recent debate (Cione et al., 2001 vs. Rossello et al., 2001).

The appearance of fluted or pseudo-fluted fishtail (Fell I) points in Patagonia and Tierra del Fuego about 11,000–10,700 BP (Flegenheimer and Zarate, 1997) signals the arrival of Clovis-descended Paleoindians around 12,900 cal bp. The earliest of the most precise (1-sigma < 100), internally consistent radiocarbon dates from good stratigraphic contexts for Fell I assemblages in southernmost South America (Miotti and Salemme, 2003) are:

Piedra Museo	11,000 ± 65 (AA-27950)
	10,925 ± 65 (OxA 8528)
Tres Arroyos	11,085 ± 70 (OxA 9248)
Cerro Tres Tetas	10,915 ± 65 (AA-22233)
Cueva Casa del Minero	10,999 ± 55 (AA-37207)
	10,967 ± 55 (AA-37208)

At first glance, these dates are as early as the best-dated Clovis sites in North America (ca. 11,050–10,950 BP) (Waters and Stafford, 2007). However, the geographic and stylistic logic of north-to-south migration makes this most obvious inference unlikely, as it leaves no time for the population movement or cultural drift implied by the Clovis-to fishtail transformation. A possibility that cannot yet be dismissed is that late glacial radiocarbon fluctuations in South America were not precisely synchronous with those recorded to the north (Kelly, 2003) (e.g., unpublished radiocarbon dates for Younger Dryas-age tree-rings sampled from a Tasmanian Huon pine log show some differences from Cariaco [and INTCAL04] dates [Barbetti, n.d.]). However, the simplest solution of this conundrum is that some Clovis sites date to about 200 or 250 years before the YD onset (i.e., about 13,200 cal bp), and the first Fell I sites date to the cusp of the YD onset event at about 12,950 cal bp (Fiedel, 2006). This solution fits the dates to the now-demonstrated late Allerød ^{14}C inversion (Kromer et al., 2004), which also seems to be present in the sequence of dated sediments in Lake Mascardi in Argentina (Hajdas et al., 2003). It also accommodates the least convoluted model of the Clovis-Fell I relationship. This solution allows some eight to twelve human generations for the migration from a Clovis staging area in Texas (e.g., the Gault site [Collins, 2002, 2003]) to the tip of South America. If it turns out that the Aubrey dates (Ferring, 2001) of ca. 11,550 BP (ca. 13,600–13,300 cal bp; see Litt et al., 2003:4) are accurate, the time available for migration and population growth could be as much as 700 years.

Apparently good associations of artifacts with megafauna bones have been reported from several South American sites. These include:

Paso Otero 5: *Megatherium americanum* (?) burned bones, 10,190 ± 120 (AA-19291) and 10,440 ± 100 (AA-39363). With fishtail point; also present, *Equus neogeus, Toxodon, Hemiauchenia, Glossotherium, Glyptodon* (Martinez, 2001)

Piedra Museo: *Hippidion saldiasi* (Alberdi et al., 2001), 10,925 ± 65, on bone (OxA-8528)

Tres Arroyos 1 Va: *Hippidion* (?) equid bone, 10,685 ± 70 (OxA-9247 AMS) (Borrero, 2003)

Pay Paso 1, Uruguay (Suarez 2003); glyptodont, equid (?) ca. 9,100 BP

Quebrada Santa Julia (Jackson et al., 2007); equid, species unidentified; dates on associated charcoal and wood, 10,920 ± 80 (Beta-194725) 11,060 ± 80 (Beta-215090), 11,090 ± 80 (Beta-215089)

Apart from these, there are two stratified sequences of ground sloth dung in caves of the Southern Cone that are analogous to the sites in the southwestern US where abrupt cessation occurs in the absence of any evidence of human activity. At Gruta del Indio, Argentina, the sloth dung, probably from a mylodont species, was deposited until about 10,500 BP (Long et al., 1998). Dates for the uppermost dung balls include: 10,200 ± 300, 10,285 ± 240, 10,610 ±

210, 10,900 ± 185, 10,950 ± 60 BP. An anomalously late date, 9,560 ± 60 BP, was obtained for dermal ossicles from a museum collection; these may have been contaminated by preservative, although the date cannot be rejected out of hand, given the ca. 9,500 BP dates for Brazilian sloths. Immediately above the dung layer are charcoal deposits interpreted as human-created hearths, although few if any artifacts were associated with them. The oldest hearth dates with acceptably small errors are: 10,530 ± 140; 10,440 + 225/-220; 10,195 ± 80; 10,170 ± 70; 10,135 ± 95. The dung and charcoal dates together suggest that the sloths disappeared here around 10,300 BP (somewhere in the range from ca. 11,900 to 12,400 cal bp).

Terminal dates for *Mylodon darwinii* at the Cueva del Milodon in Chile are broadly similar, although the large sigmas preclude a precise estimate: 10,200 ± 400, 10,400 ± 330, 10,575 ± 400 BP.

Several aspects of the South American record should be stressed:

1. A previous invasion by North American fauna after the emergence of the Isthmus of Panama ca. 3 Ma had resulted in numerous extinctions of the endemic marsupials, but not of edentate megafauna such as sloths and glyptodonts (Patterson and Pascual, 1968; Cione et al., 2003).
2. As in North America, there is no record of massive die-offs of large mammals at the onsets of previous interglacials.
3. In at least some regions where megafauna disappeared, vegetation changes seem to have been minor, and their former dietary staples did not disappear.
4. For many millennia prior to their extinction, megafauna occupied very broad ranges across the continent and adjusted flexibly to variable local ecologies by eating whatever was available; they were not specialized feeders (as shown by studies of dung composition of ground sloth [Hofreiter et al., 2000, 2003] and *Hippidion* [García et al., 2008] and isotopic signatures in teeth of Toxodonts [MacFadden, 2005] and gomphotheres [Sanchez et al., 2003]).
5. The megafaunal extinction seems slightly later in South America than farther north, even though gradual warming began much earlier there than in the north. This slight asynchrony seems to preclude the recently hypothesized bolide impact in North America (Firestone et al., 2007a) as the cause of South American extinctions.

Caribbean Extinctions

Finally, a southward glance at the Caribbean islands provides a nice analog to the late survival of mammoths on the northern island remnants of Beringia. The sloths of the Caribbean were not as gigantic as their continental relatives; the largest of the island sloths (and the largest of Caribbean terrestrial mammals) was *Megalocnus rodens*, weighing only about 200 kilograms (by the usual definition, a weight of 45 kg or more puts an animal in the megafauna category). As Steadman et al. (2005) observe, if terminal Pleistocene climate change killed off the continental sloths, one would expect to see a coeval crash of island sloths. Conversely, if human predation (or habitat disruption) was crucial, the island sloths would be expected to weather the Holocene onset unscathed, only to crash much later when humans reached the Antilles, in the mid-Holocene. In fact, the latest date for *Megalocnus rodens* in Cuba is 4,190 ± 40 BP (4,580–4,840 cal bp) (MacPhee et al., 2007). A smaller sloth, *Parocnus brownii*, has been dated in Cuba to 4,960 ± 280 BP. The oldest reliable age for human occupation of Cuba is about 5,200 BP (6,000 cal bp). The youngest sloth dates from Hispaniola (Haiti), for a small type, *Neocnus comes*, are 4,391 ± 42 and 4,486 ± 39 BP. Initial environmental disruption by human colonists on Hispaniola seems to be signaled in lake sediments by increased charcoal influx and decline of tree pollen around 5,730 BP (Steadman et al., 2005). Clearly, Caribbean sloths survived long after the onset of Holocene climate, and their demise correlates roughly with human arrival. However, the temporal overlap of first humans and last sloths seems to have been relatively long (over a thousand years) compared to that seen on the American continents (see MacPhee, Chapter 9).

Conclusion

Both an accumulating corpus of radiocarbon dates and a variety of stratigraphic data indicate that most of the megafauna of North America went extinct within 500 years of the arrival of Paleoindian hunters – by 12,700 cal bp. Extinction of most South American megafauna seems to have occurred several hundred years later. The extinction episode in North America also coincided with Terminal Pleistocene climate oscillations. However, previous climate changes of a similar scale had not caused comparable size-biased die-offs. It did not matter whether the animals were grazers or browsers, if their ranges expanded or contracted, or if their particular environments shifted from warm to cold, cold to warm, dry to wet or wet to dry conditions. In South America, a ramp-like amelioration toward warmer and wetter Holocene climate started about 20,000 BP, but megafauna survived for 8 kyr during this gradual process. In North America, proponents of climate change as the agent of mass death at least might argue that the rapid onset of the Younger Dryas, combined with opening of the ice-free corridor, subjected plants and animals to unprecedented cold winters. No such dramatic change can be postulated for South America. At present, it has not been demonstrated that the gradual climate change there reached a critical threshold at 12,500 cal bp that would have caused the disappearance of 37 or more mammalian genera within a few centuries. The ubiquity and synchrony of sudden deaths argues for a single cause, and the obvious new ecological factor that transcended all zones was a rapidly expanding human population.

For now, only about half of the extinct Pleistocene genera in North America can be demonstrated to have survived down to the time of sudden death. Even so, the simultaneous disappearance of at least 17 genera can hardly be brushed aside as a freakish coincidence (*contra* Grayson, 2007). I venture to predict that, as dates accumulate, the extinction dates of the less common Pleistocene species will come ever closer to the same ca. 11,000–10,700 BP baseline. However, we should also be prepared for surprises. If *Megaloceros* survived until 7,000 cal bp in Europe (Stuart et al., 2004), if mammoths survived on remote Siberian and Alaskan islands until about 6,000 to 4,000 cal bp, and if horse and muskox repopulated the Taimyr region from some unknown refugium after an 8 kyr absence, it is surely not unimaginable that relict populations of ground sloths, glyptodonts, or even mammoths managed to survive in isolated refugia on the American continents well into the Holocene.

Note

1. Ages are noted as 'ka,' thousands of years ago, or 'Ma,' millions of years ago, or in radiocarbon years Before Present ('BP') (that is, uncalibrated) with 1950 AD/CE considered to be the "Present." Calibrated ages (using calendar years, counting back from 1950 AD/CE) are stated as cal bp.

References

Adovasio JM, Pedler D, Donahue J, Stuckenrath R (1999) No vestige of a beginning nor prospect for an end: Two decades of debate on Meadowcroft Rockshelter. In: Bonnichsen R, Turnmire KL (eds) Ice Age peoples of North America: Environments, origins, and adaptations of the First Americans. Oregon State University Press, Corvallis, pp 416–431

Agenbroad LD (1984) New World mammoth distribution. In: Martin PS, Klein RG (eds) Quaternary extinctions: A prehistoric revolution. University of Arizona Press, Tucson, pp 90–112

Agenbroad LD (2005) North American proboscideans: Mammoths: The state of knowledge, 2003. Quat Intl 126–128:73–92

Agenbroad LD, Johnson J, Morris D, Stafford TW Jr (2005) Mammoths and humans as late Pleistocene contemporaries on Santa Rosa Island. In: Garcelon DK, Schwemm CA (eds) Proceedings of the Sixth California Islands Symposium, Ventura, CA, December 1–3, 2003. Institute for Wildlife Studies, Arcata, CA, pp 3–7

Alberdi MT, Miotti L, Prado JL (2001) *Hippidion saldiasi* Roth, 1899 (Equidae, Perissodactyla), at the Piedra Museo site (Santa Cruz, Argentina): Its implication for the regional economy and environmental reconstruction. J Archaeol Sci 28:411–419

Arnold TG (2002) Radiocarbon dates from the ice-free corridor. Radiocarbon 44(2):437–454

Arslanov Kh, Cook A, Steinar Gulliksen GT, Harkness DD, Kankainen T, Scott EM, Vartanyan S, Zaitseva GI (1998) Consensus dating of mammoth remains from Wrangel Island. Radiocarbon 40(1–2):289–294

Barbetti M (n.d.) SRT-783 tree-ring and Cariaco varve data; figure accessed online 6/28/2005 at http://www.ansto.gov.au/ainse/prorep2001/R_00_003.pdf

Beck MW (1996) On discerning the cause of late Pleistocene megafaunal extinctions. Paleobiology 22:91–103

Bement LC, Carter BJ (2003) Clovis bison hunting at the Jake Bluff site, NW Oklahoma. Curr Res Pleistocene 20:20–26

Borrero LA (2003) Taphonomy of the Tres Arroyos 1 Rockshelter, Tierra del Fuego, Chile. Quat Intl 109/110:87–93

Borrero LA, Zarate M, Miott L, Massone M (1998) The Pleistocene-Holocene transition and human occupations in the Southern Cone of South America. Quat Intl 49/50:191–199

Bradley B, Stanford D (2004) The North Atlantic ice-edge corridor: A possible Paleolithic route to the New World. World Archaeol 36:459–478

Braun A, Pfeiffer T (2002) Cyanobacterial blooms as the cause of a Pleistocene large mammal assemblage. Paleobiology 28(1):139–154

Buck CE, Bard E (2007) A calendar chronology for Pleistocene mammoth and horse extinction in North America based on Bayesian radiocarbon calibration. Quat Sci Rev 26(17–18):2031–2035

Burney DA, Robinson GS, Burney LP (2003) *Sporormiella* and the late Holocene extinctions in Madagascar. Proc Natl Acad Sci USA 100(19):10800–10805

CARD (Canadian Archaeological Radiocarbon Database, compiled by R. Morlan). (2001) Online website http://www.canadianarchaeology.ca/accessed 6/04.

Cinq-Mars J, Morlan RE (1999) Bluefish Caves and Old Crow Basin: A new rapport. In: Bonnichsen R, Turnmire KL (eds) Ice Age people of North America: Environments, origins, and adaptations. Oregon State University Press, Corvallis, pp 200–212

Cione AL, Figini AJ, Tonni EP (2001). Did the megafauna range to 4300 BP in South America? Radiocarbon 43(1):69–75

Cione AL, Tonni EP, Soibelzon L (2003) The broken zig-zag: Late Cenozoic large mammal and tortoise extinction in South America. Revista Museo Argentino de Ciencia Natural 5(1)

Clague JJ, Mathewes RW, Ager TW (2004) Environments of northwestern North America before the Last Glacial Maximum. In: Madsen DB (ed) Entering America: Northeast Asia and Beringia before the Last Glacial Maximum. University of Utah Press, Salt Lake City, pp 63–94

Collins MB (2002) The Gault site, Texas, and Clovis research. Athena Rev 3(2):24–32

Collins MB (2003) Gault site dates and findings. Newsletter of the Friends of the Texas Archeological Research Laboratory, May

Crass BA, Holmes CE (2004) Swan Point: A case for land bridge migration in the peopling of North America. Paper presented at 69th annual meeting of Society for American Archaeology, Mar. 31-Apr. 4, Montreal, Quebec

Crossen KJ, Yesner DR, Veltre DW, Graham RW (2005) 5,700-year-old mammoth remains from the Pribilof Islands, Alaska: Last outpost of North American megafauna. Paper presented at Annual Meeting of Geological Society of America, Oct. 16–19, Salt Lake City, UT

Dansie AJ, Jerrems WJ (2004) Lahontan chronology and early human occupation in the western Great Basin: A new look at old collections. In: Lepper BT, Bonnichsen R (eds) New perspectives on the First Americans. Center for the Study of the First Americans, College Station, TX, pp 55–64

Davis OK (1987) Spores of the dung fungus *Sporormiella*: Increased abundance in historic sediments and before Pleistocene megafaunal extinction. Quat Res 28:290–294

Davis OK, Shafer DS (2006) *Sporormiella* fungal spores, a palynological means of detecting herbivore density. Palaeogeogr, Palaeocl, Palaeoecol 237(1):40–50

Dent RJ (1999) Shawnee Minisink: New dates on the Paleoindian component. Poster presented at 64th Annual Meeting of Society for American Archaeology, Mar. 24–28, Chicago, IL

Dillehay TD (ed) (1989) Monte Verde: A late Pleistocene settlement in Chile, Volume 1, Palaeoenvironment and site context. Smithsonian Institution Press, Washington, DC

Dillehay TD (ed) (1997) Monte Verde, a late Pleistocene settlement in Chile, Volume 2. The archaeological context and interpretation. Smithsonian Institution Press, Washington, DC

Dillehay TD (ed) (2002) Errata, Monte Verde, a late Pleistocene settlement in Chile, Volume 2. The archaeological context and interpretation. Smithsonian Institution Press, Washington, DC

Dixon EJ (1999) Bones, boats, and bison: Archaeology and the first colonization of Western North America. University of New Mexico Press, Albuquerque

Dreimanis A (1968) Extinction of mastodons in eastern North America: Testing a new climatic-environmental hypothesis. Ohio J Sci 8(6):257–272

Dyer DL, Harington CR, Fernandez RL, Hansen MC (1986) A Pleistocene stag-moose (*Cervalces scotti*) from Licking County, Ohio. Ohio J Sci, Abstracts of 95th Annual Meeting, April 25–27, Toledo

Dyke AS, Andrews JT, Clark PU, England JH, Miller GH, Shaw J, Veillette JJ (2002a) The Laurentide and Innuitian ice sheets during the Last Glacial Maximum. Quat Sci Rev 21(1–3):9–31

Dyke AS, Moore A, Robertson L (2002b) Revised North American deglaciation maps. Paper presented at Annual Meeting Geological Society of America, Oct. 27–30, Denver, CO

Elias SA, Short SK, Nelson CH, Birks HH (1996) Life and times of the Bering land bridge. Nature 382:60–63

Fedje DW, Josenhans H (2000) Drowned forests and archaeology on the continental shelf of British Columbia, Canada. Geology 28:99–102

Ferring CR (2001) The archaeology and paleoecology of the Aubrey Clovis Site (41DN479), Denton County, Texas. Report from Center for Environmental Archaeology, University of North Texas, to US Army Corps of Engineers, Fort Worth District

Fiedel SJ (1999a) Older than we thought: Implications of corrected dates for Paleoindians. Am Ant 64(1):95–116

Fiedel SJ (1999b) Artifact provenience at Monte Verde: Confusion and contradictions. Sci Am Discover Archaeol Special Report 1(6):1–12

Fiedel SJ (2000) The peopling of the New World: Present evidence, new theories, and future directions. J Archaeol Res 8(1):39–103

Fiedel SJ (2002) Initial human colonization of the Americas: An overview of the issues and the evidence. Radiocarbon 44(2):407–436

Fiedel SJ (2006) Points in time: Establishing a precise hemispheric chronology for Paleoindian migrations. In: Morrow JE, Gnecco C (eds) Paleoindian archaeology, a hemispheric perspective. University Press of Florida, Florida, pp 21–43

Fiedel SJ (2007) Quacks in the Ice: Waterfowl, Paleoindians, and the discovery of America. In: Walker RB, Driskell BN (eds) Foragers of the terminal Pleistocene in North America. University of Nebraska Press, Lincoln/London, pp 1–14

Firestone RB, West A (2005) Evidence for the extinction of mammoths by an extraterrestrial impact event. Paper presented at Second International World of Elephants Congress, Sept. 22–25, Hot Springs, SD

Firestone RB, West A, Kennett JP, Becker L, Bunch TE, Revay ZS, Schultz PH, Belgya T, Kennett DJ, Erlandson JM, Dickenson OJ, Goodyear AC, Harris RS, Howard GA, Kloosterman JB, Lechler P, Mayewski PA, Montgomery J, Poreda R, Darrah T, Que Hee SS, Smith AR, Stich A, Topping W, Wittke JH, Wolbach WS (2007a) Evidence for an extraterrestrial impact 12,900 years ago that contributed to the megafaunal extinctions and the Younger Dryas cooling. Proc Natl Acad Sci USA 104:16016–16021

Firestone RB, West A, Revay Z, Belgya T, Smith A, Que Hee SS (2007b) Evidence for a massive extraterrestrial airburst over North America 12.9 ka Ago. Eos, Transactions of the American Geophysical Union 88(23), Joint Assembly, Supplement, Abstract PP41A-01

Flegenheimer N, Zarate M (1997) Considerations on radiocarbon and calibrated dates from Cerro La China and Cerro El Sombrero, Argentina. Curr Res Pleistocene 14:27–28

Frison GC (2000). A ^{14}C date on a late Pleistocene *Camelops* at the Casper-Hell Gap site, Wyoming. Curr Res Pleistocene 17:28–29

Funk RE, Fisher DW, Reilly EM Jr (1970) Caribou and Paleo-Indian in New York state: A presumed association. Am J Sci 268:181–186

García A (2003) On the coexistence of man and extinct Pleistocene megafauna at Gruta del Indio (Argentina) 9000 C-14 years ago. Radiocarbon 45:33–39

García A, Martínez Carretero E, Dacar MA (2008) Presence of *Hippidion* at two sites of western Argentina: Diet composition and contribution to the study of the extinction of Pleistocene megafauna. Quat Intl 180(1):22–29

Graham R (1998) Mammals' eye view of environmental change in the United States at the end of the Pleistocene. Paper presented at 63rd Annual Meeting of the Society for American Archaeology, Mar. 25–29, Seattle, WA

Graham R, Stafford T, Lundelius E, Semken H, Southon J (2002) C-14 chronostratigraphy and litho-stratigraphy of late Pleistocene megafauna extinctions in the New World. Paper presented at 67th annual meeting of Society for American Archaeology, Mar. 20–24, Denver, CO

Graham R, Stafford T, Semken H (1997) Pleistocene extinctions: Chronology, non-analog communities, and environmental change. Paper presented at symposium "Humans and other Catastrophes," American Museum of Natural History, Apr. 17–18, New York, NY

Gramly RM (2004) The Richey Clovis cache. Amateur Archaeologist 10(1)

Grayson DK (1993) The desert's past. Smithsonian Institution Press, Washington, DC

Grayson DK (2006) The late Quaternary biogeographic histories of some Great Basin mammals (western USA). Quat Sci Rev 25(21–22):2964–2991

Grayson DK (2007) Deciphering North American Pleistocene extinctions. J Anthropol Res 63(2):185–214

Grayson DK, Meltzer DJ (2002) Clovis hunting and large mammal extinction: A critical review of the evidence. J World Prehist 15:1–68

Grayson DK, Meltzer DJ (2003) A requiem for North American overkill. J Archaeol Sci 30:585–593

Grimm EC, Watts WA, Jacobson GL Jr, Hansen BCS, Almquist HR, Dieffenbacher-Krall AC (2006) Evidence for warm wet Heinrich events in Florida. Quat Sci Rev 25(17–18):2197–2211

Guthrie RD (2001) Origin and causes of the mammoth steppe: A story of cloud cover, woolly mammal tooth pits, buckles, and inside-out Beringia. Quat Sci Rev 20:549–574

Guthrie RD (2003) Rapid body size decline in Alaskan Pleistocene horses before extinction. Nature 426:169–171

Guthrie RD (2004) Radiocarbon evidence of mid-Holocene mammoths stranded on an Alaskan Bering Sea island. Nature 429:746–749

Guthrie RD (2006) New carbon dates link climatic change with human colonization and Pleistocene extinctions. Nature 441:207–209

Hajdas I, Bonani G, Moreno PI, Ariztegui D (2003) Precise radiocarbon dating of Late-Glacial cooling in mid-latitude South America. Quat Res 59:70–78

Hajduk A, Albornoz A, Lezcano MJ (2006) Levels with extinct fauna in the forest rockshelter El Trebol (northwest Patagonia, Argentina). Curr Res Pleistocene 23:55–57

Haynes CV Jr (1964) Fluted projectile points: Their age and dispersion. Science 145:1408–1413

Haynes CV Jr (1966) Elephant-hunting in North America. Sci Am 214(6):104–112

Haynes CV Jr (1991) Geoarchaeological and paleohydrological evidence for a Clovis-age drought in North America and its bearing on extinction. Quat Res 35:438–450

Haynes CV Jr (1993) Clovis-Folsom geochronology and climatic change. In: Soffer O, Praslov ND (eds) From Kostenki to Clovis: Upper Paleolithic-Paleoindian adaptations. Plenum Press, NY, pp 219–236

Haynes CV Jr (1998) Geochronology of the stratigraphic manifestation of paleoclimatic events at Paleoindian sites. Paper presented at 63rd Annual Meeting of the Society for American Archaeology, Mar. 25–29, Seattle, WA

Haynes CV Jr (2007) Nature and origin of the black mat, stratum F2. In: Haynes CV Jr, Huckell B (eds) Murray Springs. University of Arizona Press, Tucson, Appendix B, pp 240–249

Haynes CV Jr (n.d.) Clovis, pre-Clovis, climate change, and extinction. Manuscript

Haynes G (2002) The catastrophic extinction of North American mammoths and mastodonts. World Archaeol 33(3):391–416

Heaton TH, Grady F (2003) The Late Wisconsin vertebrate history of Prince of Wales Island, Southeast Alaska. In: Schubert BW, Mead JI, Graham RW (eds) Ice Age cave faunas of North America. Indiana University Press, pp 17–53

Higham TFG (2004) AMS dating of archaeological bone from Mexico: Problems and future directions. Paper presented at 2nd International Symposium, Early Man in America, Sept. 6-10, Mexico City, Mexico

Higham TFG, Jacobi RM, Bronk Ramsey C (2006) AMS radiocarbon dating of Paleolithic bone using ultrafiltration. Paper presented at 19th International ^{14}C Conference, Apr. 3-7, Oxford, UK

Hills LV, Harington CR (2003) New radiocarbon dates for Columbian mammoth and Mexican horse from southern Alberta and the Lateglacial regional fauna. Quat Sci Rev 22(14):1521–1523

Hofreiter M, Poinar HN, Spaulding WG, Bauer K, Martin PS, Possnert G, Pääbo S (2000) A molecular analysis of ground sloth diet through the last glaciation. Molecul Ecol 9:1975–1984

Hofreiter M, Betancourt JL, Pelliza Sbriller A, Markgraf V, McDonald HG (2003) Phylogeny, diet, and habitat of an extinct ground sloth from Cuchillo Cura, Neuquen Province, southwest Argentina. Quat Res 59:364–378

Huckleberry G, Beck C, Jones GT, Holmes A, Cannon M, Livingston SD, Broughton JM (2001) Terminal Pleistocene/early Holocene environmental change at the Sunshine Locality, north-central Nevada, USA. Quat Res 55:303–312

Hughen KA, Southon JR, Lehman SJ, Overpeck JT (2000) Synchronous radiocarbon and climate shifts during the last deglaciation. Science 290:1951–1954

Jackson D, Mendez C, Seguel R, Maldonado A, Vargas G (2007) Initial occupation of the Pacific Coast of Chile during late Pleistocene times. Curr Anthropol 48(5):725–731

Joyce DJ (2006) Chronology and new research on the Schaefer mammoth (?*Mammuthus primigenius*) site, Kenosha County, Wisconsin, USA. Quat Intl 142–143:44–57

Kelly RL (2003) Maybe we do know when people first came to North America. Quat Intl 109–110:133–145

Kneller M, Peteet D (1999) Late-Glacial to early Holocene climate changes from a Central Appalachian pollen and macrofossil record. Quat Res 51:133–147

Kooyman B, Hills LV, McNeil P, Tolman S (2006) Late Pleistocene horse hunting at the Wally's Beach site (DhPg-8), Canada. Am Ant 71(1):101–122

Kromer B, Friedrich M, Hughen KA, Kaiser F, Remmele S, Schaub M, Talamo S (2004) Late Glacial ^{14}C ages from a floating, 1382-ring pine chronology. Radiocarbon 46(3):1203–1209

Litt T, Schmincke HU, Kromer B (2003) Environmental response to climate and volcanic events in central Europe during the Weichselian Lateglacial. Quat Sci Rev 22:7–32

Long A, Martin PS (1974) Death of American ground sloths. Science 186:638–640

Long A, Martin PS, Lagiglia HA (1998) Ground sloth extinction and human occupation at Gruta del Indio, Argentina. Radiocarbon 40:693–700

MacFadden BJ (2005) Diet and habitat of toxodont megaherbivores (Mammalia, Notoungulata) from the late Quaternary of South and Central America. Quat Res 64(2):113–1244

MacPhee RDE (2003) The Wrangel date gap and other evidence of widespread megafaunal collapse in northern Asia during the early Holocene, Paper presented at 3rd International Mammoth Conference, May 23–29, Dawson City, Yukon

MacPhee RDE, Marx PA (1997) The 40,000-year plague: Humans, hyperdisease, and first-contact extinctions. In: Goodman S, Patterson B (eds) Natural change and human impact in Madagascar. Smithsonian Institution Press, Washington, DC, pp 169–217

MacPhee Ross DE, Tikhonov AN, Mol D, de Marliave C, van der Plicht H, Greenwood AD, Flemming C, Agenbroad L (2002) Radiocarbon chronologies and extinction dynamics of the Late Quaternary mammalian megafauna of the Taimyr Peninsula, Russian Federation. J Archaeol Sci 29:1017–1042

MacPhee RDE, Iturralde-Vinent MA, Jiménez Vázquez O (2007) Prehistoric sloth extinctions in Cuba: Implications of a new "last" appearance date. Caribbean J Sci 41:94–98

Madsen DB (2000) A high-elevation Allerød-Younger Dryas megafauna from the west-central Rocky Mountains. In: Madsen DB, Metcalf MD (eds) Intermountain archaeology. University of Utah Anthropological Papers 22, pp 100–115

Madsen DB (ed) (2004) Entering America: Northeast Asia and Beringia before the Last Glacial Maximum. University of Utah Press, Salt Lake City

Marcus LF, Berger R (1984) The significance of radiocarbon dates for Rancho La Brea. In: Martin PS, Klein RG (eds) Quaternary extinctions: A prehistoric revolution. University of Arizona Press, Tucson, pp 159–188

Martin PS (1967) Prehistoric overkill. In: Martin PS, Wright HE Jr (eds) Pleistocene Extinctions: The Search for a Cause. Yale University Press, New Haven, pp 75–120

Martin PS (2005) Twilight of the Mammoths. University of California Press, Berkeley

Martin PS, Stuart AJ (1995) Mammoth extinction: Two continents and Wrangel Island. Radiocarbon 37(1):7–10

Martinez GA (2001) 'Fish-tail' projectile points and megamammals: New evidence from Paso Otero 5 (Argentina). Antiquity 75:532–538
Massone M (2003) Fell I hunters' hearths in the Magallanes region by the end of the Pleistocene. In: Miotti L, Salemme M, Flegenheimer N (eds) Where the south winds blow. Center for the Study of the First Americans, College Station, TX, pp 153–160
McAvoy JM, McAvoy LD (1997) Archaeological investigations of site 44SX202, Cactus Hill, Sussex County, Virginia. Research Report Series No. 8, Virginia Department of Historic Resources, Richmond
McAvoy JM, Baker JC, Feathers JK, Hodges RL, McWeeney LJ, Whyte TR (n.d.) Summary of research at the Cactus Hill archaeological site, 44SX202, Sussex County, Virginia: Report to the National Geographic Society in compliance with stipulations of grant #6345–98
McDonald HG (2002) *Platygonus compressus* from Franklin County, Idaho, and a review of the genus in Idaho. In: Akersten WA, Thompson ME, Meldrum DJ, Rapp RA, McDonald HG (eds) And Whereas… Papers on the vertebrate paleontology of Idaho honoring John A. White. Volume 2, Idaho Museum of Natural History Occasional Paper 37, pp 141–149
McNeil P, Hills LV, Kooyman B, Tolman SM (2005) Mammoth tracks indicate a declining late Pleistocene population in southwestern Alberta, Canada. Quaternary Science Reviews 24(10–11):1253–1259
Mead JI, Agenbroad LD (1992) Isotope dating of Pleistocene dung deposits from the Colorado Plateau, Arizona and Utah. Radiocarbon 34:1–19.
Mead JI, Martin PS, Euler RC, Long A, Jull AJT, Toolin LJ, Donahue DJ, Linick TW (1986) Extinction of Harrington's mountain goat. Proc Natl Acad Sci 83(4):836–839
Miotti LL, Salemme MC (2003) When Patagonia was colonized: People mobility at high latitudes during Pleistocene/Holocene transition. Quaternary International 109–110:95–111.
Morgan V, Delmotte M, van Ommen T, Jouzel J, Chappellaz J, Woon S, Masson-Delmotte V, Raynaud D (2002) Relative timing of deglacial climate events in Antarctica and Greenland. Science 297(5588): 1862–1864.
Morrow J, Fiedel S (2006) New radiocarbon dates for the Clovis component of the Anzick site, Park County, Montana. In: Morrow JE, Gnecco G. (eds) Paleoindian archaeology, a hemispheric perspective. University Press of Florida, pp 123–138
Mosimann JE, Martin PS (1975) Simulating overkill by Paleoindians. Am Scientist 63:304–313
Neves WA, Pilo LB (2003) Solving Lund's dilemma: New AMS dates confirm that humans and megafauna coexisted at Lagoa Santa. Curr Res Pleistocene 20:57–60
Overstreet DF, Stafford TW (1997) Additions to a revised chronology for cultural and non-cultural mammoth and mastodon fossils in the southwestern Lake Michigan Basin. Curr Res Pleistocene 14:70–71
Patterson B, Pascual R (1968). The fossil mammal fauna of South America. Quart Rev Biol 43(4):409–451
Politis GG, Messineo PG, Kaufmann CA (2004) El poblamiento temprano de las llanuras pampeanas de Argentina y Uruguay (The early peopling of the Pampean plains of Argentina and Uruguay). Complutum 15:207–224.
Quade J, Forester RM, Pratt WL, Carter C (1998) Black mats, spring-fed streams, and Late Glacial age recharge in the southern Great Basin. Quat Res 49:129–148
Ramsey CB, Higham T, Bowles A, Hedges R (2004) Improvements to the pretreatment of bone at Oxford. Radiocarbon, 46(1): 155–163(9).
Redmond BG, Tankersley K (2005) Evidence of early Paleoindian bone modification and use at the Sheriden Cave site (33Wy252), Wyandot County, Ohio. Am Ant 70(3):503–526
Roberts DL, Solow AR (2003) Flightless birds: When did the dodo become extinct? Nature 426:245
Robinson G, Burney LP, Burney DA (2005) Landscape paleoecology and megafaunal extinction in southeastern New York. Ecol Mono 75(23):295–315
Rossello EA, Jahn BM, Liu T-K, Petrocelli JL (2001) The 4300-yr ^{14}C age of glyptodonts at Luján River (Mercedes, Buenos Aires, Argentina) and comments on 'Did the megafauna range to 4300 BP in South America' by Cione et al. Radiocarbon 43(1):77–81
Rossetti D de F, Mann de Toledo P, Moraes-Santos HM, de Araujo Santos AE Jr (2004) Reconstructing habitats in central Amazonia using megafauna, sedimentology, radiocarbon, and isotope analyses. Quat Res 61(3):289–300
Sanchez B, Prado JL, Alberdi MT (2003) Paleodiet, ecology, and extinction of Pleistocene gomphotheres (Proboscidea) from the Pampean region (Argentina). Coloquios de Paleontologia 1:617–625
Schubert BW, Graham RW, McDonald HG, Grimm EC, Stafford TW Jr (2004) Latest Pleistocene paleoecology of Jefferson's ground sloth (*Megalonyx jeffersonii*) and elk-moose (*Cervalces scotti*) in northern Illinois. Quat Res 61(2):231–240
Signor PW, Lipps JH (1982) Sampling bias, gradual extinction patterns and catastrophes in the fossil record. In: Silver HT, Schultz PH (eds) Geological implications of impacts of large asteroids and comets on the earth. Geological Society of America Special Paper 190, pp 291–296
Solow AR, Roberts DL, Robbirt KM (2006) On the Pleistocene extinctions of Alaskan mammoths and horses. Proc Natl Acad Sci USA 103(19):7351–7353
Stafford TW Jr (1994) Accelerator C-14 dating of human fossil skeletons: Assessing accuracy and results on New World specimens. In: Bonnichsen R, Steele DG (eds) Method and theory for investigating the peopling of the Americas. Center for the Study of the First Americans, Corvallis, OR, pp 45–56
Stanford D, Bradley B (2002) Ocean trails and prairie paths? Thoughts about Clovis origins. In: Jablonski NG (ed) The first Americans: The Pleistocene colonization of the New World. Cal Acad Sci Mem 27:255–271
Steadman DW, Martin PS, MacPhee RDE, Jull AJT, McDonald HG, Woods CA, Iturralde-Vinent M, Hodgins GWL (2005) Asynchronous extinction of late Quaternary sloths on continents and islands. Proc Natl Acad Sci USA 102:11763–11768
Steele J, Politis G, Pettitt P (2001) AMS radiocarbon dating of the earliest Paleoindian occupation of the southern cone of South America. Paper presented at 66th Annual Meeting of the Society for American Archaeology, Apr. 18–22, New Orleans, LA
Straus L, Meltzer DJ, Goebel T (2005) Ice age Atlantis? Exploring the Solutrean-Clovis connection. World Archaeol 37(4):507–532
Stuart AJ, Kosintsev PA, Higham TFG, Lister AM (2004) Pleistocene to Holocene extinction dynamics in giant deer and woolly mammoth. Nature 431:684–689

Suarez R (2003) First records of Pleistocene fauna for an archaeological context in Uruguay: Evidence from Pay Paso locality, site 1. Curr Res Pleistocene 20:113–116

Sugden DE, Bentley MJ, Fogwill CJ, Hulton NRJ, McCulloch RD, Purves RS (2005) Late-Glacial glacier events in southernmost South America: A blend of 'Northern' and 'Southern' hemispheric climate signals? Geografiska Annaler 82A(2):273–288

Surovell TA, Waguespack NM (2004) How many elephant kills are 14? Clovis mammoth and mastodon kills in context. Paper Presented at the 69th Annual Meetings of the Society for American Archaeology, Mar. 31–Apr. 4, Montreal, Quebec

Vartanyan SL, Arslanov KhA, Tertychnaya TV, Chernov SB (1995) Radiocarbon dating evidence for mammoths on Wrangel Island, Arctic Ocean, until 2000 bc. Radiocarbon 37(1):1–6

Waters MR, Stafford TW Jr (2007) Redefining the age of Clovis: Implications for the peopling of the Americas. Science 315: 1122–1126

Yu Z, Wright HE Jr (2001) Response of interior North America to abrupt climate oscillations in the North Atlantic region during the last deglaciation. Earth-Sci Rev 52.333–369

Zimov SA (2005) Pleistocene Park: Return of the mammoth's ecosystem. Science 308(5723):796–798

3. Estimates of Clovis-Era Megafaunal Populations and Their Extinction Risks

Gary Haynes*
Anthropology Department,
University of Nevada, Reno
Reno, Nevada 89557-0096, USA
gahaynes@unr.edu

Keywords Clovis era • megafauna • extinction risks

Introduction

In order to evaluate the contribution that Clovis-era hunting made to the end-Pleistocene extinctions, we must examine the North American empirical evidence fairly, without using models from different continents and different taxa as blueprints for the process of human hunting impacts. Before trying to decide how (or if) Clovis hunting could have had a significant effect on American megamammal extinctions,[1] a worthwhile thing to know or estimate is the size of the continental populations of megamammals during the Clovis era. Of course, no direct measure is possible, but there are some possible clues and guides in the methods employed in modern wildlife conservation practices.

Measuring the Size of Mammoth and Mastodont Populations in the Clovis Era

Here I describe two approaches to estimating the continent-wide numbers of proboscideans in the Late Glacial interval, defined as the period from 15,000 to 10,000 BP. The first approach produces a rough estimate of maximum population sizes.

Average Elephant Density and Range Size

This method involves calculating animal numbers based on total habitable land area and average density of animal per unit of land.

Mammut

Mammut americanum range in North America: I estimate this range size by adding up the apparently habitable area east of the Mississippi river (~2,860,000 km^2), plus the unglaciated or deglaciated part of Ontario and the Canadian Maritime provinces (~130,000 km^2), plus the eastern part of Texas (~130,000 km^2), then subtracting the surface area of lakes and unsuitable terrain (~260,000 km^2); total = ~2,860,000 km^2.

Mammut americanum density: I estimate this as 0.2–0.4 animals per square kilometer, which is a reasonable "average" number based on modern forest and woodland elephant densities.[2] The modern range of density for a possible analog, the Asian forest elephant, varies greatly; some Asian forest areas contain 0.12 to 0.75 animals per square kilometer, while other, more open areas have much lower densities (Eltringham, 1982; Sukumar, 1989). Waguespack and Surovell (2003) pointed out that large-mammal body size correlates with population density and cited averaged modern elephant density figures from *Walker's Mammals of the World* (Nowak, 1999), namely 0.6 Asian elephants and 1.3 African elephants per square kilometer. However, Parker and Graham (1989) showed that the highest average elephant density, present only in parts of Africa, is 0.4 to 0.8 animals per square kilometer, although some much more limited areas may have up to 5 animals per square kilometer, such as Lake Manyara (I. and O. Douglas-Hamilton, 1975), an unusual and almost ideal habitat not easily damaged by high feeding pressure. However, overall habitat variations keep the densities much lower when averaged for the entire continent of Africa.[3] I chose a density well below the upper end of the possible range because animals freely moving and able to feed in variable habitats with low predator pressure – including human pressure – usually maintain numbers well below 0.4 animal per square kilometer.

Mammut americanum population estimate in the Clovis era: 580,000, possibly as large as 1,600,000 animals. I consider this to

* Address for correspondence: gahaynes@unr.edu

G. Haynes (ed.), *American Megafaunal Extinctions at the End of the Pleistocene*, 39–53.

be a maximum estimate – the ideal or optimal numbers in all of North America, or in other words the biggest possible estimate.

Mammuthus

Mammuthus spp. range in North America: I estimate this range size by adding up nearly all of the USA (~9,206,000 sq km), plus Alaska (unglaciated) (~1,523,600 – ~910,000 km^2 glaciated/mountainous), plus eastern/central Mexico (~1,980,170 – ~910,000 km^2 of mountains and most barren land), then subtract the areas of continental mountains and pluvial lakes (~1,300,000 km^2) and thickest eastern woodlands (~1,300,000 km^2). The deglaciated parts of the Canadian Prairie Provinces should also be added, although the proportion suitable as mammoth habitat varied greatly over time. Another area that perhaps should be added is the Santarosae super-island (the California Channel Islands plus the now-submerged land connecting them during the Late Glacial period of lowered sea level). I haven't attempted to estimate these two additional areas here. The total without them is ~8,289,770 km^2.

Mammuthus spp. density: I estimate this as 0.1–0.3 animal per square kilometer, which I consider a reasonable number for elephants inhabiting open landscapes whose water sources and feeding patches are more separated than in forest and woodlands. Seasonal transhumance is very common amongst proboscideans, which means that with mammoths as with mastodonts the densities would have shifted geographically throughout the year. Hence, a single high density number cannot be appropriately applied to all areas at once; some areas would have been emptied as others were filling up. To avoid overestimating, I therefore prefer 0.1 animal per square kilometer.

Mammuthus spp. population in the Clovis era: ~828,977, possibly as large as 2,486,931. Again, I consider this to be the absolute maximum or ideal. The lower end of the range is probably more accurate. Measurable climatic stress would have cut back this number significantly during the Late Glacial interval, and therefore I suggest another method of population estimate may be more appropriate to use.

Another Method to Estimate Population Size – "Live:Dead Ratios"

This method involves comparing a range's numbers of live to dead animals in light of known ratios under different conditions in the modern world. I think this method is more appropriate to use for estimates at the end of the Pleistocene, when climatic reversals were stressing megafaunal populations, because it should provide a direct calculation rather than a potentially optimal or ideal (maximum) number.

Field biologists compare numbers of live and dead animals to get a feel for mortality increases or declines in different places and times. In modern African elephant ranges with no human hunting pressure, the number of dead animals (either fresh or skeletonized) may be 0.9% of the number of live ones in any one year. This number is based on a yearly mortality rate of about 0.5% for adults and 2.5% for unweaned animals.

Under environmentally stressful conditions, such as during the drought die-offs I documented in northwestern Zimbabwe (Haynes, 1988, 1991), the percentage may rise to 10% (or more). Other authors have estimated elephant mortality rates under conditions of high and low mortality, and the numbers are very similar (Table 3.1).

In order to calculate live numbers of mammoths and mastodonts from the numbers of fossil skeletons, we need to know how many proboscidean sites date specifically to the Clovis era. Some prehistorians (Grayson and Meltzer, 2003; also see Holliday, 2003) have cited the total number of Clovis-era proboscidean sites (archeological and nonarcheological, based exclusively on the Faunmap database) as 576, but this number is not accurate. It is an unscreened count of all of the sites in Faunmap's "Late Glacial" class, a much longer time span than the Clovis era (defined as the period from about 11,500 to 10,800 BP) (see Waters and Stafford, 2007, but also see comments by Haynes et al., 2007) The actual number of Clovis-era sites therefore must be lower, probably significantly lower. I estimate the maximum number of possible Clovis-age *Mammuthus* sites in Faunmap at about 230[4]; only about 50 mammoth sites have radiometric dates that are within or very near the Clovis era (Agenbroad, 2005),[5] and most of the others are too poorly known to be assigned any age other than 110–10 ka. Likewise I estimate the total of all *Mammut* sites of any age in the Faunmap database at only about 205.[6]

North America has produced about 185 sites with mammoth and mastodont bones reliably dated (or realistically dateable) to the Clovis era, based on Faunmap and CARD (2001) databases, abstracts from SAA meetings, brief reports published in outlets such as *Current Research in the Pleistocene*, and a literature search. But very few are actually associated with

TABLE 3.1. Estimated mortality rates in elephants. Note how similar the predictions are in both parts of the Table.

Part 1 (From Hanks, 1979)

Age class (years)	Percentage of population in this age class (GH estimate)	Percentage (and N per 100 animals of population) of age class dying with low predation pressure	Percentage (and N per 100) of age class dying with high predation pressure
< 4	25%	5% (1.25)	20% (5)
5–45	65%	1% (0.65)	4% (2.6)
>45	10%	4% (0.4)	50% (5)
	TOTAL = 100%	TOTAL = 1.715% (2.3)	TOTAL = 12.6% (12.6)

Part 2 (From Jones, 1991)

Age class (years)	Percentage of population in this age class (GH estimate)	Percentage (and N per 100 animals of population) of age class dying with low predation pressure	Percentage (and N per 100) of age class dying with high predation pressure
<3	20%	2.5% (0.5)	60% (12)
>3	80%	0.5% (0.4)	1% (0.8)
	TOTAL = 100%	TOTAL = 0.9% (0.9)	12.8% (12.8)

Clovis artifacts. The number must be considered fluid due to ambiguities in the data reported.

Therefore, if the total number of documented terminal Pleistocene mammoth and mastodont sites with highest likelihood of dating to the Clovis era is about 185 (MNI = 319) (Table 3.2), and this number of skeletons was 0.9% of the number of animals in the live population, the predicted live population was only 35,444 mammoths and mastodonts in the Clovis era. If the dead to live proportion was more like 10%, then there were barely more than 3,000 mammoths and mastodonts alive. These numbers are probably underestimated, but note that at either one of the extremes, even a very small amount of Clovis hunting of megamammals would have had a major effect on taxon survival.

If the mammoth and mastodont populations were larger, at the level of my earlier maximum estimates, then the total 319 individuals in 185 sites represents only about 0.02–0.008% of the population. The 185 sites are undoubtedly not ecologically contemporaneous, only geologically contemporary; so in fact the number of known sites of the exact same age is even smaller than 185, and hence the fossil sites are an even smaller percentage of the existing live population. What happened to all the expected death sites? Perhaps many more sites were preserved but are still undiscovered. Taphonomic factors make it likely that the number of known fossil sites is much less than the actual number created during the Clovis era.

TABLE 3.2. Part 1 *Mammuthus* sites proven or likely to be from the Clovis era.

Site name	State/Province	MNI
1. Albuquerque	NM	1
2. Aubrey	TX	1
3. Aucilla River	FL	1+
3. Avery Island	LA	1
4. Bartlett Wash	UT	1
5. Bartow	OK	1
6. Ben Franklin	TX	1
7. Bentzen Kaufman Cave	WY	1
8. Berclair Terrace	TX	1
9. Betz	MI	1
10. Blackwater Locality 1	NM	16
11. Cerros Negros	AZ	1
12. China Lake	CA	2(+)
13. Claypool	CO	1
14. Colby	WY	7
15. Dent	CO	15
16. Domebo	OK	1
17. Duewall-Newberry	TX	1
18. Dutton	CO	1
19. Elm Creek Local Fauna	OK	1
20. Escapule	AZ	1
21. Evanston	IL	1
22. Guest Mammoth	FL	2
23. Hell Gap Loess Mammuthus Locality	WY	1
24. Henry	MI	1
25. Hermit's Cave	NM	1
26. Huntington	UT	1

(continued)

TABLE 3.2. Part 1 (continued)

Site name	State/Province	MNI
27. Kassler Quad Mammoth	CO	1
28. Koehn-Schneider	KS	2
29. Lamb Spring	CO	23(+)
30. Lange/Ferguson	SD	2
31. LaPaloma Ranch	TX	1
32. Laubach Cave No. 2	TX	1
33. La Villa	MEXICO	1
34. Lehner	AZ	13
35. Leikem	AZ	2
36. Lennon	MI	1
37. Lewisville	TX	2
38. Lindsay Mammoth Kill	MT	1
39. Lubbock Lake	TX	3
40. Lucy	NM	1
41. Marion Landfill	OH	1
42. Mead	MI	1
43. Miami	TX	5
44. Mockingbird Gap	NM	1
45. Murray Springs	AZ	4
46. Naco	AZ	1
47. Navarrete	AZ	1
48. Newton	PA	1
49. Nichols	AZ	1
50. NIU-123	IL	1
51. North Sulphur River	TX	1
52. Owl Cave	ID	1
53. Page/Ladson	FL	1
54. Plainview	TX	1
55. Prillwitz	MI	1
56. Priscilla Site	FL	1
57. Professor Valley	UT	1
58. Rancho La Brea	CA	5(+)
59. Sandia Cave	NM	1
60. Santa Fe IA	FL	1
61. Santa Isabel Iztapan	MEXICO	1
62. Santa Rosa Island	CA	2(+)
63. Schaldack	AZ	1
64. Schulze Cave	TX	1
65. Seff	AZ	1
66. Selby	CO	2
67. Sloth Hole	FL	1
68. Solar One	CA	1
69. Stolle Mammoth	NM	1
70. Sun River Canyon	MT	1
71. Sweeney	MI	1
72. Tocuila	MEXICO	5(+)
73. Toyah Mammoth	TX	1
74. Tule Springs	NV	2(+)
75. Union Pacific Mammoth Kill	WY	1
76. Van Horn Mammoth	MI	1
77. Wallman Mammoth	NV	1
78. Ward Island 1	FL	1
79. Watervliet Mammoth	MI	1
80. Whitewater Draw	AZ	1

Not on this table are: Agate Basin (WY), which yielded only a mammoth-ivory artifact from the Clovis level; Bechan Cave (UT), containing mammoth-dung deposits but no bones; Brayton (IA), dated 12.2 ka; China Lake (CA), with one 18.6 ka date on ivory and other materials that are undated; Natural Trap Cave (WY), where mammoth is not in the latest Pleistocene level; and Sheaman (WY), containing one possible mammoth-ivory artifact (which may in fact be antler) but no other bone material.

TABLE 3.2. Part 2 *Mammut* sites proven or likely to be from the Clovis era.

Site name	State/Province	MNI
1. 48SW5981	WY	1
2. Adams	MI	1
3. Arborio Mastodon	NY	1
4. Aurora	IL	3
5. Avery Island	LA	1
6. Bamford Farm	IL	1
7. Ben Franklin	TX	1
8. Berclair Terrace Site 1	TX	1
9. Big Bone Lick	KY	10+
10. Boaz	WI	1
11. Bryce	MI	1
12. Burning Tree	OH	1
13. Cadmus Road	MI	1
14. Carter	OH	1
15. Chicago/Burlington/Quincy RR	IL	1
16. Coats-Hines	TN	1
17. Crystal Lake	IL	1
18. Cutler Hammock	FL	1
19. Dansville	MI	1
20. Darby Spring	FL	1
21. Deerfi eld	WI	3
22. Devil's Den	FL	1
23. Dove Springs Lignites	CA	1
24. Duncker Muskox	IN	1
25. Elkhart	MI	1
26. Evanston	IL	1
27. Evart	MI	1
28. Fairview	MI	1
29. First American Bank	TN	1
30. Four Lakes Drive	MI	1
31. Fulton	IN	4
32. Gingery Cache	FL	1
33. Glencoe	IL	1
34. Grandville	MI	1
35. Green	MI	1
36. Groleau-White Lake	MI	1
37. Hackettstown	NJ	5
38. Heisler	MI	1
39. Herrell Village	MO	1
40. Hiscock	NY	9+
41. Hollidaysburg Fissure	PA	1
42. Hot Run	VA	1
43. Hudson Mastodon	MI	1
44. Huntington Reservoir Sinkhole	UT	1
45. Ivory Pond	MA	1
46. Johnson	MI	1
47. Jolman	MI	1
48. Killin Gravel Pit	MI	1
49. Kimmswick	MO	2
50. Kolarik	IN	1
51. Kuhl	MI	1
52. La Mirada	CA	1
53. Lake Willard	OH	1
54. Latvis/Simpson	FL	1
55. Lehner	AZ	1
56. Lewis Mastodon	IN	1
57. Little Salt Spring	FL	1
58. Maurer	MI	1
59. Midland Mastodon	MI	1
60. Milwaukee	WI	1
61. New Hudson	MI	1
62. NIU-28	IL	1
63. Noble County	IN	3
64. North Sulphur River	TX	1
65. Orleton Farms	OH	1
66. Owosso	MI	1
67. Page/Ladson	FL	1
68. Parker	MI	1
69. Peccary Cave	AR	1
70. Perry Mastodon	IL	1
71. Phillips Park	IL	1
72. Pleasant Lake	MI	1
73. Powers	MI	1
74. Priscilla Site	FL	1
75. Quagaman	MI	1
76. Rancho La Brea	CA	2+
77. Rappuhn	MI	1
78. Rothbury Mastodon	MI	1
79. Rushovic	MI	1
80. Russell Farm 1 & 2	MI	2
81. Sakstrup	MI	1
82. Saltillo	PA	1
83. Sandia Cave	NM	1
84. Santa Fe IA	FL	1
85. Sebastian Canal	FL	1
86. Shaffer	MI	1
87. Sheathelm	MI	1
88. Shelton Mastodon	MI	1
89. Shine	MI	1
90. Sloth Hole	FL	1
91. Springdale	OH	1
92. Taylor	MI	1
93. Ten Mile Rock	AR	1
94. Thaller	MI	1
95. Thames River	ONTARIO	1
96. Troy Mastodon	MI	1
97. Trull	TN	1
98. Van Sickle	MI	1
99. Wakulla Springs	FL	3+
100. Warren	NY	1
101. Wattles	MI	1
102. Wells Mastodon	IN	1
103. Whitewillow	IL	6
104. Winnameg	OH	1
105. Zeller	MI	1

(continued)

TABLE 3.2. Part 2 (continued)

Grayson and Meltzer (2003) deemed 14 megamammal finds to be acceptable as killsites, which is about 8% of my best-guess total of all known sites of the Clovis era, but there's a chance that the percentage of killsites could be even higher. As many as 27 mammoth and mastodont sites may be kills (Haynes, 2002). Grayson and Meltzer (2003) did not accept these other possibilities as kills, but the debate is unfinished and all the evidence should be fairly considered. A higher count of 27 possible proboscidean killsites equals about 15% of the known sites most likely to date to the Clovis era.

This estimate (or even the smaller one of 8%) is a great surprise, especially when compared to proboscidean killsites known from Africa, another landmass which has had a much longer human occupation. The continent of Africa, three times larger than the lower 48 United States, is where hominins have probably killed elephants for hundreds of thousands of years, yet only half a dozen possible kill/scavenge sites have ever been found (Haynes, 1991). Of course less archeological prospecting has been done in Africa, and archeologists are thin on the ground, but the extreme difference in numbers of proboscidean killsites makes the North American record seem uniquely rich and numerous.

Keep in mind that any killsite made by Clovis hunters was very unlikely ever to be preserved at all, which makes it *exceptionally significant that 8–15% of all known Clovis-era proboscidean sites in North America might be cultural.*

Can We Measure Whether Mammoths and Mastodonts Were Already at Risk When Contacted by Clovis Foragers in the Latter Part of the Late Glacial?

A case can be made that Clovis foragers provided the *coup de grâce* to already doomed megafaunal populations. To evaluate the possibility, we first have to measure each taxon's extinction risks in noncultural terms. Biologists can estimate modern species' extinction risk by abstractly categorizing their *rarity* and *adaptability* (see Brown and Lomolino, 1998). The IUCN Red List (IUCN, 2001) sets out more quantitative criteria for evaluating modern species on a global scale; here I apply these criteria to genera on a continental rather than global scale.

Megafauna Rarity and Adaptability

These parameters are measured by comparing local population sizes, geographic range, and habitat specificity. Pleistocene megafauna were not found everywhere. For example, the Great Basin was home to only 19 out of the > 30 genera that became extinct (Grayson, 1993). As with modern animals, the "extent of occurrence" of megafaunal taxa – basically a line drawn around all known fossil sites – was much greater in area than the "area of occupancy" – the finer-scale mapping of only the suitable habitats.

Megafaunal taxa, while they differed in their habitat-specificity, in general probably occupied a broad variety of habitats (Harris, 1985). Thus megafauna were nonrandomly distributed in low densities in specific patches or ranges. Subpopulations may have been isolated, with little contact between them, but recolonization of most patches was possible during much of the Pleistocene. If the subpopulations became small and very isolated, they are said to have become "fragmented" and the possibility of recolonization was much reduced.

They may have been locally rare to locally abundant, and they were generally adaptable to mixed as well as zonal habitats (see, for example, the relatively unusual vegetational reconstruction for the Kimmswick mastodont site in Graham et al., 1981). Taken together, these observations lead to the conclusion that the extinction risk of the more abundant taxa (such as mammoth, mastodont, horse, and camel) would have been relatively low but perhaps rose to moderate during stressful times. In the terms used by the IUCN Red List, the extinction risk of the largest megafauna would not have been evaluated as "threatened." In order to qualify for the lowest-risk subcategory of threatened (a state called "vulnerable"), a taxon requires ten or fewer fragmented areas of occupancy of less than 2,000 km^2 within a range of under 20,000 km^2, or extreme fluctuations in numbers, or a population of less than 10,000 mature individuals fragmented into pockets of less than 1,000 each. Reasonable population estimates of mammoths and mastodonts (see above) and camels and horses (see below) do not place these taxa into the category of "vulnerable." While it is likely that a sizable percentage of the mature individuals were reproductively suppressed in each taxon, nonetheless there were probably more than 10,000 of them in the Clovis-era populations,[7] and the areas of occupancy exceeded the minimum size for elevated extinction risk. Hence, the IUCN-defined extinction risks of the better dated genera were not high during the Clovis era, even if the estimated population numbers seem low to some readers.

Johnson (2002) (also see Cardillo and Lister, 2002) found that animals with low reproductive rates are at higher risk for extinctions in the face of either major climatic changes or human hunting. Body size alone is not an especially strong predictor of risk. Animals in which reproductive rates fall below one offspring per female per year have a chance of extinction greater than 50%, regardless of body size. If climatic and vegetational changes in the Late Glacial were placing the taxa with lowest fecundity at greater risk, even the addition of very low levels of human hunting would have been enough to cause extinction (Alroy, 2001; Johnson, 2002; Mithen, 1993). If the reproductive rates of the extinct Pleistocene megafauna were low, then their extinction risks would have been elevated, either with or without human hunting. Judging on the basis of analogy with recent closely related taxa, many of the extinct taxa did indeed have low fecundity – for example, 1 offspring per 4 + years for mammoth and mastodont, or 1 offspring per 2 + years for horse and camel. Bison, on the other hand, had twice the fecundity of horse and camel – and bison did not become extinct (also see Kiltie, 1984 for related discussion).

How were megafaunal taxa distributed during the Clovis era? The Allerød chronophase corresponds to the beginning of the Clovis era, and it was a drier time than the preceding Bølling and the following Younger Dryas in a large part of the continent (Polyak et al., 2004). Decreased moisture in this interval would have stressed biotic communities, and megafaunal populations were faced with a reduction of suitable habitats within their overall distribution, which would have increased extinction risk. Thus, densities of megafauna within the continent would have declined significantly. However, the overall continental range of megafauna did not shrink, which may seem paradoxical. Graham (2003) has suggested severe range-shrinkage at the end of the Pleistocene just before extinction, but strong evidence indicates that continental distributions were not reduced compared to earlier time intervals, at least for the taxa whose fossil record is ample enough for robust spatial modeling. Agenbroad (1984, 2003, 2005) has shown that mammoth range size in North America was actually larger between 15–10 ka than it had been in any other 5,000-year span of the late Pleistocene. MacPhee et al. (2002) described a northern Asian example of a possible range bottleneck and recovery for several megafaunal taxa during deglaciation, indicating that at least some megafaunal taxa were recovering and able to expand their geographic range before they went extinct, which suggests that climate-change alone was not capable of killing off all the taxa that eventually disappeared. Yet the stability or expansion of species' extent of occurrence does not necessarily mean that megafaunal numbers were expanding, especially during the periods of climatic shifts, although it surely does imply that the animals had become even more mobile in their search for suitable remaining habitats in stressful time intervals.

To better understand how megafaunal ranges may have been affected by deglaciation climate cycles, the term metapopulation (Hanski and Simberloff, 1997) must be introduced here. A metapopulation is a composite of local subpopulations existing across fragmented landscapes where migration from one fragment to another is possible. Each subpopulation (also called deme) has independent dynamics, but migrations from time to time affect their persistence or extinction. The more persistent patches of animal population are called "sources," where population growth rates are positive. Sources are like mainlands surrounded by outlying islands of smaller populations where growth rates are low and immigration is usually negative. The interpatch migration routes usually pass through land areas called sinks, where populations do not persist or grow. I use the term "refugium" to mean source areas that seem to have persisted throughout much of the late Pleistocene, according to the fossil record.

Figure 3.1 shows some suggested refugial or source areas where megafauna were probably present in sustainable numbers during the most difficult final climatic stresses of deglaciation, as judged on the following criteria: Clovis-era dating and also earlier dates; presence of diverse carnivores in assemblages (three or more taxa); presence of more than one fossil locality and no unique taphonomic features; and availability of hydrological and botanical reconstructions showing suitable resources. Also factored in are fluted point occurrences that cluster locally. Let me note here that Meltzer and Mead (1985) proposed that the presence of some subregional fossil clusters may be due to conditions of better preservation and burial opportunities rather than to high animal numbers

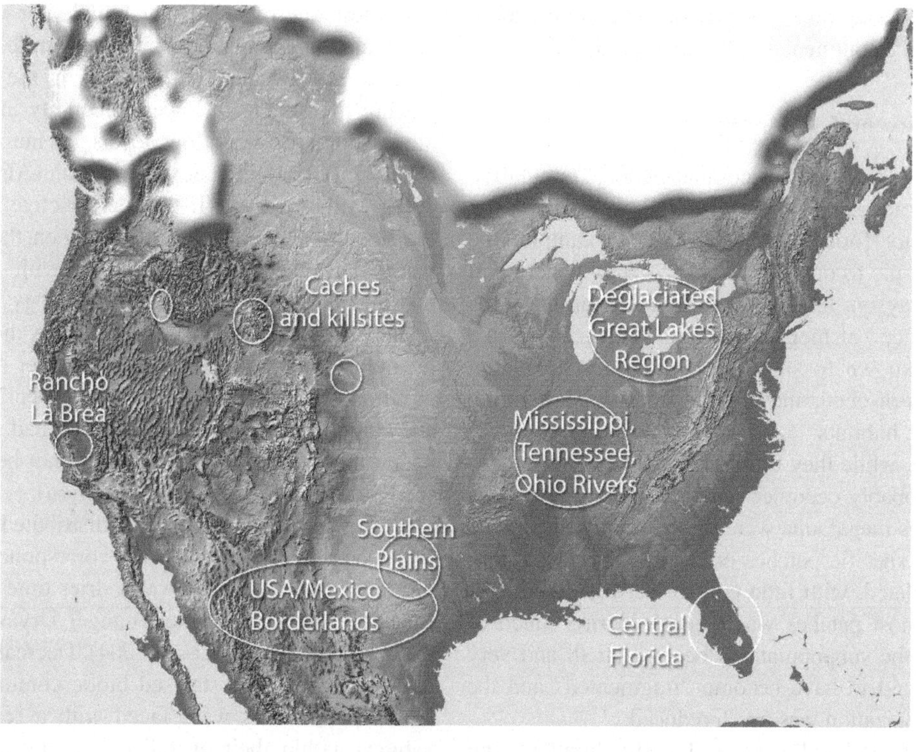

FIGURE 3.1. Possible "source" areas for megafauna and Clovis foragers, defined on the basis of relatively high numbers of fluted-point finds or megafaunal fossils. This is not a map of refugia, but rather a suggestion of the locations where multidisciplinary research can test the concept that certain parts of the continent were indeed better for megafauna (and hence for Clovis megafauna-hunting).

in those localities. But a counterargument can be made about fossil concentrations – the very conditions that make preservation so optimal also would have made the landscapes superior for live animals to aggregate within. I think that the patterns of fossil clustering in North America reflect much more than mere taphonomic luck.

Figure 3.2 shows how the distribution of fossil remains from patchy metapopulations can lead to erroneous reconstructions of larger biogeographic ranges (or "extent of occurrences") than ancient taxa actually occupied on the ground (or "areas of occupancy"). The sink areas between the sources were possibly populated by extremely low density megafaunal populations having negative growth rates. I think animal densities in the sinks were vanishingly small. Enormous overestimates of extinct megafaunal population sizes for the entire continent will result if the deglaciation-interval sinks and sources are not differentiated. Even if the sources were full of animals, the so-called sinks increased in size and number during the Late Glacial period of habitat fragmentation, which would have continually lowered continental average densities and created "dispersal sinks" (Clinchy, 1999) as well, thus more and more isolating the subpopulations.

Of course, an adequate analysis of Late Glacial sources and sinks requires knowledge of each megafaunal subpopulation's age structure, mating system, density dependence factors such as competition, variability (both stochastic and cyclic), and the local habitats' carrying capacity – all beyond the scope of this paper. No clear analysis has yet been done of potential sources and sinks in Late Glacial North America, but future research should examine each taxon's biology as well as range productivity, nutrient availability, hydrology, and fossil occurrences in possible refugia.

The special behavior seen among large terrestrial mammals when they are within such semi-isolated refugia is important to understand. Ecological and actualistic studies of megamammals such as modern elephant and rhinoceros show that in even seasonally refugial patches, large animals behave distinctly (Rachlow, 1997; Western and Martin, n.d.). Intraspecific competition for resources and especially for mates reduces the success rate among less dominant animals, and only a small proportion of males ever reproduce; aggressive encounters between younger males (and also females at times) may lead to a relatively high frequency of injury and death; smaller animals such as juveniles and the very young are killed by predators more often, due to weakening under feeding stress and inattention by stressed adults. These and other factors put local subpopulations at much greater risks for dying out, even when predation pressures are not especially high.

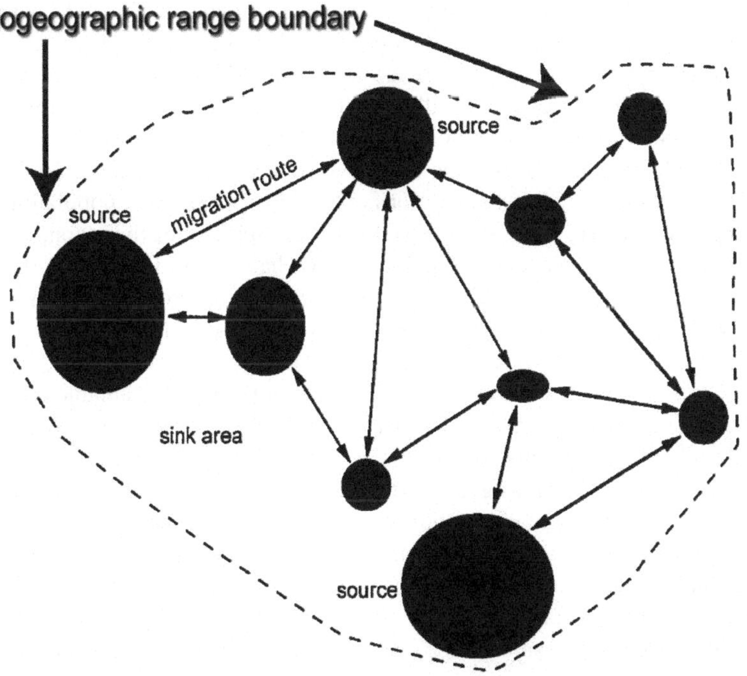

FIGURE 3.2. As shown here, the known finds of fossil bones are conveniently used to define the maximum "biogeographic range boundary" of any taxon. Yet it must be understood that animals do not exist in equal densities throughout all parts of their extent of occurrence. In some subregions ("sources") the population replacement rates equal or exceed death/emigration rates, while in others (called "sinks") the rates cancel out. It is an error to estimate total population numbers by assuming that animal densities were equal throughout the extent of occurrence.

Source-Sink Distributions and Their Implications for Clovis Hunting Decisions

There are great similarities and important differences between Paul Martin's "blitzkrieg" model and this "opportunistic" model of first-contact extinction (FCE, a term adopted in MacPhee [1999]): Both hypotheses predict rapid extinctions; both predict few archeological sites will be created, and they should be dispersed widely in the earliest stages; both predict highly mobile, exploratory human populations that leave ambiguous dietary evidence. In the blitzkrieg model, a geographic and chronological gradient marks the advance of a human wave-front through the continents (see Mosimann and Martin, 1975). Saturated human populations exist behind the front, and a high reproductive rate presses the wave to continue advancing. But in the Opportunistic model, there is no gradient because dispersals are complex, done in leapfrog, yo-yo, and directed patterns; human populations (demes) rise and fall locally; and the dispersal is stimulated by access to rich information about resources in the new landscapes, not by population pressure.

In the Opportunistic model, the extinctions probably occurred in a three-phase process. First came the foreshock which was the fragmentation of habitats and megafaunal populations due to shifting climates; in this phase the megafauna's fitness diminished as climatic reversals bottlenecked populations. Next came the actual shock phase, which was the rapid human dispersal and hunting of megafauna in the refugia and fragmented source areas; in this phase the different taxa's fitness evolved as a fluctuating pattern of population increase and collapse and frequent long-distance migrations. The amplitude of changes in reproductive rates also oscillated wildly, in what is called the classic Weibull distribution (see Lázaro et al., 2003 for an example and references). The third phase was the aftershock, when environmental alterations resulting from climate changes, human-set fires perhaps, and the removal of some taxa finally overcame many species' resistance to extinction.

Human encounter rates with megafauna would have been low in some parts of the continent and higher in others. Thus, in some regions the human foragers would have been specialists who tracked and pursued megafauna; in other times and places, such as outside the refugia and sources, the human populations would have had to behave as generalists. The search, pursuit, and processing efforts made by human foragers were variable from place to place and time to time. Because handling times influence prey choice, the optimal prey in the megafauna sinks – where megafauna were scarce – would have been medium size artiodactyls, small game, and plant foods.

The inevitable results of the Opportunistic FCE process were: foraging flexibility that changed from place to place and time to time; a high ranking of scarce prey sometimes; and a greater patchiness of megafauna that would have led human foragers to spend more time within certain patches and much less time in others.

The archeological evidence to support this model might be seen in the dispersed, low-density early archeological sites, usually lacking signs of repeat visits; the extinct fauna found in some campsites and processing sites; the rarity of actual killsites except of the very largest prey taxa; and the existence of high-ranked (usually the largest) prey remains in some sites and their absence in others which contain very diverse food remains.

Do These Calculations and Propositions Contribute Evidence that Clovis Groups Preferentially Hunted the Largest Mammals?

The preserved and known mammoth and mastodont sites (total of all cultural and noncultural) are indeed extraordinarily rare when compared to the number predicted on the basis of population estimates. Thus the subset of all sites associated with Clovis lithics (8–15% of a tiny surviving remnant) is even more amazingly high than it would seem at first glance. For whatever reasons, such as scarcity of animals, erosion of sediments after burial, or other subtractive taphonomic events, the megamammal sites did not preserve well from the Clovis era, yet a strikingly high percentage of the extremely rare ones that we have found show that widespread human hunting was done.

This empirical evidence indicates Clovis people did hunt megamammals. But could Clovis hunting deplete the megafaunal populations quickly even if the hunting was at low intensity? The short answer is yes (see Mithen, 1993). The extremely delayed maturation time of large mammals plus the long interbirth intervals[8] indicates that the large taxa were K-strategists. Breeding-age adults probably constituted less than 50% of regional populations; these animals were critical for the survival of the taxon, and any hunting of them would have had magnified effects on population viability. Within refugial areas, probably less than 20% of the population would have been actively reproducing. Clovis low-intensity hunting that targeted the adults would have reduced megafaunal reproduction to an unsustainable level fairly quickly. If there were 1,000 mammoths in a refugial area of 10,000–15,000 km^2 during the Clovis era, of which 150 were the main breeding adults, the loss of even 50 animals from hunting would have put the population at severe risk.

The very largest terrestrial mammals did not leave behind many fossil sites, and it seems inevitable that the much smaller megafaunal taxa – such as *Camelops, Equus*, and so forth – undoubtedly also would not have left behind enough fossil sites to indicate their actual abundance and density. More importantly, we should not expect to find any killsites of the smaller animals, judging from the very low likelihood of preservation suggested by the mammoth and mastodont sites.

When Did the Other Megafaunal Taxa Become Extinct, and Why is There So Little Known Evidence that People Hunted Them?

Grayson and Meltzer (2003) listed 35[9] genera that are now extinct in North America; however, five of the genera actually survived the end of the Pleistocene outside North America. These may be cases where the genera moved ("migrated" biogeographically[10]) in response to climate change or increasing predation pressure, which is the most common response in animals (Ashworth, 2003; Barnosky, 2003). These five genera could be removed from the arguments about extinction. Here I leave one in the discussion, namely *Equus*, because this genus and the entire taxonomic family of horses originated in the New World and had evolved over many periods of rapid and severe climatic changes. The complete disappearance of *Equus* from North and South America at the end of the Pleistocene was an unusual event of major importance, and seems to show that some unique factor other than climate-change must have caused the very selective extinction.

Of the remaining genera in Grayson and Meltzer's list, four are carnivores, and may never have been hunted for food. Even if Clovis-era people actively hunted them to eliminate competitors, but never butchered and ate them, no clear archeological associations ever may be found. The carcasses would have been abandoned wherever killed, without evidence of human presence such as campfires or discarded tools.

Thus, this leaves a little over two dozen genera whose extinctions most urgently need to be accounted for. Of these, as few as 8 or as many as 12 megafaunal taxa have been found in Clovis archeological sites (those sites containing typologically identifiable lithic artifacts) (see Table 3.3). Grayson and Meltzer (2003) proposed that even these occurrences are not solidly acceptable evidence of the animals having been killed by Clovis people. I don't dispute (Haynes and Stanford, 1984) that the evidence is ambiguous or equivocal, but I also think that evidence is evidence and its potential meaning cannot be selectively ignored in advance. The 8–12 occurrences should not be stubbornly brushed off just yet.

The most abundant herbivorous megafaunal taxa (see Grayson and Meltzer, 2003, whose data came from Faunmap records of fossil occurrences in the late Pleistocene) are in descending order: *Equus* [several species](horse/ass), *Mammuthus* [several species] (mammoth), *Mammut* (mastodont), *Camelops* (camel), *Bootherium* (musk-ox), *Platygonus* (peccary), *Tapirus* (tapir), *Hemiauchenia* (llama), *Megalonyx* ("great-claw" ground sloth), *Mylohyus* (long-legged peccary), *Paramylodon* a.k.a. *Glossotherium* (mylodont ground sloth), and *Nothrotheriops* (Shasta ground sloth).

Some of these taxa have spectacularly large or unusually shaped bone elements, which undoubtedly prompt much more frequent recovery and attempts to radiocarbon-date them. For example, the ground sloths, three of which are in the list of most abundant, have unusual bones and the mammoths and mastodonts have enormous bones; such taxa can be found and identified very easily, unlike the much smaller antelope-sized bones of *Stockoceros* and *Tetrameryx*, both of which have produced few records. I suggest (as have others before me) that the living mammoths, mastodonts, and ground sloths were not more abundant than contemporary camels and horses, but their bones have been discovered more often.

Twenty-three extinct genera are very scarce. For example, *Aztlanolagus* (Aztlan rabbit) is exceptionally rare (one Faunmap record) and *Brachyprotoma* (short-faced skunk) is nearly as rare (two Faunmap records). *Torontoceros* is not only rare but a bit of a mystery – it may be an ancestral caribou known only from antler material. When searching for ecological details about

TABLE 3.3. Herbivorous megafaunal taxa (other than *Mammut* and *Mammuthus*) found in Clovis-era sites. "Lithics" refers to both Clovis fluted points and other stone implements.

Site	Cultural association	Megafaunal animal(s) present
Aubrey, TX	Clovis lithics	*Paramylodon/Glossotherium*
Blackwater Draw, NM	Clovis lithics	*Smilodon, Bison, Equus, Camelops, Platygonus, Hemiauchenia*
Bull Brook, MA	Clovis lithics	*Rangifer*
Charlie Lake Cave, BC	[concave-base point, other untypable lithics]	*Bison*
Colby, WY	Clovis lithics	*Equus, Camelops, Bison*
Escapule, AZ	Clovis points	*Equus*
Gault, TX	Clovis points	*Equus, Bison*
Hiscock, NY	Clovis points	*Cervalces*
Jake Bluff, OK	Clovis points	*Bison*
Kimmswick, MO	Clovis points	*Paramylodon/Glossotherium, Mylohyus*
Kincaid Shelter, TX	Clovis points	*Equus*
Lange-Ferguson, SD	Clovis points	*Bison*
Lewisville, TX	Clovis points	*Glyptotherium, Equus, Platygonus, Camelops, Bison*
Lehner, AZ	Clovis lithics	*Equus, Tapirus, Camelops, Bison*
Lubbock Lake, TX	Clovis point	*Bison, Equus, Camelops*
Murray Springs, AZ	Clovis lithics	*Equus, Camelops, Bison, Tapirus*
Naco, AZ	Clovis points	*Bison*
Sheridan Pit (or Cave), OH	Possibly late Clovis or post-Clovis; bone point, lithics	*Arctodus, Platygonus, Castoroides*

extinction, such fossil records are hardly adequate to provide solid facts about the biology of the taxa. The uncommon taxa are represented by fossils far too scattered and poorly dated to support extinction arguments one way or another.

The most abundant nonproboscidean herbivores with adequate fossil records are *Equus*, *Camelops*, *Bootherium*, *Platygonus*, *Tapirus*, *Hemiauchenia*, and *Mylohyus*. Of these, the muskox *Bootherium* has not been found in Clovis sites and the long-legged peccary *Mylohyus*, although found at one Clovis site (Kimmswick, MO), may not have been culturally deposited, as could also be the case with *Castoroides* at Sheriden Pit (OH), and *Smilodon* and *Hemiauchenia* at Blackwater Locality No. 1 (NM).

The genus *Bootherium* is actually in doubt – it may be "congeneric with [extant] *Symbos*" (Kurtén and Anderson, 1980; Anderson, 1984), so it will not be considered further. Why would the relatively abundant megafaunal taxa *Mylohyus* and *Castoroides* not be represented in Clovis archeological sites, if Clovis people preferred hunting larger game animals, as I argue? These animals were about the size of a "small white-tailed deer" (Kurtén and Anderson, 1980: 296) in the case of *Mylohyus* or a black bear in the case of *Castoroides*. As solitary or small-group animals, and the smallest of the more abundant (well-dated) extinct megafaunal genera, their carcasses could have been stripped of meat very quickly and the skeletons left in scattered killsites where bones would have had extremely little chance of preservation. No killsites of white-tailed deer (*Odocoileus virginianus*) or black bear (*Ursus americanus*) have ever been found in North America, even though the animals were hunted for thousands of years by native peoples, so it is not logical to expect to find killsites of same-size extinct species from a much shorter timespan of 400 years or so.

What about *Equus*, *Camelops*, and *Hemiauchenia*? Of the 126 Faunmap sites of *Camelops* (all species), 39 date to the Late Glacial, and probably only 17 (MNI = 53) or fewer date specifically to the Clovis era; at least four of the 17 sites (23.5%) and as many as six (35%) are archeological (Table 3.4). Faunmap has 328 site records of *Equus* (all species, including *E. caballus*), of which 62 are Late Glacial and 27 (MNI = 70) or fewer are the likeliest Clovis-era examples; four of the Clovis-age sites (15%) are archeological (Table 3.5). There are ten possibly Clovis-age *Hemiauchenia* sites, including Blackwater Locality No. 1, but no reliable direct dating supports the possibility.

These taxa's ecological ranges differed from those of mammoths and mastodonts, as did their densities and population sizes. Faunmap (1994) records and other attempts to map taxa distribution (e.g., Martin et al., 1985) allow only a very rough estimate of range size in the lower 48 United States: *Camelops* (all species) occupied a maximum range of about 4,000,000 km^2, and I estimate a range of about 3,000,000 km^2 for *Equus* (all species). Other taxa had much more restricted ranges; for example, *Cervalces'* range was probably only about 700–800,000 km^2, and *Capromeryx* range was even smaller, about 300,000 km^2

TABLE 3.4. Camelops sites with associated radiometric dates from the Clovis era (n = 17); not all these sites contained lithics or any other artifactual items. Data from numerous sources, including Faunmap and CARD (2001), the Canadian Archaeological Radiocarbon Database http://www.canadianarchaeology.com/radiocarbon/card/card.htm.

Site name	State/Province	MNI[11]
1. Agate Basin	WY	1
2. BF Alcove	UT	1 (?)
3. Blackwater Locality Number 1	NM	(?)
4. Casper	WY	1
5. Colby	WY	1
6. Dent	CO	(?)
7. False Cougar Cave	MT	(?)
8. Galleli Pit	AB	(?)
9. Jaguar Cave	ID	1
10. La Mirada	CA	1
11. Lamb Spring	CO	8
12. Lehner	AZ	3
13. Lindenmeier	CO	1
14. Lubbock Lake	TX	1
15. Murray Springs	AZ	2
16. Sunshine Locality	NV	1
17. Wasden (Owl Caves 1 and 2)	ID	1

"(?)" in the MNI column is counted as 1, until further data are available.

TABLE 3.5. *Equus* sites with associated radiometric dates from the Clovis era (n =27); not all these sites contained lithics or any other artifactual items. Data from numerous sources, including FAUNMAP and CARD (2001), the Canadian Archaeological Radiocarbon Database http://www.canadianarchaeology.com/radiocarbon/card/card.htm.

Site Name	State/province	MNI[12]
1. Aubrey	TX	1
2. Big Bone Lick	KY	1
3. Blackwater Locality Number 1	NM	15
4. Cave Without a Name	TX	1
5. Clarke Pit	AB	1
6. Colby	WY	1
7. DgPa-VP	AB	1
8. Dry Cave	NM	1
9. Escapule	AZ	1
10. False Cougar Cave	MT	(?)
11. Galleli Pit	AB	(?)
12. Gault	TX	(?)
13. Griffin Pit	AB	(?)
14. Huntington Reservoir	UT	1
15. La Mirada	CA	2
16. Ladd's Quarry	GA	1
17. Lamb Spring	CO	2
18. Lehner	AZ	2
19. Lindoe Bluff	AB	(?)
20. Little River Rapids	FL	(?)
21. Lubbock Lake	TX	2
22. Murray Springs	AZ	3
23. Natural Trap Cave	WY	1
24. Pashley Gravel Pit	AB	1
25. Rancho La Brea	CA	24
26. Ventana Cave	AZ	(?)
27. Wilson Butte Cave	ID	(?)

"(?)" in the MNI column is counted as 1, until further data are available.

Using a variety of sources (such as Hare [2007] and Klingel [1985] on camels, and Berger [1986] on horses), I estimate the continental *Camelops hesternus* population during the Clovis era at about 55,000–60,000, and *Equus* [all species] at about 60,000 or a little more. A handy way to put these figures into perspective is to compare the total number of all public schools in the United States: 90,000. There's a public school to be found just about anywhere you go, but you won't be stumbling across one every day.

Skeletal preservation from Clovis-era camel or horse (for example, Haynes and Huckell, 2007) is much less likely than from mammoth and mastodont. No killsites of camels are known anywhere else in the world, so their absence from Clovis-era sites in North America cannot be considered an unusual fact. Horse killsites are rare in the rest of the world, with the French Upper Paleolithic site of Le Solutré being an exception. But it is unreasonable to expect to find camel, horse, and other taxa killsites from Clovis times when other periods in prehistory have left them nearly nowhere else in the world.

Perhaps the leading argument rolled out to oppose the possibility that human hunting caused the late Pleistocene extinctions is that megafaunal killsites are too rare to reflect the extent of hunting that some people think would have been necessary. To counter the case for human agency in American extinctions, some archeologists point to the massive fossil record of reindeer bones in western Europe as evidence that Paleolithic human hunting would have left behind a spectacular bone record. The implication is that if European reindeer were hunted so relentlessly for so long in the Pleistocene, and yet they survived for millennia, how could low-density stone-tool-using Clovis people ever have depleted over 30 genera of large mammals in North America and left behind almost no identifiable killsites?

The European reindeer sites are very different from the Clovis-era sites, and the differences are all-important. The reindeer-dominated sites are often found in relatively steep-sided European valleys that would have funneled reindeer herds during migrations, thus concentrating game animals near the sites; people could have intercepted and ambushed reindeer herds over and over again for centuries at these locales, accounting for the abundance of bones. Clovis-era sites have not been found in such settings. Also, the reindeer-rich European sites are mostly in limestone caves and grottos that offered people long-term shelter and that preserved bones extraordinarily well in a fixed number of locales, unlike the very rare and scattered open-air Clovis sites, most of which have no organic preservation at all. And perhaps most importantly the European sites were the results of repeated visits over long time intervals – sometimes tens of thousands of years – whereas Clovis-era sites with animal bones in them were single or limited occupations, with the continental sample of sites spanning no more than about 400 years. The point to be made here is that Clovis killsites are extraordinarily rare, and campsites with prey bones are even rarer. In the grand scheme of archeological knowledge, Clovis sites do not measure up against sites elsewhere in the world, and therefore comparing them is not a useful exercise.

O'Connell et al. (1992:322) reported that in Africa most Hadza killsites were "single event locations, marked on abandonment by one or two small scatters of bone debris." Scavenging carnivores removed almost all epiphyses and even some shaft fragments from the small sample of killsites visited 24 h after abandonment by Hadza groups. Only long bone shaft fragments were minimally disturbed by the scavengers, a very slim assemblage indeed to represent Hadza hunting success. Nearly no long-term follow-up studies have been done of human hunters' killsites, but neotaphonomic studies of comparably sized large-mammal taxa in North America, Africa, and Australia (Haynes, 1981, 1991) and a survey of information from ethnographic publications and the eHRAF database (Malinky, 2003) show how extremely unlikely it is that sub-mammoth-sized species ever would have become fossilized, both in cultural and noncultural sites.

Ideally, transported body parts of sub-proboscidean-sized prey animals might be found in Clovis-era campsites or central places, but there are very few of them known with good bone preservation. The Aubrey and Lubbock Lake sites in Texas had areas interpreted as camps, and indeed both sites have yielded bones of extinct megafauna. In spite of the unlikelihood that bones from animals smaller than proboscideans would have been fossilized, nearly all of the non-rare extinct genera are represented in Clovis archeological sites, notably those interpreted as camps or processing sites adjacent to kills. The non-rare genera that are not present or not securely associated with the Clovis occupation, such as *Platygonus*, often had the smallest skeletons. An examination of Clovis archeological sites that have any organic preservation at all demonstrates (Table 3.6) that nearly two-thirds of the adequately published Clovis-lithic sites that yielded plant and animal remains contained the bones or teeth of extinct megafauna.

Grayson and Meltzer (2003), who suggested megafauna were rarely hunted, did not deem most of the sites in Table 3.6 to be behavioral associations of megafaunal bones and artifacts. Critical doubt about the human-animal associations should also be extended to the smaller fauna and to the botanical remains as well, whose main indication of human use often is their proximity to purported but frequently undocumented features in the sites or undocumented indications of burning. Possible burning or spatial proximity to possible fire features can not automatically be considered solid proof of human behavioral association.

The evidence for preferred subsistence use of smaller fauna, as proposed by Cannon and Meltzer (2004) and Banks (2001), among others, is not as strong as it should be. For example, turtle shells are said to have been stacked around a hearth at Blackwater Locality 1, but only a personal communication unspecific about the exact number of specimens (cited by Johnson, 1977 and repeated in Johnson, 1987) and no maps or photographs of indisputable cutmarks have been provided as support. Jeff Saunders (2007 personal communication) has

TABLE 3.6. Clovis-era sites that contained floral or faunal remains associated with diagnostic lithics. The designation "Clovis" refers to fluted bifaces such as Clovis or Gainey types.

Site	Organic material recovered	Cultural material and approx. age (BP)
Aubrey, TX	Bones/teeth of deer, bison, rabbit, muskrat, fishes, birds, turtles, rodents, ground sloth	Clovis lithics
Blackwater Draw, NM	*Mammuthus columbi* bones, MNI=6; also bones/teeth of bison, horse, camel, box turtle, carnivores, rodents, plus teeth only of antilocaprid, extinct paleoleama, flatheaded peccary	Clovis lithics; 11,040–11,630
Bull Brook, MA	Bones/teeth of caribou, beaver	Clovis lithics
Charlie Lake Cave, BC	Bones/teeth of bison, lagomorph, muskrat, rodent, fish, bird	[concave-base point]
Colby, WY	*Mammuthus columbi* bones, MNI=7; also bones/teeth of pronghorn, camel, hare, ass, bison	Clovis lithics; 11,220; 10,864
Dent, CO	*Mammuthus columbi* bones, MNI=15	Clovis lithics; 11,200; 10,980–10,670
Domebo, OK	*Mammuthus columbi* bones, MNI=1	Clovis lithics; ~11,000
Dutton, CO	*Mammuthus columbi* bones, MNI=1	Clovis lithics; <11,710
Escapule, AZ	*Mammuthus columbi* bones, MNI=1; also horse	Clovis lithics; no date
Hiscock, NY	*Mammut americanum* bones, MNI=10; also bones/teeth of caribou, stag-moose, *long-nosed peccary,* California condor, pied-billed grebe, small unidentified mammal	Clovis [Gainey] points; 9,205 ± 50 to 11,450 ± 50
Holcombe, MI	Bones/teeth of caribou	Clovis lithics
Jake Bluff, OK	Bones of *Bison*	Clovis lithics
Kimmswick, MO	*Mammut americanum* bones, MNI=2; also bones/teeth of micromammals (mainly rodents)	Clovis lithics; no date
Kincaid Shelter, TX	Bones/teeth of alligator, slider and box turtle, armadillo, badger, raccoon, mice	Clovis lithics
Lange-Ferguson, SD	*Mammuthus columbi* bones, MNI=2; also micromammals, deer, bison	Clovis lithics; no direct date
Lehner, AZ	*M. columbi* bones, MNI=13; also at least 11 taxa, incl. micromammals, and horse (teeth), camel, bison	Clovis lithics; 10,900
Leikum, AZ	*M. columbi* bones, MNI=2	Clovis lithics; no date
Lewisville, TX	Bones/teeth of horse, peccary, camel, deer, small mammals, amphibians, reptiles (and reptile eggs), mud-dauber larvae, hackberry seeds in hearth features	1 Clovis point, hearth features
Lubbock Lake, TX	*M. columbi* bones, MNI=2(?); also horse, peccary, camel spp., short-faced bear	Clovis lithics; 11,100
Miami, TX	*M. columbi* bones, MNI=5	Clovis lithics; no date
Murray Springs, AZ	*M. columbi* bones, MNI=2; also bones/teeth of numerous taxa, incl. micro-mammals, and horse (teeth), camel, bison	Clovis lithics; 10,900
Naco, AZ	*M. columbi* bones, MNI=1; also bison	Clovis lithics; no date
Navarette, AZ	*M. columbi* bones, MNI=1	2 Clovis points, no date
Sloth Hole, Aucilla River, FL	*Mammut americanum* bones, MNI=1+; also other mammalian taxa	Fluted point variants, lithics, bone tools, 33 ivory points; no direct dates
Shawnee Minisink, PA	Calcined bones of unidentified fish, micromammals, and reptiles; also 76+ botanical specimens, including examples of *Chenopodium*, blackberry, carbonized *Crataegus* (hawthorn) seeds, *Physalis* (Cape gooseberry), *Acalypha*-like, *Vitis* (wild grape) seeds, *Celtis* (hackberry) seeds, 1 Clovis point, many scrapers, debitage	1 Clovis point, other lithics, hearth/fire floor, ~10.9 ka
Sheridan Pit (or Cave), OH	Bones/teeth of snapping turtle, caribou, flatheaded peccary, giant beaver, plus microfauna (noncultural?)	Possibly late Clovis or post-Clovis (Holcombe-like point), bone point, lithics
Udora, Ontario	Bones/teeth of cervid (including caribou), hare/rabbit, arctic fox	Gainey/Clovis lithics, calcined bones (possibly hearth sweepings)
Whipple, NH	Bones/teeth of caribou	Clovis lithics
One site in Israel River valley, NH	1 charred water lily seed in an excavated feature	?

noted at least three terrapin shells from Blackwater Locality 1 that had punctures interpreted as killing marks, and which were apparently butchered. "Indeterminate turtle" shell (presumably fragments) outnumber all but the bone fragments of "indeterminate mammal – large" in the Camp B area of the Aubrey Clovis site in Texas (Yates and Lundelius, 2001: 106–108), but no data indicate whether there were many turtles represented or just a couple with comminuted shells. None were noted as cut-marked. Turtle also dominated Aubrey's Area A pond deposits; a small proportion is said to have been burned, but no fire pit features were found. Neotaphonomic studies have shown (e.g., Gary Hurd's online website http://medtsta.med.utah.edu/kw/

osteo/hurd/apage.htm) that wildfires can differentially burn bones, charring some specimens and calcining others, thereby deceptively appearing to represent burning patterns created by nonrandom clustering around cultural fire features. Sites that produce burned bones, even when the burning is apparently "patterned," do not necessarily provide good, clear evidence of a turtle-heavy Clovis diet.

Some of Aubrey's Area A rodent remains are from burrowers and may be intrusive (Yates and Lundelius, 2001), but a great many of the rodent bones were burned, more than any other taxon's in Area A pond strata. Is this evidence of rodents cooked on open fires? These kinds of data are interesting and cannot be dismissed, but they are not qualified to be uncritically considered as "strong evidence of Early Paleoindian subsistence use" (Cannon's and Meltzer's [2004] words).

Conclusions

The significant association of artifacts and megafauna might have resulted from collector bias (the biggest taxa's bones are easiest to find), or from taphonomic accidents (the biggest bones are best preserved), or from a real Clovis preference for hunting megafauna in some parts of the continent. The hypothesis that Clovis foragers were small-game hunters and plant-food collectors who rarely or never tried to kill megafauna has been proposed, but this story is no more strongly supported than the hypothesis that Clovis foragers were opportunistic gatherers of all edible plant products and hunters of all animals including megafauna, and they deliberately chose to exploit the largest animals *under certain conditions* (Haynes, 2002) set up by shifting climate at the end of the Pleistocene.

In this chapter I have suggested that the continental populations of the most frequently found extinct taxa (mammoth, mastodont, horse, and camel) were not large, teeming, or especially abundant everywhere. I also suggested that human hunting left an empirical record that is underappreciated by archeologists who unrealistically expect to find abundant bone piles throughout the continent – this is simply not how the fossil record ever would have formed in the geologically brief interval of the Clovis era. There are few killsites of the largest megafaunal genera, mammoth and mastodont, but these are still a very significant proportion of all fossil sites dating from this time interval, in my view reflecting the locale-specific emphasis that Clovis foragers put on the hunting of proboscideans. Most of the other extinct megafaunal taxa are too poorly known and too rarely dated to be used in arguments against the possibility that human hunting impacted the populations. But the best dated and more frequently found megafaunal taxa are represented in some of the extremely few Clovis campsites that have been excavated, and this fact should keep us from overlooking the potential paleoecological impacts of even low-level megafaunal hunting by opportunistic Clovis foragers.

Notes

1. I only briefly discuss *Bison* here, because the genus never became extinct in North America. Note, however, that there was a bison species replacement at the beginning of the Holocene, quite possibly caused by a combination of factors including human hunting. Drummond et al. (2005), among others recently, noted a severe population bottleneck affecting Beringian bison at around the time of the first human presence.
2. Density is geographically variable (see Berger, 1986:238–242). Under conditions where animals can forage freely, animal density is not fixed, but adjusts to the quality of habitats. If the better habitats are limited, the density will be high, but part of the population also may be forced into low-quality areas, where density would be lower. In species where some individuals such as territorial males monopolize the better areas, the densities may be fairly consistent in the better areas, but higher or lower outside them.
3. In Douglas-Hamilton's (1980) census of Africa's elephants, the average continental density was 0.18/km^2 in 35 countries. However, this figure is the result of several unnatural forces – habitat removal and destruction by humans, intensive hunting in certain regions, and crowding of elephants into protected areas. Densities in the protected areas themselves varied from > 4.5 to 0.09/km^2.
4. There are small differences between the online and the diskette versions of the Faunmap database.
5. Not all these dates were derived directly from mammoth bone samples, and in some cases the mammoth bones are associated with a range of dates older than Clovis.
6. Furthermore I propose that there are not *ca.* 2,000 "Clovis-age localities" containing all the taxa of extinct megafauna in them, as has been stated (e.g., Holliday, 2003).
7. Here I use the term "population" to refer to all individuals of a taxon and not to just the mature ones as the IUCN Red List does (IUCN, 2001). A "subpopulation" is a geographically distinct group that has little exchange with other groups.
8. For example, 22 month gestation plus four-year interbirth interval in proboscideans, or 11 to 14 month gestation in horses and camels, respectively, versus nine months in bison.
9. The number differs from author to author, depending on the preferred taxonomy and the authors' decision to include genera that survived on other continents.
10. In this case, "migrated" does not necessarily mean the animals each physically moved; biogeographic migration refers to a species changing range, accomplished either through movement and local extirpation behind an advancing front, or extirpation within all the outlying ranges.
11. The number is unsettled because not all the bones are thought to be culturally associated in each site.
12. The number is unsettled because not all the bones are thought to be culturally associated in each site.

References

Agenbroad LD (1984) New World mammoth distribution. In: Martin PS, Klein RG (eds) Quaternary extinctions: A prehistoric revolution. University of Arizona Press, Tucson, pp 90–112

Agenbroad LD (2003) Post-30,000 BP mammoth distribution in North America. Paper presented at 3rd International Mammoth Conference, May 24–29, Dawson City/Whitehorse, Yukon

Agenbroad LD (2005) North American proboscideans: Mammoths: The state of knowledge, 2003. Quat Intl 126–128:73–92

Alroy J (2001) A multispecies overkill simulation of the end-Pleistocene megafaunal mass extinction. Science 292:1893–1896

Anderson E (1984) Who's who in the Pleistocene: A mammalian bestiary. In: Martin PS, Klein RG (eds) Quaternary extinctions: A prehistoric revolution. University of Arizona Press, Tucson, pp 40–89

Ashworth AC (2003) The Quaternary record for extinction and speciation in the Coleoptera. Paper presented at the XVI INQUA Congress, July 23–30, Reno, NV

Banks NT (2001) Review of evidence for Clovis subsistence. Ph.D. Dissertation, The University of Texas, Arlington

Barnosky AD (2003) Effects of Quaternary climatic change on speciation in mammals. Paper presented at the XVI INQUA Congress, July 23–30, Reno, NV

Berger J (1986) Wild horses of the Great Basin: Social competition and population size. University of Chicago Press, Chicago

Brown JH, Lomolino MV (1998) Biogeography (2nd ed) Sinauer, Sunderland, MA

Cannon MD, Meltzer DJ (2004) Paleoindian foraging: Examining the faunal evidence for large mammal specialization and regional variability in prey choice. Quat Sci Rev 23:1955–1987

CARD (Canadian Archaeological Radiocarbon Database, compiled by R. Morlan) (2001) Online website <http://www.canadianarchaeology.ca/> accessed 6/04

Cardillo M, Lister A (2002) Death in the slow lane. Nature 419:440–441

Clinchy M (1999) Does immigration "rescue" populations from extinction? Ph.D. Dissertation, University of British Columbia, British Columbia

Douglas-Hamilton I (1980) IUCN/WWF/NYZS elephant survey and conservation programme: The African Elephant action plan [excerpts]. Elephant 1(4):47–68

Douglas-Hamilton I, Douglas-Hamilton O (1975) Among the elephants. Viking Press, New York

Drummond AJ, Rambaut A, Shapiro B, Pybus OG (2005). Bayesian coalescent inference of past population dynamics from molecular sequences. Mol Biol Evol 22(5):1185–1192

Eltringham K (1982) Elephants. Blandford Press, Dorset

Faunmap Working Group (1994) Faunmap: A database documenting Late Quaternary distributions of mammal species in the United States. Illinois State Museum Scientific Papers, Vol. XXV, No. 1 and No. 2

Graham RW (2003) Environmental models for late Pleistocene megafaunal extinctions and MIS 5E. Paper presented at the XVI INQUA Congress, July 23–30, Reno, NV. Abstract published in XVI INQUA Congress Programs with Abstracts, p 89

Graham RW, Haynes CV Jr, Johnson DL, Kay M (1981) Kimmswick: A Clovis-mastodon association in eastern Missouri. Science 213:1115–1117

Grayson DK (1993) The desert's past: A natural prehistory of the Great Basin. Smithsonian Books, Washington, DC

Grayson DK, Meltzer DJ (2003) A requiem for North American Overkill. J Arch Sci 30:585–593

Hanks J (1979) The struggle for survival. Mayflower Books, New York

Hanski I, Simberloff D (1997) The metapopulation approach, its history, conceptual domain, and application to conservation. In: Hanski IA, Gilpin ME (eds) Metapopulation biology. Academic Press, San Diego, pp 5–27

Hare J (2007) *Camelus ferus*. In: IUCN 2007. 2007 IUCN Red List of Threatened Species. <www.iucnredlist.org>. Downloaded 2 January 2008

Harris AH (1985) Late Pleistocene vertebrate paleoecology of the West. University of Texas Press, Austin

Haynes CV Jr, Huckel BB (eds) (2007) Murray Springs: A Clovis site with multiple activity areas in the San Pedro Valley, Arizona. Anthropology Papers Number 71, University of Arizona

Haynes G (1981) Utilization and skeletal disturbances of North American prey carcasses. Arctic 35(2):266–281

Haynes G (1988) Longitudinal studies of African elephant death and bone deposits. J Arch Sci 15:131–157

Haynes G (1991) Mammoths, mastodonts, and elephants: Biology, behavior, and the fossil record. Cambridge University Press, New York

Haynes G (2002) The early settlement of North America: The Clovis Era. Cambridge University Press, Cambridge

Haynes G, Anderson DG, Ferring CR, Fiedel SJ, Grayson DK, Haynes CV Jr, Holliday VT, Huckell BB, Kornfeld M, Meltzer DJ, Morrow J, Surovell T, Waguespack NM, Wigand P, Yohe RM II (2007) Comment on "Redefining the age of Clovis: Implications for the peopling of the Americas," by Waters MR, Stafford TW Jr. Science 317, DOI: 10.1126/science.1141960

Haynes G, Stanford D (1984) On the possible utilization of *Camelops* by Early Man in North America. Quat Res 22:216–230

Holliday VT (2003) Where have all the mammoths gone? [Review of The Early Settlement of North America by G. Haynes]. Science 300:1373–1374

Hurd G (2005) (accessed online 22 February 2005) Faunal taphonomy. Online text and image site <http://medtsta.med.utah.edu/kw/osteo/hurd/apage.htm>

IUCN (The World Conservation Union) (2001) IUCN Red List Categories and Criteria, Version 3.1. Prepared by the IUCN Species Survival Commission, Approved by the 51st Meeting of the IUCN Council, Gland, Switzerland

Johnson CN (2002) Determinants of loss of mammal species during the Late Quaternary 'megafauna' extinctions: Life history and ecology, but not body size. Proc Roy Soc Lond B 269:2221–2227

Johnson E (ed) (1977) Paleo-Indian lifeways. Mus J 17 (Texas Tech University)

Johnson E (ed) (1987) Lubbock Lake: Late Quaternary studies on the southern high plains. Texas A&M University Press, College Station, TX

Jones M (1991) Unpublished aerial census data and files on elephants and other large mammals in North-West Matabeleland. (Notes and copies in author's possession)

Kiltie RA (1984) Seasonality, gestation time, and large mammal extinctions. In: Martin PS, Klein RG (eds) Quaternary extinctions: A prehistoric revolution. University of Arizona Press, Tucson, pp 299–314

Klingel H (1985) Soziale Organisation des Dromedars. Verhandlungen der Deutsche Zoologische Gesellschaft 78:210

Kurtén Bj, Anderson E (1980) Pleistocene mammals of North America. Columbia University Press, New York

Lázaro E, Escarmis C, Pérez-Mercader J, Manrubia SC, Domingo E (2003) Resistance of virus to extinction on bottleneck passages: Study of a decaying and fluctuating pattern of fitness loss. Proc Natl Acad Sci 100(19):10930–10835

MacPhee RDE (ed) (1999) Extinctions in near time: Causes, contexts, and consequences. Plenum, New York

MacPhee RDE, Tikhonov AN, Mol D, de Marliave C, van der Plicht H, Greenwood AD, Flemming C, Agenbroad L (2002) Radiocarbon chronologies and extinction dynamics of the late Quaternary mammalian megafauna of the Taimyr peninsula, Russian Federation. J Arch Sci 29:1017–1042

Malinky B (2003) Discard patterns in ethnographic hunter-gatherer large game killsites. Manuscript

Martin LD, Rogers RA, Neuner AM (1985) The effect of the end of the Pleistocene on man in North America. In: Mead JI, Meltzer DJ (eds) Environments and extinctions: Man in Late Glacial North America. Center for the Study of Early Man (University of Maine at Orono), Orono, pp 15–30

Meltzer DJ, Mead JI (1985) Dating late Pleistocene extinctions: Theoretical issues, analytical bias, and substantive results. In: Mead JI, Meltzer DJ (eds) Environments and extinctions: Man in Late Glacial North America. Center for the Study of Early Man (University of Maine at Orono), Orono, pp 145–173

Mithen S (1993) Simulating mammoth hunting and extinction: Implications for the late Pleistocene of the central Russian Plain. In: Peterkin GL, Bricker HM, Mellars P (eds) Hunting and animal exploitation in the Later Palaeolithic and Mesolithic of Eurasia. Anthropol Pap Am Anthropol Assoc 4:163–178

Mosimann JE, Martin PS (1975) Simulating overkill by Paleoindians. Am Scient 63:304–313

Nowak RM (1999) Walker's mammals of the World, 6th edn. Johns Hopkins University Press, Baltimore

O'Connell JF, Hawkes K, Blurton-Jones NG (1992) Patterns in the distribution, site structure and assemblage composition of Hadza kill-butchering sites. J Arch Sci 19:319–345

Parker ISC, Graham AD (1989) Men, elephants and competition. In: Jewell P, Maloiy GMO (eds) The biology of large African mammals in their environment: Proceedings of a symposium held at the Zoological Society of London, 19–20 May 1988. Symp Zool Soc Lond 61:241–252

Polyak VJ, Rasmussen JBT, Asmerom Y (2004) Prolonged wet period in the southwestern United States through the Younger Dryas Geology 32(1):5–8

Rachlow JL (1997) Demography, behavior, and conservation of White Rhinos. Ph.D. Dissertation, University of Nevada, Reno

Sukumar R (1989) The Asian elephant: Ecology and management. Cambridge University Press, Cambridge

Waguespack NM, Surovell TA (2003) Clovis hunting strategies, or how to make out on plentiful resources. Am Ant 68(2):333–352

Waters MR, Stafford TW Jr (2007) Redefining the age of Clovis: Implications for the peopling of the Americas. Science 315:1122–1126

Western D, Martin PS (n.d.) A fragmentation-debilitation model of Pleistocene extinctions. Manuscript

Yates B, Lundelius E (2001) Vertebrate faunal remains from the Aubrey Clovis site. In: Ferring CR (ed) The archaeology and paleoecology of the Aubrey Clovis site (41DN479), Denton County, TX. Center for Environmental Archaeology, University of North Texas, Denton, pp 103–119

4. Paleobiology and Extinction of Proboscideans in the Great Lakes Region of North America

Daniel C. Fisher*
Museum of Paleontology and Department of Geological Sciences
University of Michigan
Ann Arbor, Michigan 48109-1079, USA
dcfisher@umich.edu

Keywords Mastodon • mammoth • tusk growth • season of death • life history

Introduction

I would like to address the topic of this chapter in calm reflection on a mature body of data, representing a balanced sampling of the empirical record and unhurried evaluation of its possible interpretations. I would also like to be 5 – no, 10 – years further along in the very labor-intensive process of compiling that empirical record! For now, however, I will have to settle for a status report on a series of ongoing investigations designed to assess the nature of late Pleistocene proboscidean occurrences and evaluate aspects of proboscidean paleobiology that have the potential to yield insights concerning the ecological stresses encountered by these animals during the centuries and millennia leading up to the time of their ultimate extinction.

This book focuses on the broad problem of late Pleistocene losses of megafaunal taxa across the Americas, which is itself a geographically, taxonomically, and temporally restricted subset of the larger problem of worldwide losses of megafaunal diversity. In contrast, my title carves out an even smaller region (and set of taxa) as the domain for my analysis. Work in progress actually involves proboscideans from more diverse regions of the Americas and from Siberia as well, and it has involved a variety of aspects of proboscidean paleobiology, but only for the Great Lakes region of North America are there enough data in hand at this time to warrant a summary of trends that offer evidence of the cause of extinction.

* Address for correspondence: dcfisher@umich.edu

G. Haynes (ed.), *American Megafaunal Extinctions at the End of the Pleistocene*, 55–75.

The strategy on which this work is based is to investigate the behavior, physiology, and life history of proboscidean victims of the late Pleistocene extinction. My emphasis on victims is in part because generations of scholars working before me have already shed much light on factors that might have been responsible for this extinction – essentially, the "cause" end of the chains of cause-and-effect relationships that brought about the extinction. We of course know the ultimate "effect" end of the chains, the extinction itself, but what remains to be identified and evaluated are intermediate effects that might constitute evidence that one or another causal factor was at work. The advantage of seeking these intermediate effects within the biology of victim species is that only an effect documented there, ideally as a change in "state" that is temporally associated with, and plausibly premonitory to, the extinction event, can be said to have really "completed" one of the hypothesized chains of cause-and-effect relations. Changes wrought in victim species, beyond serving as evidence for identifying causes, also demonstrate, as nothing else can, that the identified cause had an *impact* on the victim (Fisher, 1996a, 2001a).

Since we have no direct, observational access to most of the behavior and physiology that we think is important for understanding late Pleistocene events, we would have little chance of following the strategy described above were it not for the remarkable records of growth and life history that are encoded in the tusks of mastodons and mammoths (most observations presented here involve *Mammut americanum*, but some deal with *Mammuthus primigenius*, *M. columbi*, and specimens referred to as *M. jeffersonii*, which I have suggested may represent hybrids between *M. primigenius* and *M. columbi* (Fisher, 2001a; Hoyle et al., 2004). Other victims of the late Pleistocene extinction offer dental records that might prove tractable in certain respects, but no other animal offers a structure quite like a tusk, capable of recording virtually the entire life.

Tusks are enlarged incisors that grow continuously, without remodeling. In tusks of older individuals, especially males, the earliest part of the record may be missing due to tip fracture or abrasion, but typically the middle and late portions of life are preserved in their entirety. Periodically formed laminae within tusk dentin provide internal temporal control, and changes in dentin composition provide clues to aspects of life history and environment. In some cases, parts of this record may help to constrain causes of death, but an even more general outcome is that the record clarifies circumstances under which animals *lived*. After all, most of the tusk is a record – many years long – of life and growth; only the last layers pertain to the time of death.

Morphologic and Taphonomic Contexts for Tusk Studies

Tusk analyses that are most informative about environment and life history involve fine structural and compositional details that could be documented and interpreted on an essentially microscopic scale, without explicit reference to the larger geometry of the tusk, let alone the rest of the animal or its conditions of preservation. However, a tusk always has some larger-scale morphological configuration, and it is often found associated with additional parts of the animal, within a particular depositional setting and taphonomic context. Each of these successively larger contextual scales comes with its own potential for recording information that can be relevant for interpreting even the finest details of tusk structure and composition.

Sexual Dimorphism of Tusks

Previous workers have remarked on the pronounced dimorphism of proboscidean tusks, in which tusks of adult males are typically longer and much larger in diameter, while those of females are shorter and more slender (Osborn, 1936; Vereschchagin and Tikhonov, 1999; Fig. 4.1A). Still, distinguishing sex from tusk size and shape is not trivial when a sample includes tusks of younger individuals. In particular, tusks of a young, incompletely grown male can be difficult to distinguish from tusks of adult females. However, Elder (1970) recognized that male African elephants of intermediate and larger size have pulp cavities that are longer, extending distally past the alveolar margin, while females have shorter pulp cavities that end proximal to the alveolar margin. This same difference has been observed in both mastodons and mammoths (Sher and Fisher, 1995; Smith and Fisher, 2007; Fisher, 2008, and unpublished data; 2001b on *M. primigenius*), although pulp cavities for both sexes get shorter in the oldest individuals, as tusk diameters also begin to decrease. More informative than single extremal measures such as tusk length, maximum diameter, or pulp depth, are series of measurements (e.g., diameter or girth) made at a sequence of positions relative to the tusk tip (Fig. 4.1B). These comprise a "profile" or a time series (uncalibrated to time-in-life, if the independent variable is *distance* from the tusk tip) for the measurement in question. Profiles of tusk girth vs. length for nine males and eight females in Fig. 4.1B trace out largely non-overlapping trajectories. Only one young male (the Heisler mastodon; all specimens referred to here are listed in Table 4.1) occupies the

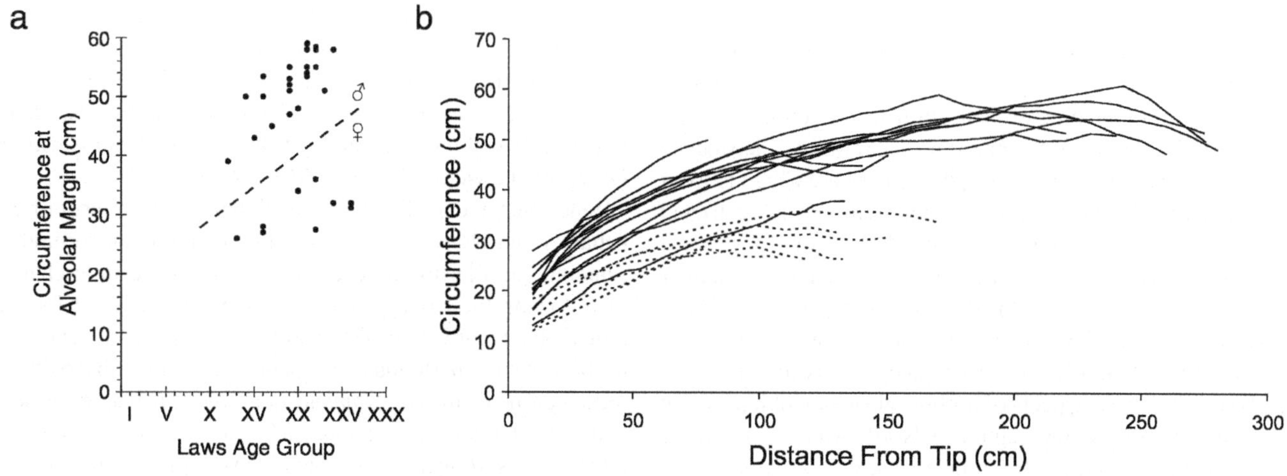

FIGURE 4.1. Sexual dimorphism in tusks of *Mammut americanum* (after Fisher 2008). a. Graph of tusk circumference (averaged when both tusks are present) near the alveolar margin vs. Laws' Age Group, based on cheek tooth dentition, for all specimens listed in Table 4.1 (Krugler has no associated molars, but is given an approximate Laws' age group, based on epiphysis fusion; Fisher, 2008). Tusk circumference increases with age for both male and female mastodons, but males of a given age, or stage of molar development, show greater tusk girth than do same-stage females. Dashed line (placed by eye) follows separation between putative male and female morphs. b. Tusk girth profiles measured relative to distance from the tusk tip. Solid lines, inferred males (Hyde Park, Buesching, Burning Tree, Cohoes, Farview, Grandville, Heisler, Parker, Pleasant Lake); dashed lines, inferred females (Alma, Laur, Miller, North Java, Owosso, Powers, Sheathelm, Shelton).

TABLE 4.1. American mastodon specimens referred to in this work. Institutional abbreviations: AC, Alma College, Alma, MI; CIS, Cranbrook Institute of Science, Bloomfield Hills, MI; INSM, Indiana State Museum, Indianapolis, IN; KPMNH, Kanagawa Prefectural Museum of Natural History, Yokahama, Japan; MSU, Michigan State University, East Lansing, MI; NYSM, New York State Museum, Albany, NY; PMGR, Public Museum of Grand Rapids, Grand Rapids, MI; PRI, Palaeontological Research Institution, Ithaca, NY; RMSC, Rochester Museum & Science Center, Rochester, NY; UM, University of Michigan, Ann Arbor, MI. Sex (M, F) inferred from data presented in Fisher (2008). LAG, Laws' (1966) Age Groups (assignments in brackets based on estimated number of years in tusk, rather than cheek teeth); TG, tusk girth (cm) near alveolar margin (values in brackets estimated from partial tusks or alveoli).

Specimen name	Specimen number	Sex	Material available	State	LAG	TG
Hyde Park	PRI	M	Partial skeleton	NY	XXII	55.0
Cohoes	NYSM V101	M	Partial skeleton	NY	XX	48.0
East Bloomfield	RMSC	M	Partial skeleton	NY	XXII	58.1
Farview	RMSC	M	Partial skeleton	NY	XXI	53.4
Bloomfield Hills	UM 11308	M	Partial skeleton	MI	XVI	50.0
Pleasant Lake	UM 57705	M	Partial skeleton	MI	XXI	58.0
Brennan	UM10627	M	Partial skeleton	MI	XVI	53.4
Darling	UM 22273	M	Cranium	MI	XXI	59.0
Kuhl	UM 59936	M	Partial skeleton	MI	XXIV	58.0
Johnson	UM 57648	M	Partial skeleton	MI	XXII	58.5
Smith-Running	UM 10934	M	Partial skeleton	MI	XV	43.0
Parker	AC	M	Partial skeleton	MI	XXI	54.0
St. Johns	UM 12306	M	Palate	MI	XXI	55.0
McAlpin	UM 11731	M	Partial skeleton	MI	XIX	55.0
Quagaman	UM 24240	M	Cranium	MI	XIX	52.0
Heisler	UM 61888	M	Partial skeleton	MI	XII	39.0
Russell Farm I	UM 37811	M	Partial skeleton	MI	XVII	45.0
Krugler	UM 16303	M	Tusk, misc.	MI	[XXII]	[58]
Grandville	PMGR	M	Partial skeleton	MI	XXIII	51.0
Striker	UM 3489	M	Cranium	MI	XIX	47.0
Winnameg	UM 11230	M	Partial skeleton	OH	XIX	[51]
Burning Tree	KPMNH	M	Partial skeleton	OH	XIX	53.0
Buesching	INSM 71.3.261	M	Partial skeleton	IN	XXI	55.0
Elkhart	UM 34302	M	Cranium	IN	XIV	50.0
North Java	PRI	F	Tusk	NY	[XXVI]	31.2
Shelton	CIS	F	Partial skeleton	MI	XIII	26.0
Sakstrup	UM 54910	F	Cranium, misc.	MI	XXIV	32.0
Sheathelm	MSUVP 1355	F	Cranium, tusk, mand.	MI	XVI	28.0
Owosso	UM 23498	F	Partial skeleton	MI	XXII	36.0
Alma	AC	F	Tusk	MI	[XXII]	30.2
Eldridge	UM 58075	F	Partial skeleton	MI	XX	34.0
Laur	UM 16190	F	Tusk	MI	[XXII]	27.5
Powers	UM 13971	F	Partial skeleton	MI	XXVI	32.0
Miller	UM 16191	F	Tusk, molars	IN	XVI	27.0

portion of the morphospace in which females fall. It falls below the plots for most male tusks because it has not experienced appreciable tip breakage (making it longer, relative to its girth), but it matches males, and is distinguished from females, by its rapid rate of increase of girth with length.

Skeletal Evidence of Sexual Dimorphism and History of Injuries

As informative as tusks can be, we would be remiss to ignore the larger anatomical system of which they are but one element. Part of the value of associated skeletal material, when it is available, is its ability to refute or confirm tusk-based determinations of sex. In all extant proboscideans and in woolly mammoths where preserved genitalia permit unambiguous sex determination, adult males are significantly larger than adult females (Vereshchagin and Tikhonov,1986; Haynes, 1991). Body size dimorphism is also conspicuous in mastodons (Fisher, 2008) and Columbian mammoths (Lister and Agenbroad, 1994). Using bone size for sex determination requires some comparative framework, but this can be provided by comparing bone size among individuals and by referencing bone size to stages of epiphysis fusion on the same or other bones (Roth, 1984; Haynes, 1991) or to stage of molar eruption and wear.

The clearest and most easily understood example of sexual dimorphism in bone shape involves the innominates, or pelvis. As shown by Deraniyagala (1955), Haynes (1990), and Lister (1996; Coope and Lister, 1987; Lister and Agenbroad, 1994), females have a broad (and high) pelvic aperture relative to the width of the shaft of the ilium (a useful measure of size and robustness, even when there is peripheral damage to other parts of the innominate), giving them a large birth canal relative to their body size. In addition, recent work on mastodon pelves (Fisher, 2008) has shown that there is striking dimorphism in the ischia, replicating a pattern seen in other mammals as well (Warwick and Williams, 1973). In posterior aspect, looking along the axis of the birth canal, the ischial tuberosities of females form a broad, U-shaped trough (with outwardly directed limbs, forming an angle ≥ 90°), while in males, they form an acutely angled V-shape. In addition, the ventral extremity of the ischia of the male pelvis protrudes anteriorly as a prominent buttress for the closely adherent corpora cavernosa (Fisher, 2008).

Skeletal material also offers the prospect of gaining information on events during the life of an animal, via evidence of injury or pathology, and on postmortem events, including bone modification by human or nonhuman agents. Although analyses of this sort represent a distinct line of investigation from tusk studies, there is rich potential for complementary perspectives. Notable cases of injury in mastodons include only two females (Powers [Garland and Cogswell, 1985] and Eldridge [Kapp et al., 1990]) but are much more common among males. Another categorization relevant to injuries is between those that had healed (without returning to a normal configuration) prior to death and those that did not heal. Cases in the former group reflect to some extent conditions of life, but those in the latter group may be associated with the cause of death. On the other hand, they may be difficult to distinguish from postmortem bone modification caused by any of a number of agents. Further descriptions of cases of injury are given below, where they can be presented in conjunction with details of taphonomic context.

Recovery and documentation of skeletal material took on new urgency following the sale of the Burning Tree mastodon, an important specimen on which a great deal of work had already been done (Lepper et al., 1991; Fisher et al., 1994). This specimen was purchased by the Kanagawa Prefectural Museum of Natural History in Yokahama, Japan, and although the work interrupted by this sale could be resumed, the logistics are now far more complex. In retrospect, the sale of this specimen both resulted from, and contributed to, the growing commercialization of fossils. In the short run, it led to overestimation of the commercial value of mastodon material, and this misreading of the market complicated negotiations over the disposition of subsequently excavated specimens almost as much as if the market had in fact been more vigorous. One of the important specimens lost to commercial interests in the wake of Burning Tree was the Manitou Beach mastodon, discussed below. The combination of excavators trying to reach firm agreements with landowners before investing a great deal in sites and landowners looking for opportunities for commercial gain has made for challenging times. With the loss of some opportunities to document sites thoroughly, it has become even more important to do so whenever the chance arises. This has led to redoubled efforts to encourage placement of specimens into the public trust and to document osteological material as thoroughly as possible. Products of these efforts now nearing completion include complete photographic series (six standard anatomical views of each bone) and complete sets of research-quality molds and casts of two relatively complete, adult males, the Buesching and Hyde Park mastodons. In addition, the Buesching mastodon and parts of several others have been digitized to produce 3D models that are being prepared for use in comparative and biomechanical studies and development of mapping protocols for documenting newly excavated sites (e.g., Fisher, 2005, 2008).

Taphonomic Patterns and Their Association with Sex, Age, and Season of Death

Two decades have passed now since taphonomic patterns of Great Lakes region mastodon sites were systematically reviewed (Fisher, 1987). Some aspects of the pattern of occurrences remain stable, but, not unexpectedly, new sites have brought some new perspectives.

Two Site Types Recognized in Fisher (1987)

In Fisher (1987) one series of sites was distinguished as showing patterns of bone modification suggestive of carcass processing (butchery) by humans. Mastodons recovered at these sites were predominately males, with a younger age distribution than was observed in a complementary group of specimens interpreted as not showing evidence of human association, and they all turned out to have died in autumn or earliest winter (based on analysis of their last-formed dentin). All of these specimens came from wetland depositional settings, with bones occurring in marl or peat, and were interpreted as meat-caches consisting of minimally butchered carcass parts submerged in shallow ponds or along lake margins to promote preservation and extended access to the meat (Fisher, 1995). Some of these caches had apparently been abandoned without utilization, while others probably were utilized, in some cases after retrieval through a hole in ice (evidence for this includes burned wood, thought to represent a fire built on the frozen surface of the pond, associated with the bones in utilized caches, but not with unutilized caches at the Heisler site [Fisher, 1987]; other sites preserve cobble- to boulder-sized rocks that may represent a strategy [Fisher, 1995] for "passive-solar-ice-entry" ... let them melt their own way through as far as they will go and then punch through the remainder). These sites included some for which contextual data were available and others, excavated by previous University of Michigan staff, for which minimal contextual data had been recorded. However, all were united by occurrence of shared patterns of bone modification.

The complementary series of sites interpreted as not butchered included relatively few for which detailed contextual data were available. One of these few was the Johnson mastodon, a relatively complete skeleton of an old male preserved in a quasi-articulated condition within a small-scale fluvial setting, with the most conspicuous disarticulation being downstream displacement of small foot bones (Fisher, 1984a). Most of the other specimens in this group were categorized with Johnson because of the absence of bone modification like that observed on the specimens interpreted as butchered, though in some cases patterns of presence and absence of bones different from those seen in the butchered cases also played a role. These not-butchered cases showed an "even" sex ratio, an older age distribution than in the other group, and season of death ranging from the end of winter to late spring, with one midsummer death. Given the disparate sex, age, and season of death traits shown by these two groups, I proposed an argument framed in Bayesian terms that the butchered animals were most likely procured via hunting.

I had intended to document each of these assemblages in greater detail in the following years, but a steady stream of new sites was reported and took precedence until they could be secured and accessioned. These efforts, coupled with increased emphasis on tusk analysis and initiation of the digitizing project that would make site descriptions more effective, delayed completion of assemblage and site reports, though the first of many of these is now complete (Fisher, 2008). With this steady growth in the number of sites, my initial two-fold classification has become increasingly inadequate, while not completely breaking down.

New Sites and New Observations Complicate the Picture

One of the first sites that showed differences from the patterns reported in Fisher (1987) was the Eldridge mastodon, described briefly by Kapp et al. (1990). This was an adult female with evidence of butchery, but she turned out to be a spring death. Unfortunately, she had lost one tusk in life, and the other was not recovered at the site (season of death was determined from a molar), so a full life history was not available (though much might be learned from further analysis of cheek teeth). She also showed a complex array of healed injuries on her facial region, including several large perforations of parts of the skull that would normally be solidly ossified. We were unable to determine what caused her facial injuries or whether the loss of the tusk during life was related, but both of these events probably predated death by more than a year. Since most other spring deaths had shown no evidence of butchery, it seemed most parsimonious to interpret Eldridge as a natural death, of unknown cause, and therefore to consider the butchery as representing an instance of human scavenging of a natural death.

Another female previously interpreted as not butchered (Fisher,1987) is the Owosso mastodon. This interpretation was based on examining the full skeletal mount displayed in the University of Michigan Exhibit Museum. Although it was clear that a few parts of the skeleton had been restored, the apparent absence of butchery damage on exposed areas provided no basis for interpreting the specimen as associated with humans. However, when a skeletal cast of the adult male Buesching mastodon was recently mounted alongside the adult female Owosso mastodon, the latter had to be partly dismounted to reconfigure its permanent base to match the style of mobile base used for Buesching. In cleaning and remounting Owosso foot bones, it became clear that they showed types of damage previously observed only on butchered specimens. We know that the Owosso skeleton occurred in peat, in a lacustrine setting, but the skeleton was excavated by the neighbor of the landowner (leading to a dispute over ownership), and no data are available now on the spatial structure of the site.

Two of the sites interpreted as not butchered in Fisher (1987), Sheathelm and Quagaman, were unusual in preserving only heads, and each was encountered in nearly pristine condition (though Sheathelm was damaged by the landowner's efforts to extract it). These were interpreted as non-butchered because the hypothesis of "no human association" was treated as a null hypothesis, and at the time, there were no grounds for rejecting it, even though I had no good explanation for why a mastodon head might occur in isolation. Both were recovered from sediments suggestive of shallow ponds, as were the sites interpreted as meat caches, but this was not considered sufficient reason to suggest human association. However, my perspective on these sites changed with recovery of the St. Johns mastodon, an adult male represented only by a palate (with upper cheek tooth dentition) and basicranium (Fisher, n.d.). This too was from a lacustrine setting (marl) comparable to those of meat caches, but the season of death was spring, as had been observed for Sheathelm and Quagaman, and different from all prior meat caches (except Eldridge). Though severely broken, the specimen was essentially unweathered. We expected to find other bone fragments nearby, representative of other parts of the highly pneumatized skull typical of proboscideans, but a major recovery effort supported by the Michigan Department of Transportation produced nothing. A "fresh"-looking skull remnant was even more difficult to account for out in a shallow lake than had been the heads of Sheathelm or Quagaman, the intact condition of which might permit them to be interpreted as having floated out into deeper water from a carcass on the pond margin that ultimately failed to be preserved.

Heightening the sense that a new pattern was emerging, UM collections contained two additional specimens that were almost indistinguishable from St. Johns (UM 3488 and UM 3489) and more that probably were similar at the time of discovery (e.g., UM 11308), but had been restored with plaster, making them difficult to compare. These were also spring deaths. I subsequently proposed (Fisher, n.d.) that "heads-alone" represent a distinct type of occurrence.

Although I cannot rule out the possibility that these animals were hunted, I regard them, conservatively, as natural deaths scavenged by humans, where the main tissues that warranted caching were those associated with the head, especially the brain and extensive nasal mucosa lining the diplöe of the greatly expanded system of cranial sinuses. In any death due to non-human causes, humans might be lucky enough to be first on the scene and scavenge an entire carcass, as in the case of Eldridge, but often other scavengers might be expected to find the carcass first, such that when humans arrived, many of the most accessible parts might already have been eaten. However, even in such cases, important fat reserves inaccessible to most other scavengers remain in the brain and nasal mucosa, and the nutritional importance of lipids for humans subsisting on high-protein diets (Speth and Spielman, 1983) may have made these tissues attractive. More to the point, spring deaths might often have involved fat-depleted animals in which much of the meat was lean enough to impose a calorie deficit on humans eating it, but this would not apply to brain tissue, which retains its lipid content even in starving animals (Speth and Spielman, 1983). Heads cached, but never retrieved, might account for pristine skulls such as those of Sheathelm and Quagaman. In contrast, specimens consisting of a palate+basicranium would be ones that were cached subaqueously, retrieved in winter (when the need for lipids was greatest) through a hole in the iced-over surface of the pond, "harvested" by breaking away upper portions of the skull, where the brain and diplöe are located, and then abandoned on the ice surface after the lipid-rich tissues attached to the broken fragments had been gathered and returned to a camp where they could be rendered. After limited exposure on the surface of the frozen pond, the palate+basicranium would have melted through the ice and been preserved on the pond bottom (if breakup of the skull did not occur on the ice, it is hard to explain how this dense part of the skull got out into a central area of a pond, since without diplöe, it could not have floated). An Inuit practice of storing heads of game underwater and returning to harvest the brain and nasal mucosa (Taylor, 1969) was part of the ethnographic support for the hypothesis of subaqueous meat-caching in the first place (Fisher, 1995), and deer heads that I stored in ponds and bogs at the E.S. George Reserve (prior to legs of lamb and a draft horse) retained brain tissue over summer and through the next winter. This hypothesized behavior has another parallel in the winter "head-collecting" documented for Neanderthals by Stiner (1991).

Another type of site for which we now have additional examples is the multiple death site, where more than one individual is preserved. Russell Farm I and II were discussed in Fisher (1987), and at least one additional mastodon was discovered at the Johnson site. There is no reason to assume that all individuals preserved at a site represent the same taphonomic history, but in any given case, there may be evidence supporting such a conclusion. The Bothwell site, excavated by R. Richards (Indiana State Museum) in northern Indiana and under collaborative study with K. Smith (UM), yielded 13 tusks and additional cranial and postcranial remains, all apparently representing adult females (Smith and Fisher, 2007). We know nothing yet about site formation processes in this case, but determining season of death for all individuals will provide a good starting point.

Another interesting multiple death site is Manitou Beach (UM 18288 and 18289), where an adult female mastodon and a young calf were found. The calf was represented by only a left malar (= jugal) and several ribs, but from their size, age at death was estimated as 1–2 years. Measurements of annual increment lengths on a tusk of the adult female at the site, interpreted as in Fisher (1996a) to present a record of calving intervals, suggested that she had given birth to a calf about 1.5 years prior to death. The adult female was originally donated to the University of Michigan, but just as detailed analysis was about to begin (and after a great deal of fieldwork and lab work cleaning and stabilizing the specimen), the landowner demanded its return and sold it to a private collector (though several small samples were at least left to document the find in part). Through a complicated arrangement involving neighbors of the landowner who had assisted with much of the fieldwork (H. and D. Hoppe), the remains of the calf were acquired and (except for one rib) donated to the University of Michigan. Analysis of these specimens has been deferred in the hope that access to the remains of the adult might again be arranged, but a provisional interpretation of this site is that it represents a mother and her own calf, probably autumn deaths, and probably hunted and cached for later use.

An Important Cause of Natural Death for Adult Males

Another new development has been recognition of what may have been an extremely important cause of natural death for (mainly) adult males – death as a victim of musth battle. "Musth" refers to the hormonally mediated season of fasting, heightened aggression, and nearly exclusive focus on mating that is well known in extant Asian and African elephants (Poole and Moss, 1981). Musth has been recognized as an important cause of death in adult male elephants (Buss, 1990), but from a behavioral and life history point of view, this is only part of its significance. The frequency and duration of musth is related to dominance rank of a male, with larger and/or more highly ranked males remaining in musth longer (up to three months) and coming into musth at the optimum time of year (given the 22-month gestation period of extant elephants and the optimum season for calf birth). Less dominant males are inhibited from coming into musth in "prime-time," showing shorter musth intervals, timed either earlier or later in the year (Poole, 1987). Independent of size, non-musth males typically give way to musth males, and females in estrous typically consent to mate with a musth male in preference to any non-musth male. Musth behavior and the physiological ability to enter and maintain a musth episode are thus prime determinants of reproductive success. Musth battles in elephants may be brief and are generally non-lethal when disparity in size and power is clear, but cases in which two well matched musth males come into contact can lead to serious injury or death (Buss, 1990).

The first mastodon musth victim identified was the Cohoes mastodon, but its fate might not have been recognized had I not decided to test an earlier tusk-based season-of-death determination by repeating the analysis for one of the molars (lower left quadrant). To my surprise, the molar yielded a season of death about one month later! Both showed a spring death, but the tusk had stopped forming dentin before the molar. This refocused attention on a small puncture in the bone of the lateral aspect of the left temporal fossa, near the growing end of the tusk that yielded the earlier time of death. The fragments of bone surrounding this puncture were still attached along their margins and had all been rotated toward the skull interior, forming a round hole (Fig. 4.2A). I now interpret this hole as a bone puncture formed by the tip of an opponent's tusk, after plunging through much of the masticatory musculature on this side of the head. This would have caused a great deal of blood loss, and perhaps an infection that "killed" the tusk physiologically, soon after the injury was sustained. The animal itself, however, did not die immediately, but slowly starved. Its injury precluded mastication on the left side of its mouth, and ironically, the right side of its dentition had already been incapacitated by a developmental anomaly that I trace to an earlier injury that was probably incurred in an adolescent precursor to a full-fledged musth battle (Fisher and Fox, 2007b). Unable to eat, the Cohoes mastodon may have nursed a fever until he finally succumbed, about a month after his injury. Coincidentally, there was also evidence of butchery and meat-caching, implying that his carcass was scavenged by humans who must then have perceived some residual value in it. With additional analyses of state of health at death, we might make sense of some of the contrasting fates of different carcasses (i.e., whether they are scavenged or not), but it seems most conservative for now to attribute such outcomes (without claiming independent evidence at this time) to variability in timing of access to the carcass and in the condition of remaining tissue when a carcass was encountered by humans.

Unfortunately, circumstances of access to the Cohoes mastodon (assisting with its remounting in the New York State Museum) did not allow for detailed study of all postcrania. However, two other parallel cases were recognized in quick succession – the Buesching and Hyde Park mastodons – and the first trait linking them all, beyond the fact that they were all adult males that died in spring, was the occurrence of remarkably similar, unhealed puncture wounds in the same anatomical position as the puncture on Cohoes. Both Buesching and Hyde Park show evidence of old, healed injuries like those sustained by extant elephants in musth battles (ribs broken by impact of the ventrolateral surface of a tusk, proximal caudal vertebrae broken and/or dislocated, probably by a tusk blow to the rump, as an unsuccessful competitor beat a hasty retreat, and zygapophysial remodeling following transient dislocation of thoracic vertebrae by ramming a tusk into a victim's flank). In addition, Hyde Park (which has been studied in greatest detail) shows evidence of fresh tusk blows to the lateral flank and associated vertebral dislocation and an injury from a tusk tip driven between two thoracic vertebrae from a dorsolateral direction, an indication that the Hyde Park mastodon was already recumbent when it sustained this injury (Fisher, 2008).

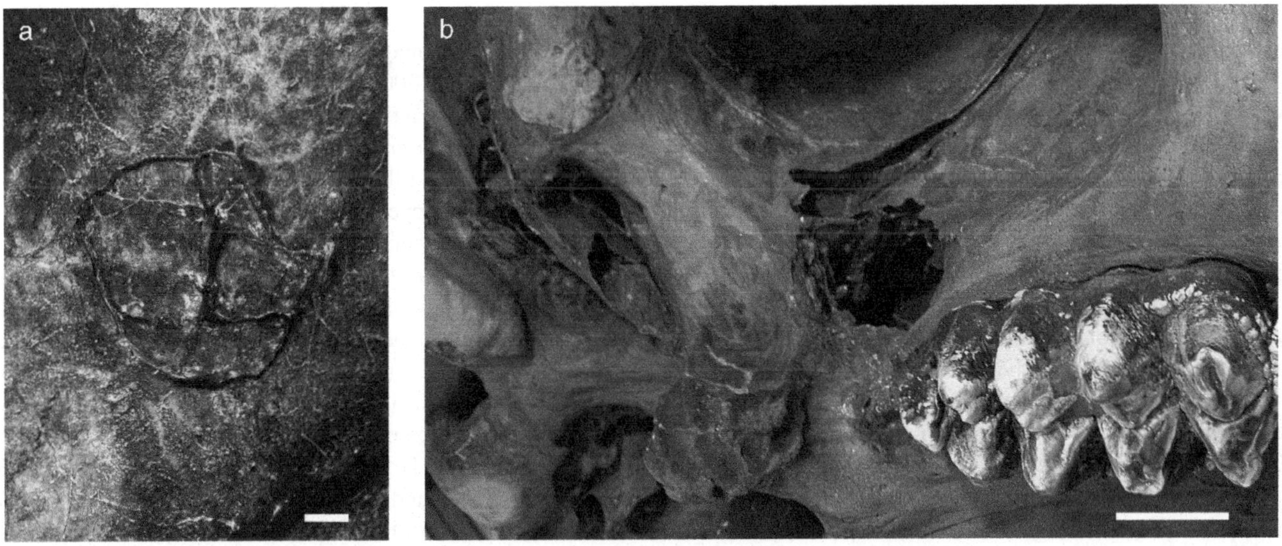

FIGURE 4.2. Bone punctures in the alisphenoid region (inner wall of temporal fossa) on adult male victims of musth battles. a. Cohoes mastodon, left temporal fossa; circular puncture restored by rotating fragments back into plane of bone surface (all remained partly attached); scale = 1 cm. b. Buesching mastodon, right temporal fossa; circular puncture left as found (for orientation, tooth is the right M^3, foramen magnum is dark opening at left of image, and the puncture is just below the alisphenoid canal; scale = 5 cm.

I interpret repeated occurrences of wounds implying tusk penetration of masticatory musculature and puncture of the lateral aspect of the skull bordering the temporal fossa as a result of a stereotyped mode of fighting in which opponents engaged each others' tusks, twisting their necks and driving forward to try to catch the other off balance. The key "move" would have come when one of the adversaries sensed even a momentary advantage of orientation or balance, dropped his head and tusks suddenly, and then vigorously thrust them upward and toward his opponent, on whichever side seemed vulnerable for a strike. The upward and inwardly turned tusk tip in the opportune position would then have caught the opponent in the cheek, or even entered his mouth, tearing through the masseteric and pterygoideus musculature, driving upward and toward the midline, and penetrating the alisphenoid bone, which formed the bony wall of the temporal fossa. This damage was usually unilateral, perhaps because the recipient of such a blow either withdrew from battle immediately or was sufficiently incapacitated by the blow that no comparable opportunity would likely be presented again. However, Quagaman, though identical in all other respects (except that as an isolated head, it lacked associated postcrania), shows identical damage bilaterally. This damage was observed during review of specimens prior to Fisher (1987), but it did not fit patterns of bone modification associated with butchery, and without the parallel examples of Cohoes, Bueschfing, and Hyde Park, it was assigned to an uninterpreted residue of peri- or postmortem damage that did not (fortunately) compromise interpreting Quagaman as a natural death.

The skull damage described above does not replicate the style of damage seen on extant elephant victims of musth battles, but this may be a consequence of different tusk geometry in these taxa. Wounds inflicted by elephant tusks are described as perforations of the temporal margins of the skull (or of other parts of the body) that suggest thrusting of their much straighter tusks in a direct "jab" at the opponent's head or body (Buss, 1990). The "upper cut" style of motion implied by mastodon tusk wounds is functionally similar in that the tusk is moved along a trajectory parallel to its tip axis, but kinematically distinct, in association with the greater tusk curvature of mastodons. In addition, the alveolar surface of the dorsolateral portion of adult male mastodon premaxillary bones shows a localized specialization implying hypertrophy of the periodontal ligament precisely in the area needed to serve as a shock-absorber for the stresses of impact of an upthrust tusk tip (Fisher, 2008). Finally, as discussed below, use of the tusk in this way leaves a signature style of damage in dentin and cementum along the outer curve of the tusk (described below) such that individual bouts of fighting are visible in the record of tusk growth.

The only case in which I now feel reservations about a specimen interpreted in Fisher (1987) as probably hunted is the Heisler mastodon. This specimen has a small perforation of the alisphenoid region on the right side of the skull, which could be a tusk wound. As with Quagaman, this was noted before, but these two cases alone were not enough for me to recognize the pattern. If Heisler was killed in a musth battle, two points are unusual. First, his autumn death would be the only exception to the otherwise exclusive association between musth battles and spring. Second, as a young male only about 16 years old at death, Heisler seems too young to have yet experienced onset of typical musth. This did not occur in Hyde Park until an age of about 23 and is commonly delayed in extant elephants until even later (Poole and Moss, 1981). One interpretation of these two anomalies is that as a young male in musth for one of his first times, Heisler may have been inhibited from going into musth in spring, when older, larger males were in musth, and may instead have delayed his novice musth episode until autumn. Even so, it is not clear that musth in this young a male is plausible, unless the local population was characterized by notably few adult males (Slotow et al., 2000). A second possibility is that perhaps this was not really a full-blown musth episode at all, but an example of the kind of fighting in young males that occurs throughout adolescence and that, bout by bout, prepares them for the onset of typical musth. The alisphenoid perforation on Heisler does suggest a very sharp tusk tip, like that of Heisler's own tusks, so perhaps he was just an unlucky pre-musth teen. This may be resolved by closer analysis of Heisler's tusk record, but this work is not yet complete. However Heisler died, the site still presents a compelling case of butchery and meat-caching.

Since it is now clear that some adult male mastodons met extremely violent ends in musth battles, and since some musth victims have been interpreted as scavenged and butchered (e.g., Cohoes, Bueschfing), it is worth asking whether some of the bone modification interpreted as due to butchery could be reinterpreted as damage sustained in musth battles. This might be difficult to assess without comparative observations from a musth victim that was not butchered, but this is exactly what we encountered in the Hyde Park mastodon (Fisher, 2008). Despite the damage evident on his skeleton, none of it resembles the types of damage interpreted as indicative of butchery (Fisher, 1984a, b, 2008; Fisher et al., 1994). This makes him, like Johnson, an important reference for documenting the contrasting patterns of "butchered" and "non-butchered" carcasses. We also still have cases such as Pleasant Lake and Burning Tree that appear to have been butchered but show no evidence of perimortem musth battle.

A final note on musth battles is that they constitute the first positive indication of a cause of death (normally applicable only to adult males) unrelated to human activity. In principle, studies of dentin increments and isotope profiles might allow recognition of death due to drought and/or nutritional stress (recorded as thin increments and elevated oxygen and/or nitrogen isotope values; Heaton et al., 1986; Hobson et al., 1993; Koch, 1998; Fisher and Fox, 2003), but we have not yet studied cases that appear to fit such patterns. Rather, as explained in Fisher (1987), most deaths now attributed to "natural" causes are identified as such because we have made the conservative choice to treat natural death as a null hypothesis that has to be rejected before an interpretation of hunting can be put forward.

I would not want to defend this as necessarily the best choice in all cases; for example, Fisher and Fox (2003) note as "suspicious" that three, or possibly four, of four females analyzed at the Hiscock site show no evidence of nutritional stress or severe seasonality, but all died between winter and early summer, each apparently with a yearling calf. As new evidence emerges, we may well want to revisit default "natural death" interpretations, but in the context of a debate where skeptics regard hunting as an extraordinary claim requiring extraordinary evidence, I am willing to maintain a conservative stance.

Additional Findings at New Sites

The presence and certain aspects of the behavior of mastodons are documented by footprints as well as bones. At the Brennan site, while excavating remains of a butchered, cached mastodon, footprints were discovered that record passage of a solitary adult male and, sometime after, a group that appears to have been composed of two females and at least one calf (Fisher, 1994). We followed the trackway of the male for about 60 m, paralleling a lake margin at about 1 m paleodepth and exposed a profusion of additional mastodon footprints just at and beyond the lake margin. Subsequently, with more of an idea of what to look for, abundant mastodon footprints were exposed just shoreward of the pond margin at the Heisler site. These footprints, in conjunction with the finely laminated character of the peaty marl of the bone horizon within the pond sediments, suggest that mastodon behavior was characterized by an appropriate avoidance of the soft, yielding substrates encountered within the pond itself.

The typical circumstances that bring new sites to our attention (receipt of a call from a landowner who has discovered large bones) represent a strong bias favoring animals as large as mastodons, and against specimens in a body size range that would permit them to be interpreted by landowners as simply domestic livestock buried by some prior property owner. Nevertheless, once working at a site, we sometimes find remains of taxa other than proboscideans. Perhaps the most notable case was recovery of limited remains of two *Cervalces scotti* at the Brennan mastodon site. One of these was represented by a partial cranium, with antlers broken off, that could plausibly be a cached Scott's moose head retrieved for recovery of its brain. Likewise, at the Dempsey site, in what may be a Holocene assemblage, white-tailed deer (*Odocoileus virginiana*) crania with antlers girdled by stone tools, before being snapped off at the base, were recovered from lacustrine marl. Muskox crania in our collection also show damage that could reflect subaqueous caching and subsequent recovery for harvesting of brain and nasal mucosa.

Discussion of Morphologic and Taphonomic Context of Tusk Studies

The sites reviewed by Fisher (1987) still warrant more detailed analysis, but only one site interpreted then as reflecting hunting is now viewed conservatively as a result of scavenging (Heisler). Slightly more change in interpretation has affected sites thought to show no evidence of human association, with one (Owosso) now interpreted as a scavenged natural death and two others (Sheathelm and Quagaman) now interpreted as cached heads scavenged from natural deaths. Even with these changes, sites without human association comprise a set with a more even distribution of sexes and an older age distribution. With respect to season of death, specimens that show evidence of human association are still mostly autumn deaths, a season not represented at all among specimens without human association. This means they cannot all be readily interpreted as scavenged natural deaths. Instead of suggesting that all human-associated carcasses were probably hunted, I would now treat them as including both cases of hunting and cases of scavenging, but the inclusion of scavenging is mostly in the direction of considering scavenging for specimens that were formerly not recognized as involving humans at all.

The picture of human subsistence adaptation that emerges from these patterns is consistent with a generalist, opportunistic strategy in which megafauna are hunted or scavenged as circumstances allow, and in which proboscideans, predominantly mastodons in the Great Lakes region, represent a source of protein, fat, and possibly other materials that is at least seasonally important. The occurrence of these remains in the context of resource stores suggests some risk-reduction strategy, implying that humans in the Great Lakes region during the latest Pleistocene may have had to deal with some uncertainty in access to food, especially to lipid-rich dietary components. This would have been a human-scale problem, affecting human nutrition, and it would be a mistake to generalize this without further evidence to other contemporary mammals, especially proboscideans. The nature of proboscidean response to environmental conditions will be best determined from records of proboscidean growth and life history in tusks.

Elements of Tusk Structure and Their Interpretation

Tusk and cheek tooth structure in mastodons and mammoths has been reviewed previously (e.g., Fisher, 1987, 1988, 1996a, 2001a; Fisher and Fox, 2003), but each of these treatments introduced elements that significantly change our understanding, and the pace of new observations is such that I doubt the last of these revisions is behind us. Because enamel is present on tusks only near their tips and is removed from cheek teeth by occlusal attrition, it does not tend to provide a long and continuous record for either. Cementum on tusks is subject to information loss due to abrasion following eruption and is generally not thick enough in any one area, especially on cheek teeth, to provide adequate temporal resolution. Tusk cementum does figure in studies of adult male mastodons, but except for this, I focus here on dentin. To keep the scope of this discussion manageable, even some unique aspects of proboscidean dentin (such as the "Schreger" pattern; Espinosa and Mann, 1993; Trapani and Fisher, 2003) must give way to an exclusive focus on the patterns of layering that record the appositional history of this tissue.

Proboscidean Dentin: Themes and Variations

The most basic aspect of proboscidean dentin structure is the existence of a hierarchical organization of laminar features, such that laminae at coarser scales are composed (typically) of a relatively fixed number of laminae at the next finer spatial scale. In tusk dentin there appear to be at least three orders of laminar features, and cheek teeth show two clearly (and probably three, but we do not usually use magnifications high enough to resolve the finest; Fisher, 1987, 1988, 1996a). The largest-scale layering in dentin is the repetition of so-called "first-order features" reflecting an annual periodicity, the cycle of seasons. This is expressed in variation in dentin color, density, the spacing of the next-finer laminar elements, and often the topography of the dentin-cementum junction. The next finer scale of lamination is the repetition of "second-order" features, reflecting a different period in different contexts, as discussed below. Second-order features are bounded, in general, by more strongly marked versions of the same type of structural discontinuities that demarcate the next-finer, or third-order, laminae. Third-order laminae are couplets of more and less dense dentin, where the less dense layers are essentially zones of vacuities, left in intertubular dentin as the mineralizing front passes a given locus of apposition. Third-order laminae appear to recur with daily periodicity and may be some reflection of circadian physiological rhythms.

First- and third-order features thus have at least plausible physiological causes, but second-order features are more variable and less well understood. Their apparent period in mastodon tusks is one fortnight, or two weeks, repeating about 26 times per first-order unit. However, in mastodon molars, I have observed about 13 second-order features per first-order feature, suggesting a period close to a lunar month. Both of these periods are different in mammoths, where second-order features in tusks recur at about one-week intervals, with about 52 per first-order unit, in mammoths living at temperate latitudes (Fisher et al., 2003). Mammoth molars, in contrast, show second-order features with a two-week period (Fisher and Fox, 2007a). Thus, each taxon shows a different period (related by a factor of 2) in tusks and molars, and controlling for the type of tooth, there is another factor-of-2 difference between mastodons and mammoths. At this point, insufficient work has been done on cheek teeth to say much about variability, but enough tusks have been analyzed to show that there can be variation within and between individuals. Some of this is attributable to the uncertainty in where exactly to say a given first-order feature begins. Seasonal changes are, after all, gradual on some scale; spring, for example, can come early or late in any given year. However, cases where one mastodon, for example, shows 25 second-order increments per first-order increment for several years running (Fisher et al., 2008) imply real variation from the more common value of 26.

It is worth pausing here to note that the periodicities in mammoth tusk dentin are almost identical to those of human tooth enamel and dentin (Dean and Scandrett, 1996) and that other mammals show comparable hierarchical patterns (Klevezal, 1996). What I refer to as second-order features figure in literature on primate dental anatomy as "long-period striations," while third-order features are "short-period striations." Without dwelling on terminology, reference to "orders" implicitly addresses the hierarchical character of the system, acknowledges existence of annual as well as shorter-term features, and promotes the concept that relative position within the hierarchy may be more relevant for some problems than the actual duration of a period.

The interpretation of these patterns of tusk lamination is more complex than can be treated fully here. However, an overview of my working hypothesis is that second-order laminae are a reflection of a "beat-frequency" (see also Newman and Poole, 1974) driven by interaction of circadian rhythms and some other rhythm, possibly a spatiotemporal rhythm of dentin mineralization that is endogenous to the tusk itself (and likewise, but with different parameters, to cheek teeth). This physical model allows me to calculate the frequency and period of this hypothesized rhythm, even though its actual identity remains unresolved. As long as circadian rhythms are entrained to environmental light-dark cues, cycling with a 24-h period, as will always obtain at temperate (or equatorial) latitudes, the beat-frequency retains its period of about 7 days. However, for animals living north of the Arctic Circle, a portion of the year (winter and summer) lacks 24-h light-dark cues, and under those conditions, the circadian rhythm shifts to its endogenous "free-running" frequency, for which the period is generally longer (or shorter) than 24 h. This induces a change in the pattern of lamination for mammoths living above the Arctic Circle (Fisher, 2001b, 2007), but since these populations are not the subject of this chapter, I defer additional discussion of this phenomenon. In any case, the relevance of this model for mastodons and mammoths of the Great Lakes region (not to mention other instances of hierarchical dentin lamination) is that it provides insight into the cause of second-order lamination and supports treating second-order laminae as periodic features that can be used to extract data on short-term variation in rates of dentin apposition.

Isotope Evidence of Annual Nature of First-Order Features

Analyses of light stable isotope compositions of mineralized tissues of organisms have recently become a major source of new perspectives on paleobiology and paleoenvironments. The isotope systems we have used for working on North American mastodons and mammoths are carbon and oxygen from structural carbonate of hydroxyapatite, phosphate oxygen from hydroxyapatite, and carbon and nitrogen from collagen.

The systematics of these isotopes have been reviewed on many occasions (e.g., Koch et al., 1994; Koch, 1998; Cerling and Harris, 1999; Kohn, 1996; Fisher and Fox, 2003). Factors that contribute to variation in isotope composition are broadly understood, and many aspects of such variation are highly predictable. However, the nature of isotope records, depending as they do on the actual fluxes of isotopes in and out of the body and all the environmental, physiological, and behavioral factors that control these, is highly contingent. This is the basis for the strength of isotope records as proxies for real variation in causal factors, but it is also a source of potential problems because, especially in the case of a large, mobile animal such as a proboscidean, we are monitoring a system in which relevant factors do not always maintain constant relations to one another.

For example, the most fundamental aspect of dentin lamination that has been addressed using isotope composition is the annual nature of first-order features. These highly regular, but not invariant, largest-scale features of dentin lamination could in principle have been under some hypothetical system of spatial or structural control, producing layers of roughly constant thickness but with no consistent temporal dimension. However, this interpretation has been effectively ruled out by finding patterns of variation in isotope composition matching those expected for seasonal variation on an annual cycle in nearly constant association with particular parts of the structural cycle. This was first established using oxygen from structural carbonate in hydroxyapatite, in the first study of intra-annual isotopic variation in dental remains (Koch et al. 1989), but it has been replicated many times over using all the isotope systems noted above (e.g., Fisher and Fox, 2003, 2007a; Fisher, 2001a; Hoyle et al., 2004). The fact that the expected patterns are not observed in every case is explained by variations in behavior and context. For instance, under temperate-latitude conditions, patterns of seasonal variation in meteoric water composition (Gat, 1980; Rozanski et al., 1993) and the calculated lag (reservoir effect) due to gradual shift in body water composition relative to inputs from drinking water and food water (Koch, 1989) lead us to expect the lowest oxygen isotope values ($\delta^{18}O$) near the winter-spring boundary and the highest values in late summer or early autumn (Stuart-Williams and Schwarcz, 1997). This is seen often enough that we are confident that first-order features really are annual; yet the expected pattern of variation is based on the assumption that drinking water and food water are dominated by meteoric sources that display the usual seasonal pattern of compositional change. In work on four Hiscock mastodon females, from western New York, Fisher and Fox (2003) found a limited annual range of variation in oxygen isotope ratios (< 3‰ in most tusk-years), no consistent pattern of intra-annual variation either within or between individuals, and little correspondence to the typical pattern of variation in meteoric water composition. Given the highly regular patterns of oxygen isotope variation observed for other sites, failure to observe the same patterns here cannot be treated as refuting the annual nature of first-order features. Based on where the Hiscock mastodons died, they probably lived near glacial meltwater sources that were relatively stable isotopically and different in composition from meteoric waters. The most plausible interpretation of these animals' oxygen isotope profiles may therefore be that they reflect largely stochastic, individual histories of switching between water sources. If so, their profiles carry little information on local seasonality though they may still, to some degree, "reflect behavior."

The contingent nature of compositional profiles also means that the observed pattern of variation depends on the pattern of environmental change in the locale inhabited by the animal under study. Temperate-latitude settings in North America may show a simple, sinusoidal pattern of change in oxygen isotope values, one cycle per year, as seen today in meteoric water values in the same regions, but at lower or higher latitudes qualitatively different patterns may emerge. In a Florida mastodon, for example, there were two cycles of variation in oxygen values per first-order feature, interpreted as reflecting a more "tropical" pattern of two wet and two dry seasons per year (Fisher and Fox, 2006). On the other hand, woolly mammoths in Chukotka (northeastern Siberia) show an annual cycle that appears to depend entirely on seasonal shifts in the vapor source and rain-out history of air masses (Fox et al., 2007).

In many studies of intra-tooth variation in isotope composition, samples provide a time series of compositional data only in the sense that measured values are ordered in time; the amount of time represented by the whole series may be estimated if a large enough fraction of an annual cycle is recorded, but there is typically little independent control on the amount of time represented in each sample. In contrast, sampling proboscidean dentin, where first-order features are often marked by color banding or repeating patterns of accentuation or spacing of second-order features, provides an opportunity for greater precision in relating samples to time in the life of the animal. This also lets us express compositional time series in either the spatial domain (as in distance from the pulp surface at the time of death) or the temporal domain (as in weeks or fortnights since the last winter-spring boundary). Thus far, this temporal resolution has served mainly to control the sampling process itself and insure that we achieve comprehensive but minimally overlapping recovery of dentin from consecutive intervals of time, but we could in principle evaluate rates of change in composition as well as the general pattern of change.

Just as the pattern of oxygen isotope variation over the course of a year depends on the local hydrologic cycle (and an animal's behavioral interaction with it), the pattern of variation of other isotope systems is a function of behavioral and physiological interactions with locally available plants and environmental conditions. One interesting example of this may become an increasingly important pattern in studies of temperate-and high-latitude ecosystems. We are used to consulting oxygen isotope profiles to monitor seasonal changes in environment, seeing only moderate changes seasonally in carbonate carbon profiles and interpreting them as indicative of subtle changes in diet (switching between C_3 and C_4 plants) or habitat use (habitats that are more or less arid, or closed-canopy; e.g., Koch, 1998). However, in the northern portions of temperate-latitude ranges, C_4 plants become rare or drop out of floras completely (Teeri and Stowe, 1976). Although this might be expected to subdue seasonal patterns in carbon isotope profiles, profiles of carbonate carbon isotope composition recovered from the North American Great Lakes region and from Siberia often show a strong seasonal signal (Fisher and Fox, 2003; Gohman et al., submitted). Especially in Siberia, C_4 photosynthesis would have been functionally absent, and there would have been no closed-canopy habitats in which more ^{12}C-enriched isotope values would have been generated.

What we suspect is going on here is a seasonal cycle between fat utilization and fat formation, usually expressed most clearly in carbonate carbon, which reflects the carbon isotope composition of bulk diet (Ambrose and Norr, 1993). Because lipids tend to be depleted in ^{13}C (DeNiro and Epstein, 1981), when late winter caloric deficits are balanced by metabolizing stored fat reserves, the body's carbon reservoir (and thus the carbonate carbon in newly mineralized dentin) is shifted toward lighter values (Polischuck et al., 2001). The opposing phase of this cycle, showing heavier values in late summer and fall, could reflect in part simply the opposite effect; that is, as caloric intake exceeds immediate needs, and excess energy is stored as fat, preferentially sequestering ^{12}C in adipose tissue, the body's carbon reservoir is shifted toward heavier (more ^{13}C-enriched) values. However, other carbon fluxes are important as well (P. Koch, 2007 personal communication), so additional investigation of this cycle is needed. In some cases, these patterns in carbonate carbon isotope composition may provide a clearer seasonal signal than corresponding patterns in oxygen isotope composition, especially where oxygen is influenced by vapor source and rainout history more than local temperature of condensation.

Age Determination

A prerequisite for quantitative studies of life history is a means of determining individual age, ideally at critical points during life as well as at the end of life. Counting first-order features in tusk dentin satisfies much of this requirement, providing a minimum constraint on age at any given point in the development of the recovered part of a tusk. However, it leaves unaccounted for the number of years missing from the tusk tip due to breakage and abrasion, and these "missing years" may comprise a larger portion of life the longer an individual lives. In certain cases, useful information may be gained by correlating between teeth of the same individual, but by the time a tusk tip is broken, the cheek tooth that was forming during the corresponding part of life may have been worn down and lost long before.

The most effective response to this problem is to develop a composite reference frame (i.e., based on multiple individuals) for interpreting age relative to temporal reference features involving tusk composition and/or geometry. Rountrey et al. (2007a) have shown that compositional time series through the tusks of juvenile mammoths can track seasonal changes in the relative importance of milk and plant components of the diet and the longer-term trends associated with weaning. We expect the timing of weaning to be environmentally dependent itself, but it should be less variable than the number of years missing from a tusk tip. We are now documenting weaning in a number of calves that have lost little or no material from their tusk tips, and in some cases, we observe a record of birth itself via recognition of a neonatal line (Rountrey et al., 2007b, submitted). There may also be other transitions in juvenile tusk development, such as shifts in the composition of milk, that can be used to define reference features within tusks.

Tusk geometry is also important for building this composite reference frame. In both mastodons and mammoths (and in extant elephants; Elder, 1970), tusks develop as conical structures with proximally increasing circumference (except for parts of the tusk formed late in adult life) at the dentin-cementum boundary and externally, on the outside of the cementum. Distal-to-proximal gradients of increasing dentin radius (distance from the axis to the dentin-cementum boundary at a given radial position and distance from the tip) appear to be less variable among individuals of the same sex than between individuals of different sexes, possibly even starting with the earliest stages of tusk formation. In addition, the thickness and length of first-order dentin increments display sexual dimorphism during later tusk growth (Fisher, 2008; Fisher et al., 2008), and possibly in early years. Differential provisioning of male and female calves by mothers (Lee and Moss, 1985) may provide the nutrient influx critical for supporting higher tusk growth rates in males. Additional aspects of geometry include the transverse cross section of increments, which changes from elliptical to a more nearly circular configuration along both distal to proximal and axial to external gradients.

Although breakage and abrasion remove material from the tip and dynamically reconfigure it to yield a secondarily acute appearance, loss of material can always be recognized and to some degree estimated. In the first place, material loss is almost always asymmetrical, such that the structural axis of dentin increments is displaced from the morphologic tip of the tusk. This asymmetrical loss can usually also be seen by following the outcrop of the dentin-cementum boundary, which is typically displaced farther proximally on the outside (ventral) curve of the tusk (Fisher, 2008). Wherever cementum has been lost from the tusk surface, it is likely that some dentin has been removed too, but as long as even traces of cementum are left, the dentin radius at that position cannot yet have been altered. Measurements of unaltered dentin radii can then be used to correlate between tusks with different degrees of tip loss.

Male Life History

The most striking feature of male life histories, viewed from the perspective of records of tusk growth, is eviction of males from their matriarchal family units upon sexual maturation. Observations of extant elephant males undergoing this transition show that this is a time of increased stress and risk of mortality (Moss, 1988). This is partly because recently evicted males waste much time and energy "appealing" the sentence pronounced against them by the matriarch, and partly because they simply have not yet acquired the knowledge of local resources from which they benefited while in the company of their matriarch. Even records of tusk growth rate compiled at an annual scale show a pronounced drop in year length (i.e., first-order increment length measured along the dentin-cementum junction) that has been interpreted as indicative of eviction, and records of second-order increment thicknesses (Fisher, 1996a, 2008, and unpublished work on an African elephant tusk) show an abrupt, mid-year drop in tusk growth rate that is sustained, with only gradually moderating effects, for a period of 2–3 years. Annual records of tusk growth rate based on year thickness show a similar drop, though usually not as pronounced, as first-order increment thickness is inherently less variable over the whole ontogeny than first-order increment length (Fisher, 2001a). For the Hyde Park mastodon, examination of both types of annual increment profile has been augmented by pair-wise comparison of each of the 32 complete years in the tusk. Year 12 in the tusk, interpreted as year 15 in life, is the year that matches criteria for eviction on annual increment profiles, and it also stands out as the most distinctive year in the tusk from the perspective of Kolmogorov-Smirnov tests of both the temporally ordered sequence of second-order increment thicknesses and the temporally unordered distribution of second-order increment thicknesses (Fisher, 2008). We have not yet had a chance to "ground-truth" these interpretations by analyzing tusks from male elephants with known life histories, but the appearance of this depression in tusk growth is so consistent in character in all male proboscidean tusks we have examined that no other interpretation seems plausible. On any given tusk, we see no more than one depression in tusk growth rate matching the pattern interpreted as eviction, and it always occurs between the ages of 9 and 20 years, the same range attributed to eviction in extant elephants (Sikes, 1971).

Attainment of physiological sexual maturity in male elephants is the first but not the only threshold leading to full reproductive function. At a more advanced age, often in the later 20s or even 30s, comes the onset of musth, beginning with short, irregular periods of fasting and aggressive behavior lasting days or weeks, and later developing into longer and more regular episodes. In mature males, musth may recur annually as long as physical condition is sufficient to sustain the fasting and high levels of aggressive interaction associated with the state (Poole, 1987). In less mature males, even the presence of more mature males within the population tends to inhibit musth or displace its timing to less opportune parts of the year. Comparisons between local populations where older bulls are and are not present have made it clear that the timing of musth is not programmed into individual development but is rather a reversible, inducible response to social interaction (Slotow et al., 2000).

For *Mammuthus columbi*, the most dramatic evidence of musth is presented by the Crawford mammoths (University of Nebraska State Museum 2448 and 2449), a pair of mature males that died with their tusks interlocked; in the end, both were victims of a musth battle from which neither could retreat. Recent studies (Fisher, Rountrey, and Voorhies, unpublished) of dentin increments and carbon and oxygen isotope compositions of hydroxyapatite carbonate show that both died in mid- to late-spring, roughly 22 months from what would have been an optimum early spring calving season. This same season of death and, implicitly, musth was observed for the Brooks mammoths (University of Nebraska State Museum 1384, 1952, 1953, 1954), an assemblage of four tusks that shows every indication of being analogous to the Crawford mammoths, but without such spectacular preservation of the rest of the skeletons (Fisher, 2004b). Musth at this time of year is also indicated by intervals of slow tusk growth (thin second-order increments) induced by the musth fast in years prior to the time of death, recorded in the tusks of all of these individuals.

Mammut americanum offers no pairs of gladiators like the Crawford mammoths, but we do have inferred solitary victims of musth battles such as the Cohoes, Hyde Park, and Buesching mastodons. Again, death and musth were in the mid- to late-spring, implying a roughly 22-month gestation period if the calving season was in early spring, as seems optimal. As with mammoths, we also have indications of earlier musth episodes in the form of intervals of reduced rate of dentin apposition, presumably induced by the musth-fast. In addition, the Hyde Park mastodon shows a recently recognized type of tusk record that is in its own way almost more dramatic than pairs of victims. As argued at length elsewhere (Fisher, 2008), forceful upthrusting of its tusks during musth battles displaced its tusks within their alveoli to the extent that the proximal, growing margin was fractured along an arcuate tract on the ventral aspect (outer curve) of the tusk. After each such battle, tusk growth resumed, but odontoblasts and cementoblasts along the damaged locus produced anomalous dentin and cementum for many years after, leaving a visible record (externally and in thin section) of each significant battle in the life of this individual. The first recorded battle was at an age of 23, and more or less severe battles recurred in every subsequent year, always in mid- to late spring, until the death of this animal in a musth battle at an estimated age of 36. The combination of healed injuries and fresh injuries attributable to the final battle make it clear that such battles were brutal even when not lethal, taking a mounting toll on the physical condition of the animal. Nonetheless, annual increments of dentin thickness (rate of apposition) maintain roughly constant values through most of life, and there is no long-term diminution in the length or severity of musth episodes, suggesting that this was simply the lifestyle to which mastodons were adapted, and which they successfully maintained within late Pleistocene environments. Patterns of tusk growth throughout life suggest that the seasonal cycle was quite variable from year to year, but in general, neither winter cold nor summer drought were harsh enough to impact tusk growth adversely. Clearly this work needs to be replicated on other individuals, but it promises to bring the lives and fates of individual mastodons into spectacular focus.

Beyond descriptive detail concerning the lives of individuals, comparative treatment of comparable records of maturation and musth battles may provide demographic evidence that is ordinarily extremely difficult to extract from the fossil record. Other mastodon males have not yet been studied in as great detail as Hyde Park, but specimens such as the Buesching and Pleasant Lake mastodons appear to have been involved in fewer severe musth battles. Those that are recorded show the same seasonal timing, but they do not occur in every year of these animals' adult lives. I suspect this means that the populations in which they lived were characterized by lower densities of adult males, leading to encounters that were less regular, though no less severe when they did occur. This suggests a social context in which young males would be expected to mature at younger ages. As noted below, eviction does seem to have occurred earlier for Pleasant Lake than for Hyde Park, but more work is needed to make this determination for Buesching.

Concerning the possibility of human-induced mortality (i.e., hunting), the evidence is still indirect. Roughly half of the mastodons recovered in the Great Lakes Region are young adult males that died in autumn (Fisher, 1987). If there was ever any suspicion that these were musth deaths, the emerging record of mid- to late spring musth argues against such an interpretation. The sole exception to this pattern, as noted above, is the Heisler mastodon, who, if he was even in musth at all at the time of his death, was probably a case of early onset of musth, at a completely inopportune time of year. Most other males dying at this time of year are not nearly this young and show no evidence of having died in musth battles.

Female Life History

Tentative interpretations of tusk records of female mastodon life histories were proposed in Fisher (1996a) and discussed further in Fisher (2001a). Although all details require further evaluation, a recent comprehensive analysis of a tusk of the North Java mastodon (Fisher et al. 2008) corroborates earlier suggestions and provides much additional data, including comparisons and contrasts with male life history. There is some age uncertainty due to fracture of the tusk tip, but our best estimate of age for the North Java mastodon suggests first conception at 9–10 years and a lifetime total of 6–7 calves, most at 3–4-year intervals (based on cycles of variation in annual increments of tusk length and dentin thickness), prior to death at an age of about 40. To carry the analysis beyond simple profiles of variation in annual increment dimensions, pair-wise comparisons between each year in the tusk (evaluated, as for Hyde Park, by Kolmogorov-Smirnov tests of both the temporally ordered sequence of second-order increment thicknesses and the temporally unordered distribution of second-order increment thicknesses) showed a pattern of similarities and differences between years that was dramatically different from the pattern seen in an adult male (Hyde Park; Fisher, 2008) and reflective of the inferred calving cycles. Results of these comparisons make it clear that there are objective differences in the growth histories of these tusks. North Java's tusk growth record is compatible with conception in mid- to late spring (equivalent to the timing of musth in males) and calving in early spring, roughly 22 months later. The pattern of seasonal variation in tusk growth rate, compared year-to-year, suggests that conditions for growth were highly variable but generally good. Slow-growth intervals in winter were not protracted and in most cases not extreme (Fisher et al., 2008). The emphasis of new studies of female life histories, beyond simply replicating prior analyses on additional individuals, is to identify additional structural and compositional factors in tusk growth that might provide independent evaluation of the calving cycles inferred from tracing changes in tusk growth rate (Fisher and Fox, 2003).

One pattern noted in Fisher and Fox (2003) and Fisher et al. (2008) is that female mortality does not seem to be associated with the time – either the season or the particular year – of calving, but rather with the time of mating. This needs to be addressed with larger samples, but it raises the question of what factors are responsible for most deaths of females. Neither tusk growth rates nor nitrogen isotope values of dentin collagen (evaluated for four females from the Hiscock site; Fisher and Fox, 2003) give any hint (e.g., elevated $\delta^{15}N$ values; Hobson et al., 1993) that nutritional stress is involved.

Whether or to what extent there was human hunting of adult female mastodons and/or mastodon calves is not yet clear. The Manitou Beach mastodons are suggestive, but we have not even completed a formal season-of-death analysis on this material. Likewise, the Hiscock mastodon females analyzed by Fisher and Fox (2003) were difficult to explain as natural deaths, though terminal growth histories extracted from isolated tusks do not provide much to go on.

The possibility of a completely different source of mortality was proposed by Fisher et al. (2008) based on injuries on the Powers and Eldridge mastodons that are similar to injuries caused in tusk battles. Although this is "all wrong" in terms of normal proboscidean behavior (i.e., musth males might fight with each other, but ordinarily not with the females with whom they should be mating), it is reminiscent of the misdirected aggression of young male elephants showing premature onset of musth in populations where poaching has removed most or all of the older bulls (Slotow et al., 2000). This mechanism does not address the issue of how the density of older bulls was reduced, but it does represent one possible behavioral pathology that could emerge in populations that display, for whatever reason, this type of demographic structure.

At present, much less has been done on life histories of North American mammoth females, but some information on tusk growth rates was presented in Fisher (2001a), and oxygen isotope time series for the last year or more of dentin formation in three of the Dent mammoths (Fisher and Fox, 2007a) suggest that this site (with at least 15 mammoths, all either young or adult females; Brunswig, 2007) preserves remains of individuals that died on at least three (and very likely more) occasions. Ironically, although this may strike some readers as less dramatic than Saunders' (1980) "herd confrontation hypothesis", in which hunters are envisioned as killing an entire matriarchal family unit on one occasion, the seasonal selectivity and site fidelity that our analysis revealed (also compatible with results of Hoppe, 2004) seem to us to strengthen the inference of human agency in site formation. Finally, the autumn season of death of one of the Colby mammoths (an adult female; Fisher, 2001a) is compatible with Frison and Todd's (1986) analysis of that site and with other evidence that autumn deaths may reflect hunting.

Life History Changes as Tests of Hypothesized Causes of Extinction

Changes in life history traits are not necessarily useful for resolving every question regarding mechanism of extinction. For example, it is not clear that any prediction can be made about how life history should respond to "hyperdisease" (MacPhee and Marx, 1997) or to an extraterrestrial impact (Firestone et al., 2007). This is not to say that there would *be* no life history response to such events, but rather that the details of any response would depend on too many local factors and environmental contingencies to predict an overall response. In contrast, climate change as a mechanism of extinction, whether affecting animal populations directly or through mediation of a vegetational response, works by depressing organisms' capacity for growth and reproduction. If it was the principle driver of extinction, we should see evidence of reduced growth rates and retarded reproductive schedules. Whether or not this is a general prediction, it seems quite secure for a large-bodied herbivore with long gestation time, single births, and a social organization involving dominance hierarchies based largely on body size (Fisher, 1996a, 2001a, 2004a). Lower growth rates are essentially a direct response to environmental degradation, and delays in reproductive schedules tend to mitigate the effects of resource limitations. The fundamental trade-offs between growth and reproduction (e.g., Clutton-Brock et al., 1982) are probably especially strong for proboscideans, where females incur the substantial energetic cost of producing, carrying, and provisioning large-bodied young, and males endure the rigors of the musth fast, not to mention the risks of encounters with dominant adversaries among older males. Enduring lower growth rates without a delay in reproductive schedules implies reproduction at smaller body size. This is bad enough for females, as it would tend to lead to reduced calf size, reduced survivorship (perhaps for both mother and calf), and reduced status for both mother and calf (to the extent that this is related to body size). However, it is even worse for males, for whom body size is a strong predictor of reproductive success (Hollister-Smith et al., 2007). In the absence of significantly elevated risk of adult mortality, delay of reproductive schedules is thus usually the best strategy. Proboscideans must have operated under this regime throughout much of their evolutionary history, and they have evolved behavioral and physiological traits that serve to regulate their life history response to environmental stress on ecological time scales. It is thus no surprise that under conditions of resource limitation, extant elephants show delayed maturation and prolonged calving intervals (Douglas-Hamilton, 1973; Laws et al., 1975).

In cases of moderate resource limitation, moderate delay of reproductive schedules might allow maturation at nearly the ancestral body size, but this is not to say that any amount of environmental stress can be accommodated in this way. If only because lifespan itself is finite, a proboscidean living under conditions of chronic resource limitation must eventually commit to reproduction, even if at a less than optimal body size. If further delay of reproductive schedules is not likely to lead to realized gains in fitness, then the "time-value" of earlier reproduction will begin to dominate the calculus of tradeoffs that control life histories (Clutton-Brock et al., 1982). This is the type of scenario generally called upon to explain the phenomenon of island dwarfing that has occurred so many times within probosocidean history, though to be sure, the developmental mechanisms behind this phenomenon have not yet been documented to the degree that would be possible with tusk-based studies of life history. Similarly, King and Saunders (1984) have appealed to resource limitation within the continental setting to explain reduction in mastodon body size during the late Pleistocene. Although this is certainly plausible, without actual data on ages and reproductive schedules, we cannot distinguish this cause of body size reduction (essentially a nutritionally induced neoteny, or retardation of growth and maturation) from an alternative scenario according to which smaller body size is viewed as a consequence of accelerated reproductive schedules (essentially progenesis). To understand why this might occur, we have but to consider hunting.

Hunting tends to reduce population densities and relax constraints on growth and reproduction imposed by resource limitations. Females that are in adequate condition to go into estrous do so, and those that lose a calf for any reason typically cycle again with little delay. Losses from any portion of the spectrum of age classes remove potential competitors for reproductive opportunities and diminish any expected benefits of delaying reproduction until greater size can be attained. Under this regime, when survival to successful reproduction is itself less certain, the "time-value" of early reproduction (like compound interest) assumes greater importance than any "strategic" delay of reproductive effort. This effect may be especially strong for males, who are normally capable of reproduction long before the social hierarchy affords them significant opportunities. When that social hierarchy is "thinned out" by elevated mortality rates, onset of musth and reproductive function may be accelerated (Slotow et al., 2000). A relevant example of this phenomenon, though not involving elephants, was documented by Carrick and Ingham (1962) in southern elephant seals. On South Georgia Island, where hunting had been allowed for years, females matured about a year earlier, and males 2–3 years earlier, than on Macquarie Island, where no hunting had been allowed. Again, because these responses are mediated by behavioral and physiological mechanisms, they maintain a dynamic equilibrium with the social and environmental circumstances encountered by individual organisms.

It would be unreasonable not to consider what might happen under the combined assault of environmental deterioration and hunting pressure, but because these ecological forces have such diametrically opposed effects, a situation in which they were precisely matched would be expected to yield little change in life histories. In a less precisely balanced scenario, the stronger force would likely prevail, and we would see evidence of its action in the direction of change of life history traits. Before we get caught up, however, in worrying about still more complex scenarios (e.g., temporally varying combinations of opposing forces), we should remember that our role as empiricists is not so much to predict what would happen under any and all circumstances, but rather to document the patterns that history actually presents for our evaluation and correctly assess their differential likelihood under (and thus their differential support for) competing causal hypotheses.

As a proxy for somatic growth rates, two different measures of tusk growth rate (rate of dentin apposition, measured by thickness of increments, and rate of tusk extension, measured by length of annual increments) were considered in Fisher (2001a). Annual length increments may be easier to observe (as periradicular features near the growing end of the tusk, especially under conditions of high seasonality), but studies of their pattern of variation throughout the lifespan (e.g., Fisher, 2008; Fisher et al., 2008) make it clear that there is a strong ontogenetic component in both sexes and pronounced sex differences, making appropriately controlled comparisons critical. In contrast, thicknesses of annual dentin increments are more nearly constant throughout the adult portions of tusks (though they still differ between males and females) and can be compared more easily. Both sets of tusk growth rates reported by Fisher (2001a), for North American mammoths showed higher values toward the end of the Pleistocene than for times earlier in the Pleistocene. I do not yet have this much time depth for mastodons, but their tusk growth rates are on par with those of late Pleistocene mammoths.

As noted above, calving intervals are tentatively recognized based on relatively regular cycles of variation in tusk growth rate in adult female mastodons. These patterns differ from what is seen in immature females and males of any age (Fisher, 1996a). Annual increments within such time series have been corroborated by oxygen and carbon isotope profiles (Fisher and Fox, 2005), and the pattern taken as characteristic of adult females has been documented at high temporal resolution and with interannual statistical comparisons (Fisher et al., 2008). We still need more individuals to be analyzed in these ways, and we especially need to trace structural and/or compositional traits that provide independent indications of calving cycles. Despite these reservations, all indications in hand now point toward late Pleistocene calving cycles in mastodons as lasting an average of 3–4 years, equivalent to those of African elephants under the most benign environmental conditions (Douglas-Hamilton, 1973; Laws et al., 1975).

Age of maturation is the life history trait for which some of the best data exist currently. For males, this is the time of eviction of adolescent males from the matriarchal family unit, and for females, it is the time of first conception. Males in particular are not fully mature until later, on or after the onset of musth, but eviction shows up clearly in the tusk record as a year of sharply reduced tusk growth rate, and it is at least one important point of reference. Some of the first data on age of maturation in mastodons were presented in Fisher (1996a), and a graph like Fig. 4.3, showing age of maturation vs. radiocarbon years before present, has been shown publicly on multiple occasions since then, starting with Fisher (1996b). In each case, these data have been described as preliminary: not yet based on sufficiently detailed tusk analyses, not yet fully accounting for loss of years from the tusk tips, and some requiring additional radiocarbon dating. Improving on these deficiencies is a priority, but unfortunately, it has at times had to be deferred to address newly discovered specimens and ephemeral opportunities for progress on other critical issues. For these reasons, the data themselves remained unpublished for over a decade, until my colleague Michael Foote convinced me they were useful enough as an example of the kind of argument being envisioned that they should be reported (Foote and Miller, 2007). Having now given permission for the graph to be published, I feel obliged to present the data on which it is based, and this is done in Table 4.2. In the intervening years, new data have been obtained, some of which are actually more highly resolved (Fisher, 2008; Fisher et al., 2008), and at the risk of mixing generations of results of differing inherent quality, these are added (where radiocarbon dates are available). All the caveats listed above still apply, and I would certainly entertain the

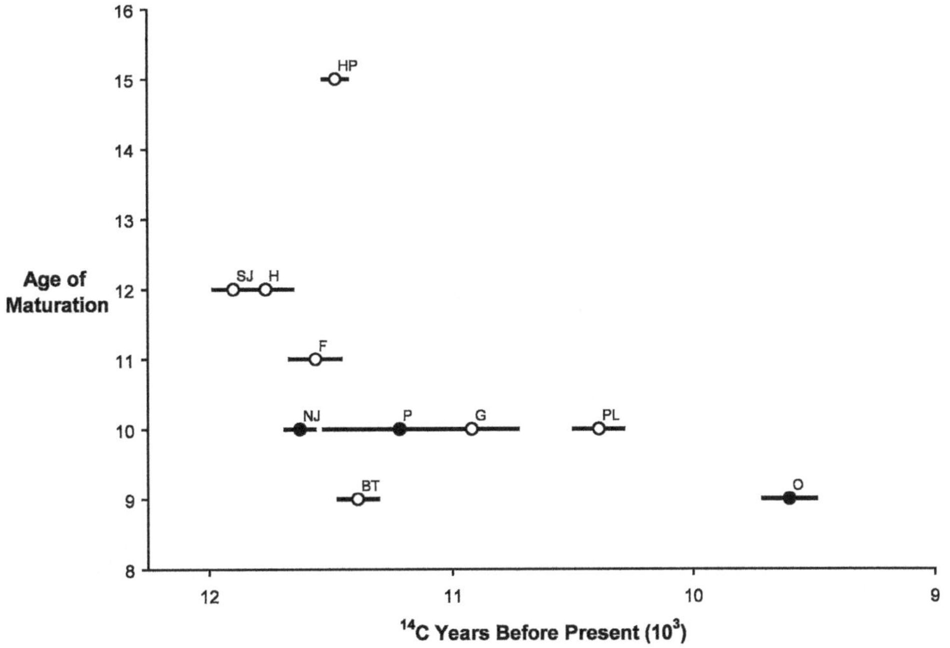

FIGURE 4.3. Graph showing age of maturation, inferred from annual increment measurements in tusk, relative to radiocarbon age estimates; data provided in Table 4.2. Males are shown as open circles, females as solid circles, labels as the initials of the site name, and reported uncertainties on age estimates as solid bars.

TABLE 4.2. Radiocarbon age estimates and ages of maturation for Great Lakes Region mastodons; sex of each individual is listed in Table 4.1.

Specimen name	Radiocarbon age estimate	Laboratory number	Age of maturation
St. Johns	11,900 ± 80	Beta 78626	12
Heisler	11,770 ± 110	AA 6979, NSRL 282	12
North Java	11,630 ± 60	Beta 176928	10
Farview	11,565 ± 105	AA 7397, X 359	11
Hyde Park	11,480 ± 50	Beta 141061	15
Burning Tree	11,390 ± 80	AA 6980, NSRL 283	9
Powers	11,220 ± 310	Beta 9482	10
Grandville	10,920 ± 190	Beta 15265	10
Pleasant Lake	10,395 ± 100	Beta 1388	10
Owosso	9600 ± 110	Beta 74159	9

possibility that something on the order of two years should be added to many of these ages to account for time missing from the tip. However, I think it unlikely that as many as 5 or 7 years should be added, and that is what would be required to make these data match the ages of maturation of elephants under conditions of environmental stress. Moreover, even this would leave unexplained the pronounced decline in age of maturation toward the end of the Pleistocene. I keep an open mind, but I see strong indications that age of maturation for Great Lakes Region mastodons declined as they neared extinction.

Discussion: Extinction Scenarios

This chapter is not intended as a general discussion of the merits, relative or absolute, of all scenarios under consideration for explaining late Pleistocene megafaunal extinctions. Other recent discussions address a wider range of factors than I can treat here (e.g., Alroy, 2001; Haynes, 2002; Barnosky et al., 2004; Martin, 2005). I have attempted to show, however, that if we focus on resolving the relative merits of climate change and human hunting as potential causes of extinction, life history data derived from tusk analyses of Great Lakes Region mastodons run counter to expectations based on climate change models, but follow closely the patterns expected for hunting. Interestingly, the duration of human-proboscidean interaction implied by Fig. 4.3 is much longer than suggested for Martin's (1967) "Blitzkrieg" style of overkill and suggests rather a more protracted attrition of proboscidean populations. Much of the hunting appears to have focused on adult males (Fisher, 1987), but agents of mortality for females are not well resolved.

Each determination in Table 4.2 and Fig. 4.3 deserves its own discussion, but this would take us into more detail than is possible here. One maturation age that stands out from the rest is the 15-year age for Hyde Park. Although this is higher than other maturation ages, this is also the animal that shows annual musth battles in his tusk record, implying a dense local population of adult male adversaries. Delayed maturation (though this is still not "late" by African elephant standards) is the expected response in this social setting. No other male in this data set shows a comparable frequency of musth battles, which could mean that other males are derived from populations that had already experienced some reduction in the density of adult males.

Likewise, one date that stands out is the figure of 9,600 rybp for the Owosso mastodon. This is the individual interpreted in Fisher (1996b) as having lost a large number of calves prior to weaning. At the time, this seemed suggestive of human predation, and subsequent discovery of what could be a mother-calf pair at the Manitou Beach site enhances this impression, without of course yet providing firm evidence. This date for Owosso is later than is generally accepted for mastodons in North America (the youngest currently accepted date being 10,395 rybp for the Pleasant Lake mastodon; Fisher, 1984a; Meltzer and Mead, 1983) and it needs to be replicated, but it raises the possibility of a late phase of human-mastodon interaction in which hunting of females and calves became more common. As if this were not enough, the possibility (discussed above) that misdirected aggression of novice males in premature musth was responsible for some deaths of adult females adds an even darker note to the last days of mastodons in North America.

Data on mammoth life histories in North America are not yet plentiful enough to resolve a pattern as clear as that for mastodons, but they are compatible with the same broad trends. We simply need more analyses of more tusks with a wider sampling of regions and time intervals. Parallel studies are underway on life histories of Siberian woolly mammoths (*Mammuthus primigenius*), which offer the promise of large numbers of exquisitely well preserved specimens of male and female adults (e.g., Fisher et al., 2007; Gohman et al., 2007) and juveniles of both sexes (e.g., Rountrey et al., 2007a, b). Whether woolly mammoths show life history changes like those of North American mammoths and mastodons is much too soon to tell, but we are eagerly at work deciphering their tusk records.

Acknowledgments. I am grateful to Gary Haynes for inviting this contribution and to Paul Koch and an anonymous reviewer for comments that greatly improved its clarity. Scott Beld has provided able assistance during much of the work reported here; Paul Koch and David Fox remain valued colleagues in multiple ongoing projects; and Adam Rountrey and Kathlyn Smith have already taken on significant roles in moving this work forward. We have also benefited from many undergraduate assistants and volunteers in fieldwork and laboratory phases of documenting mastodon skeletons. Finally, Larry Agenbroad, Warren Allmon, Bruce Bourque, Robert Brunswig, George Frison, Elizabeth Garland, Alan Holman, Gary Hoyle, Ronald Kapp, Jude Kirkpatrick, Richard Laub, Bradley Lepper, George McIntosh, Norton Miller, Ron Richards, Jeffery Saunders, Ed Smith, Dennis Stanford, and Mike Voorhies have provided access to many important specimens and participated in rewarding collaborations on North American material.

References

Alroy J (2001) A multispecies overkill simulation of the end-Pleistocene megafaunal mass extinction. Science 292:1893–1896

Ambrose SH, Norr L (1993) Experimental evidence for the relationship of the carbon isotope ratios of whole diet and dietary protein to those of bone collagen and carbonate. In: Lambert JB, Groupe G (eds) Prehistoric human bone: Archaeology at the molecular level. Springer, Berlin, pp 1–37

Barnosky AD, Koch PL, Feranec RS, Wing SL, Shabel AB (2004) Assessing the causes of late Pleistocene extinctions on the continents. Science 306:70–75

Brunswig RH (2007) New interpretations of the Dent mammoth site. In: Brunswig RH, Pitblado BL (eds) Frontiers in Colorado Paleoindian archaeology: From the Dent site to the Rocky Mountains. University Press of Colorado, Boulder, CO, pp 87–121

Buss IO (1990) Elephant life: Fifteen years of high population density. Iowa State University Press, Ames

Carrick R, Ingham SE (1962) Studies on the southern elephant seal, *Mirounga leonina* (L.). I. Introduction to the series. Commonwealth Scientific and Industrial Research Organization Wildlife Research 7, 89–101

Cerling TE, Harris JH (1999) Carbon isotope fractionation between diet and bioapatite in ungulate mammals and implications for ecological and paleoecological studies. Oecologia 120:347–363

Clutton-Brock TH, Guiness FE, Albon SD (1982) Red Deer: Behavior and ecology of two sexes. University of Chicago Press, Chicago

Coope GR, Lister AM (1987) Late-glacial mammoth skeletons from Condover, Shropshire, England. Nature 330:472–474

Dean MC, Scandrett AE (1996) The relation between long-period incremental markings in dentine and daily cross-striations in enamel in human teeth. Arch Oral Biol 41:233–241

DeNiro MJ, Epstein S (1981) Influence of diet on the distribution of nitrogen isotopes in animals. Geochimica et Cosmochimica Acta 45:341–351

Deraniyagala PEP (1955) Some extinct elephants, their relatives, and the two living species. National Museum of Ceylon, Colombo

Douglas-Hamilton I (1973) On the ecology and behavior of the Lake Manyara elephants. East Afr Wildlife J 11:401–403

Elder WH (1970) Morphometry of elephant tusks. Zoologica Africana 5:143–159

Espinoza EO, Mann MJ (1993) The history and significance of the Schreger pattern in proboscidean ivory characterization. J Am Institute Conserv 32:241–248

Firestone RB, West A, Kennett JP, Becker L, Bunch TE, Revay ZS, Schultz PH, Belgya T, Kennett DJ, Erlandson JM, Dickenson OJ, Goodyear AC, Harris RS, Howard GA, Kloosterman JB, Lechler P, Mayewski PA, Montgomery J, Poreda R, Darrah T, Que Hee SS, Smith AR, Stich A, Topping W, Wittke JH, Wolbach WS (2007) Evidence for an extraterrestrial impact 12,900 years ago that contributed to the megafaunal extinctions and the Younger Dryas cooling. Proc Natl Acad Sci 104:16016–16021

Fisher DC (1984a) Mastodon butchery by North American Paleo-Indians. Nature 308:271–272

Fisher DC (1984b) Taphonomic analysis of late Pleistocene mastodon occurrences: Evidence of butchery by North American Paleo-Indians. Paleobiology 10:338–357

Fisher DC (1987) Mastodont procurement by Paleoindians of the Great Lakes Region: Hunting or scavenging? In: Nitecki MH, Nitecki DV (eds) The evolution of human hunting. Plenum, New York, pp 309–421

Fisher DC (1988) Season of death of the Hiscock mastodonts. In: Laub RS, Miller NG, Steadman DW (eds) Late Pleistocene and Early Holocene paleoecology and archeology of the Eastern Great Lakes Region. Bull Buffalo Soci Natural Sci 33:115–125

Fisher DC (1994) Late Pleistocene mastodon trackways in pond-margin sediments in southeastern Michigan. J Vert Paleontol 14(Suppl to 3):25A

Fisher DC (1995) Experiments on subaqueous meat caching. Curr Res Pleistocene 12:77–80

Fisher DC (1996a) Extinction of proboscideans in North America. In: Shoshani J, Tassy P (eds) The Proboscidea: Evolution and palaeoecology of elephants and their relatives. Oxford University Press, Oxford, pp 296–315

Fisher DC (1996b) Testing late Pleistocene extinction mechanisms with data on mastodon and mammoth life history. J Vert Paleontol 16 (Suppl to 3):34A

Fisher DC (2001a) Season of death, growth rates, and life history of North American mammoths. In: West DL (ed) Mammoth site studies: Proceedings of the first international conference on mammoth site studies. Publications in Anthropology, University of Kansas, Lawrence 22, 121–135

Fisher DC (2001b) Entrained vs. free-running physiological rhythms and incremental features in the tusks of woolly mammoths. J Vert Paleontol 21(Suppl to 3):49A–50A

Fisher DC (2004a) Mastodons, mammoths, and humans in the North American midcontinent. In: Lepper BT, Bonnichsen R (eds) New perspectives on the first Americans. Center for the Study of the First Americans, College Station, TX, pp 81–86

Fisher DC (2004b) Season of musth and musth-related mortality in Pleistocene mammoths. J Vert Paleontol 24(Suppl to 3):58A

Fisher DC (2005) Osteology, taphonomy, and life history of two late Pleistocene American mastodons (*Mammut americanum*). In: Agenbroad LD, Symington RL (eds) 2nd World of elephants congress: Short papers and abstracts. Mammoth Site of Hot Springs, South Dakota, pp 44–46

Fisher DC (2007) Life history analysis of the Yukagir mammoth. In: Boeskorov GG, Tikhonov AN, Suzuki N (eds) The Yukagir mammoth. Institute of Applied Ecology of the North [Yakutsk], St. Petersburg, pp 142–156

Fisher DC (2008). Taphonomy and paleobiology of the Hyde Park mastodon. In: Allmon WD, Nester PL (eds) Mastodon paleobiology, taphonomy, and paleoenvironment in the late Pleistocene of New York State: Studies on the Hyde Park, Chemung, and Java sites. Palaeontographica Americana 61:197–289

Fisher DC n.d. (1997) Recovery and analysis of the St. Johns mastodon, Clinton County, Michigan. Bureau of Transportation Planning, Environmental Section, Michigan Department of Transportation, 91 pp

Fisher DC, Beld SG, Rountrey AN (2008) Tusk record of the North Java mastodon. In: Allmon WD, Nester PL (eds) Mastodon paleobiology, taphonomy, and paleoenvironment in the late Pleistocene of New York State: Studies on the Hyde Park, Chemung, and Java sites. Palaeontographica Americana 61:417-463

Fisher DC, Fox DL (2003) Season of death and terminal growth histories of Hiscock mastodons. In: Laub RS (ed) The Hiscock Site: late Pleistocene and Holocene paleoecology and archaeology of western New York state. Bulletin of the Buffalo Society of Natural History 37, 83–101

Fisher DC, Fox DL (2005) Calving histories of female mastodons (*Mammut americanum*). J Vert Paleontol 25 (Suppl to 3):57A

Fisher DC, Fox DL (2006) Five years in the life of an Aucilla River mastodon. In: Webb SD (ed), The First Floridians. Springer, Dordrecht, pp 343–377

Fisher DC, Fox DL (2007a) Season of death of the Dent mammoths: Distinguishing single from multiple mortality events. In: Brunswig RH, Pitblado BL (eds) Frontiers in Colorado Paleoindian archaeology: From the Dent site to the Rocky Mountains. University Press of Colorado, Boulder, CO, pp 123–153

Fisher DC, Fox DL (2007b) Life history and unilateral loss of molar function in the Cohoes mastodon: A case study in nutritional stress? J Vert Paleontol 27 (Suppl to 3), 74A–75A

Fisher DC, Fox DL, Agenbroad LD (2003) Tusk growth rate and season of death of *Mammuthus columbi* from Hot Springs, South Dakota, USA. In: Reumer JWF, De Vos J, Mol D (eds) Advances in mammoth research (Proceedings of the Second International Mammoth Conference, Rotterdam, May 16–20 1999). Deinsea 9:117–133

Fisher DC, Lepper BT, Hooge PE (1994) Evidence for butchery of the Burning Tree mastodon. In: Dancy WS (ed) The first discovery of America: Archaeological evidence of the early inhabitants of the Ohio area. Ohio Archaeological Council, Columbus, pp 43–57

Fisher DC, Rountrey AN, Beld SG, Fox DL, Gohman S, Tikhonov AN, Mol D, Buigues B, Boeskorov G, Lazarev P (2007) Life history of the Yukagir mammoth. Abstracts, International Mammoth Conference IV:129–130

Foote M, Miller AI (2007) Principles of Paleontology. Third Edition. WH Freeman, New York

Fox DL, Fisher DC, Vartanyan S, Tikhonov AN, Mol D, Buigues B (2007) Paleoclimatic implications of oxygen isotopic variation in late Pleistocene and Holocene tusks of *Mammuthus primigenius* from northern Eurasia. Quat Intl 169–170:154–165

Frison GC, Todd LC (1986) The Colby mammoth site: Taphonomy and archaeology of a Clovis kill in northern Wyoming. University of New Mexico Press, Albuquerque

Garland EB, Cogswell JW (1985) The Powers mastodon site. Michigan Archaeolo 31:3–39

Gat JR (1980) The isotopes of hydrogen and oxygen in precipitation. In: Fritz P, Fontes JC (eds) Handbook of environmental isotope geochemistry, 1, The terrestrial environment. Elsevier, Amsterdam, pp 21–47

Gohman S, Fox DL, Fisher DC, Vartanyan S, Tikhonov AN (2007) Paleodietary and environmental implications of carbon and nitrogen isotopic variation in late Pleistocene and Holocene tusks of *Mammuthus primigenius* from northern Eurasia. J Vert Paleontol 27 (Suppl to 3):82A

Gohman SCL, Fox DL, Fisher DC, Vartanyan SL, Tikhonov AN, Mol D, Buigues B (Submitted) Paleodietary and paleoenvironmental implications of carbon and nitrogen isotopic variations in late Pleistocene and Holocene tusks of *Mammuthus primigenius* from northern Eurasia. In: Boeskorov GG, Lazarev PA (eds) Proceedings of the 4th International Mammoth Conference, Yakutsk. 8 ms. pp., 4 figures

Haynes G (1990) The mountains that fell down: Life and death of heartland mammoths. In: Agenbroad LD, Mead JI, Nelson LW (eds) Megafauna and man: Discovery of America's heartland. The Mammoth Site of Hot Springs, South Dakota, Hot Springs, pp 22–31

Haynes G (1991) Mammoths, mastodonts, and elephants: Biology, behavior, and the fossil record. Cambridge University Press, Cambridge

Haynes G (2002) The early settlement of North America: The Clovis era. Cambridge University Press, Cambridge

Heaton THE, Vogel JC, von la Chevallerie G, Collett G (1986) Climatic influence on the isotopic composition of bone nitrogen. Nature 322:822–823

Hobson KA, Alisauskas RT, Clark RG (1993) Stable-nitrogen isotope enrichment in avian tissues due to fasting and nutritional stress: Implications for isotopic analyses of diet. Condor 95:388–394

Hollister-Smith JA, Poole JH, Archie EA, Vance EA, Georgiadis NJ, Moss CJ, Alberts SC (2007) Age, musth and paternity success in wild male African elephants, *Loxodonta africana*. Animal Behaviour 74(2):287–296

Hoppe KA (2004) Late Pleistocene mammoth herd structure, migration patterns, and Clovis hunting strategies inferred from isotopic analyses of multiple death assemblages. Paleobiology 30:129–145

Hoyle BG, Fisher DC, Borns HW, Jr, Churchill-Dickson LL, Dorion CC, Weddle TK (2004) Late Pleistocene mammoth remains from coastal Maine, USA. Quat Res 61, 277–288

Kapp RO, Cleary DL, Snyder GG, Fisher DC (1990) Vegetational and climatic history of the Crystal Lake area and the Eldridge mastodont site, Montcalm County, Michigan. Am Midland Natural 123:47–63

King JE, Saunders JJ (1984) Environmental insularity and the extinction of the American mastodont. In: Martin PS, Klein RG (eds) Quaternary extinctions: A prehistoric revolution. University of Arizona Press, Tucson, pp 315–339

Klevezal GA (1996) Recording structures of mammals: Determination of age and reconstruction of life history. Revised and updated edition (Mina MV, Oreshkin AV, trans.). A.A. Balkema, Rotterdam

Koch PL (1989) Paleobiology of late Pleistocene mastodonts and mammoths from southern Michigan and western New York. Ph.D. Dissertation, University of Michigan

Koch PL (1998) Isotopic reconstruction of past continental environments. Ann Rev Earth Planet Sci 26:573–613

Koch PL, Fisher DC, Dettman D (1989) Oxygen isotope variation in the tusks of extinct proboscideans: A measure of season of death and seasonality. Geology 17:515–519

Koch PL, Fogel ML, Tuross M (1994) Tracing the diets of fossil animals using stable isotopes. In: Lajtha K, Michener RH (eds) Stable isotopes in ecology and environmental science. Blackwell, Oxford, pp 63–92

Kohn M (1996) Predicting animal $\delta^{18}O$: Accounting for diet and physiological adaptation. Geochimica et Cosmochimica Acta 60:4811–4829

Laws RM (1966) Age criteria for the African elephant, *Loxodonta a. africana*. East African Wildlife J 4:1–37

Laws RM, Parker ISC, Johnstone RCB (1975) Elephants and their habitats: The ecology of elephants in North Bunyoro, Uganda. Clarendon Press, Oxford

Lee PC, Moss CJ (1985) Early maternal investment in male and female African elephant calves. Behav Ecol Sociobiol 18:353–361

Lepper BT, Frolking TA, Fisher DC, Goldstein G, Wymer DA, Sanger JE, Ogden JG, III, Hooge PE (1991) Intestinal contents of a late Pleistocene mastodont from midcontinental North America. Quat Res 36:120–125

Lister AM (1996) Sexual dimorphism in the mammoth pelvis: An aid to gender determination. In: Shoshani J, Tassy P (eds) The proboscidea: Evolution and palaeoecology of elephants and their relatives). Oxford University Press, Oxford, pp 254–259

Lister AM, Agenbroad LD (1994) Gender determinations of the Hot Springs mammoths. In: Agenbroad LD, Mead JI (eds) The Hot Springs mammoth site: A decade of field and laboratory research in paleontology, geology, and paleoecology. The Mammoth Site of Hot Springs, SD, Hot Springs, pp 208–214

MacPhee RDE, Marx PA (1997) The 40,000-year plague: Humans, hyperdisease, and first-contact extinctions. In: Goodman S, Patterson B (eds) Natural change and human impact in Madagascar. Smithsonian Institution Press, Washington, DC, pp 169–217

Martin PS (1967) Prehistoric Overkill. In: Martin PS, Wright HE, Jr (eds) Pleistocene extinctions: The search for a cause. Yale University Press, New Haven, pp 75–120

Martin PS (2005) Twilight of the mammoths: Ice Age extinctions and the rewilding of America. University of California Press, Berkeley

Meltzer DJ, Mead JI (1983) The timing of late Pleistocene mammalian extinctions in North America. Quat Res 19:130–135

Moss CJ (1988) Elephant memories: Thirteen years in the life of an elephant family. William Morrow, New York

Newman HN, Poole DFG (1974) Observations with scanning and transmission electron microscopy on the structure of human surface enamel. Arch Oral Biol 19:1135–1143

Osborn HF (1936) Proboscidea: A monograph of the discovery, evolution, migration and extinction of the mastodonts and elephants of the world. Vol. 1: Moeritherioidea, Deinotherioidea, Mastodontoidea. The American Museum of Natural History, New York, 802 pp

Polischuck SC, Hobson KA, Ramsay MA (2001) Use of stable-carbon and -nitrogen isotopes to assess weaning and fasting in female polar bears and their cubs. Can J Zool 79:499–511

Poole JH (1987) Rutting behaviour in African elephants: The phenomenon of musth. Behaviour 102:283–316

Poole JH, Moss CJ (1981) Musth in the African elephant, *Loxodonta africana*. Nature 292:830–831

Roth VL (1984) How elephants grow: Heterochrony and the calibration of developmental stages in some living and fossil species. J Vert Paleontol 4:126–145

Rountrey AN, Fisher DC, Vartanyan S, Fox DL (2007a) Carbon and nitrogen isotope analyses of a juvenile woolly mammoth tusk: Evidence of weaning. Quat Intl 169–170:166–173

Rountrey A, Fisher DC, Mol D, Lazarev P, Boeskorov G (2007b) Prenatal to early juvenile woolly mammoth life history as revealed by structural and compositional analyses. J Vert Paleontol 27 (Suppl to 3):136A

Rountrey AN, Fisher DC, Mol D, Buigues B, Tikhonov AN, Lazarev PA, Boeskorov GG (Submitted). Stable isotope time-series in juvenile mammoth tusks. In: Boeskorov GG, Lazarev PA (eds) Proceedings of the 4th International Mammoth Conference, Yakutsk. 7 ms. pp., 2 figures

Rozanski K, Araguás-Araguás L, Gonfiantini R (1993) Isotopic patterns in modern global precipitation. In: Swart PK, Lohmann KC, McKenzie J, Savin S (eds) Climate change in continental isotopic records. Am Geophys Union Geophys Mono 78:1–36

Saunders JJ 1980 A model for man-mammoth relationships in late Pleistocene North America. Can J Anthropol 1:87–98

Sher TM, Fisher DC (1995) Sexual dimorphism in *Mammut americanum*. J Vert Paleontol 15(Suppl to 3):53A

Sikes SK (1971) The natural history of the African elephant. Weidenfeld & Nicolson, London

Slotow R, van Dyk G, Poole J, Page B, Klocke A (2000) Older bull elephants control young males. Nature 408:425–426

Smith K, Fisher DC (2007) Sexual dimorphism in tusks of Great Lakes-region American mastodons (*Mammut americanum*). J Vert Paleontol 27 (Suppl to 3):149A

Speth JD, Spielman KA (1983) Energy source, protein metabolism, and hunter-gatherer subsistence strategies. J Anthropol Archaeol 2:1–31

Stiner MC (1991) Food procurement and transport by human and non-human predators. J Archaeol Sci 18:455–482

Stuart-Williams HLeQ, Schwarcz HP (1997) Oxygen isotopic determination of climatic variation using phosphate from beaver bone, tooth enamel, and dentine. Geochimica et Cosmochimica Acta 61:2539–2550

Taylor JG (1969) William Turner's journeys to the caribou country with the Labrador Eskimos in 1780. Ethnohistory 16: 141–164

Teeri JA, Stowe LG (1976) Climatic patterns and the distribution of C_4 grasses in North America. Oecologia 23:1–12

Trapani J, Fisher DC (2003) Discriminating proboscidean taxa using features of the Schreger pattern in tusk dentin. J Archaeol Sci 30:429–438

Vereshchagin NK, Tikhonov AN (1986) [A study of mammoth tusks.] Issledovaniya bivnyey mamontov. In: Vereshchagin NK, Kuz'mina, IYe (eds) Mlekopitayushchiye chetvertichnoy fauny SSSR. Trudi Zoologicheskovo Instituta 149:3–14

Vereschchagin NK, Tikhonov AN (1999) Exterior of the mammoth. Cranium 1:1–93

Warwick R, Williams PL (1973) Gray's anatomy, 35th British Edition. WB Saunders, Philadelphia

5. Human Prey Choice in the Late Pleistocene and Its Relation to Megafaunal Extinctions

Todd A. Surovell* and Nicole M. Waguespack
Department of Anthropology
University of Wyoming
1000 East University Avenue
Laramie, WY 82071, USA
surovell@uwyo.edu, nmwagues@uwyo.edu

Keywords Early Paleoindian diet • large-game predation

Introduction

Like many dimensions of human behavior during the early phases of New World occupation, interpretations of Early Paleoindian subsistence practices are highly contentious. Different researchers examining the same faunal record have arrived at opposing conclusions regarding what Early Paleoindians were hunting, collecting, and eating. Some argue that Early Paleoindians were quintessentially "large game specialists;" others see a pattern of "generalized foraging." This debate has important implications for evaluating possible causes of Pleistocene extinctions. While at the core of the issue is a fundamentally simple question – "What did Early Paleoindians hunt?" – the interpretation of direct human involvement in the demise of multiple species of animals is clouded by larger issues concerning hunter-gatherer economics and climate change. Our concern is with the former, and we examine Early Paleoindian hunting from an ethnographic, zooarcheological, and behavioral ecological standpoint.

Why Hunt Big Animals, Especially Really Big Ones?

Kurtén and Anderson (1980) estimated a woolly mammoth (*Mammuthus primigenius*) to have weighed over six metric tons, roughly 100 times the weight of an average human. The magnitude of this size difference is unprecedented among predatory mammals and their prey. Even a wolf capturing an adult bull moose is killing an animal only about eight times its body weight. Yet, interpretations of Early Paleoindian subsistence frequently portray foraging societies as specialized hunters of Pleistocene megafauna. While the image of mammoth hunters is compelling, it remains difficult to conceive of megafaunal hunting as a habitual component of the food quest. This scenario challenges the limits of hunter-gatherer economic diversity as it is ethnographically known and more generally presents a unique predator-prey dynamic in prehistoric ecosystems.

Whether or not Pleistocene foragers of the Americas were capable of killing mammoths and other megafauna is not in question. Historic and recent hunter-gatherers are proficient at killing elephant sized game using hand propelled weapons (Hodgeson, 1926; Johnson et al., 1980; Duffy, 1984; Fisher, 1993) and replicative experiments indicate that Early Paleoindian weaponry is suitable for megafauna predation (Stanford, 1987; Frison, 1989). While we can comfortably assume that Pleistocene foragers of the Americas could hunt extremely large-bodied prey, we are left to explore the complex issue of why they would or would not have chosen to do so (e.g., Haynes, 2002a:198–200). In addition, questions regarding the potential human involvement in Pleistocene extinction necessitates that the degree of megafauna exploitation also be addressed. Exploring the personal motivations of Paleoindian hunters is well beyond our interpretive capacities. But the more general question of how foragers make economic decisions regarding prey selection can be addressed.

Given various options, how do hunter-gatherers choose which prey to pursue? Models and concepts derived from foraging theory, such as diet breadth ranking systems, predation risk, and variance, provide one means for exploring prey choice. According to the diet breadth model, a hunter can maximize net

* Address for correspondence: surovell@uwyo.edu

G. Haynes (ed.), *American Megafaunal Extinctions at the End of the Pleistocene*, 77–105.

return rates by focusing on taxa whose post-encounter returns exceed the average environmental return rate (Charnov, 1976; Stephens and Krebs, 1986). Individual prey species can then be ranked according to their potential caloric returns relative to other prey. In its simplest form, the optimal diet is comprised of the highest ranked resources. Taking lower ranked taxa only serves to lower overall return rates, an unproductive activity from an evolutionary standpoint. Focusing one's subsistence efforts on high ranked items ensures worthwhile expenditure of a forager's time and energy. High ranked prey are often larger than lower ranked species because they provide greater caloric returns per unit of resource (Winterhalder, 1983; Winterhalder et al., 1988; Ugan, 2005). Although roughly comparable, prey rank and prey size are not necessarily the same. For instance, a mule deer and giant tortoise may be similar in weight but fundamental differences in how these animals are captured and butchered may result in disparate handling costs and hence alter their caloric returns and relative rank. For many predators, excessively large- and small-bodied prey (relative to predator size) may provide so little caloric return due to the difficulty involved in their capture that their rank is far lower than predicted by body size alone (Griffiths, 1980; Byers and Ugan, 2005).

The "zero-one rule" of the diet breadth model predicts that a species will either always be taken or always be ignored upon encounter (Stephens and Krebs, 1986:20–21). From an optimality standpoint an individual predator should always pursue high ranked prey upon encounter and never pursue low ranked items, assuming that the ranking of resources remains constant relative to a fixed average environmental return rate. For as long as high ranked prey species are frequently encountered, low ranked prey are not profitable. Rarely, if ever, can the rank of available prey be considered constant. In a risk-sensitive model, that allows for variation in environmental and prey-specific return rates (e.g., Winterhalder et al., 1999), the optimal diet will consist of a more diverse mix of high and low ranked prey. For example, if a forager encounters a low-ranked prey item in circumstances in which handling costs are minimal, the effective return rate for that animal is enhanced, and it should be exploited (e.g., Madsen and Schmitt, 1988). Temporary scarcity of high-ranked prey could also cause low-ranked prey to be included in the diet (Krebs and McCleery, 1984). In addition, optimal diet breadth should vary for different segments of a foraging population with respect to age, skill, mobility patterns, time constraints, and, for humans, access to hunting technology. By allowing for contingencies such as fluctuations in prey encounter rate, capture efficiency, and capture success rate, variation in the relative ranking of available prey more realistically accounts for the diversity of prey likely to be pursued and captured by a demographically variable group over a period of time.

Thus, although the diet breadth model would generally predict use of the largest encountered prey, it does not imply that the highest ranked resources are always the largest available or that small prey are universally low ranked. Studies of diet breadth in both ethnographic and prehistoric human contexts do commonly indicate that large prey are high ranked and pursued when encountered (Winterhalder, 1981; Hill et al., 1985; Hawkes, 1991; Smith, 1991). However what constitutes "large" prey depends on the species available within the hunting environment and the inherent behavioral/physiological properties of the hunter (Hill et al., 1985; Bailey, 1991; Bird and Bird, 2000; Walker et al., 2002). Although subject to ecological, individual forager, and cultural conditions, prey size remains an important attribute of non-human and human predation strategies. Large prey can provide great economic payoffs, whether these are measured in total weight, calories, or nutrients procured. Given the choice between procuring a 1,000 kg animal and a 20 kg animal with comparable handling costs, why not choose the larger?

The potential caloric benefits of large animals are frequently matched by social (Testart, 1986; Condon et al., 1995; Cox et al., 1999; Gurven et al., 2000; Wiessner, 2002) and/or reproductive benefits (Hawkes, 1991; Kaplan et al., 2000; Marlowe, 2001, 2003) enjoyed by hunters capable of procuring large game in many hunter-gatherer societies. The ability to reliably procure large-bodied game, animals with returns which exceed the immediate caloric needs of the procurer and their dependants, requires successfully capturing prey that are likely to be encountered far less frequently than smaller prey. The combination of providing excess resources and procuring comparatively rare prey animals often brings enhanced social prestige upon the hunter. The larger the resource package, the greater the opportunity to share and exchange hunted prey with others. Large prey animals provide a commodity, utilized among hunter-gatherer societies across the globe, as a widely shared and exchanged resource conferring social benefits (in the form of social capital, economic security, and mating/marriage opportunities) to successful hunters. Thus there are clear economic and social incentives for capable hunters to actively pursue the largest game species available to them. If such social mechanisms were in place among early foragers of the Americas, Pleistocene megafauna presented an obvious medium for attaching social value to hunted prey. The combination of social and economic payoffs provides theoretical impetus for taking seriously the possibility that Pleistocene foragers consistently included megafauna in their subsistence economy.

The Drawbacks of Hunting Really Big Prey

While the potential windfall of calories and social advantages resulting from large animal predation is well documented among subsistence hunting societies from across the globe, it must also be acknowledged that such behavior has its drawbacks and may present an unfeasible economic strategy in many socio-environmental contexts. Negative aspects of large game predation include specific costs incurred to the hunter (in the form of risk and reduced post-encounter caloric returns when hunted prey is widely shared), costs distributed among cooperative individuals (variance in return rates), and more general issues concerning the abundance and processing costs associated with extremely large game.

Hunting large prey involves a certain degree of risk. A hunter must be willing/able to absorb the potential risk of failing to successfully capture encountered prey and the consequences of coming home empty-handed as well as the potential risk of injury involved in pursuit and capture (i.e., not coming home at all). Risk is generally interpreted to increase with prey size (Griffiths, 1980; Jochim, 1981; Bird et al., 2002) and undoubtedly the successful pursuit and capture of Pleistocene sized megafauna clearly entailed the risk of being fatally stomped-on, gored, kicked, and other "unsuccessful" outcomes. However, a certain degree of risk is associated with hunting regardless of the prey being pursued. Assuming equal search times, you are just as likely to be bitten by a snake or trip and break your leg while hunting rabbits as you are hunting mammoth. Fatal attacks on human foragers by other animals such as jaguars and snakes occur more commonly while hunting than engaging in other activities (Treves and Naughton-Treves, 1999), suggesting that time spent hunting is also time a hunter is at higher risk for personal injury regardless of the prey pursued. But it must also be acknowledged that there is a greater probability of being injured or killed by a mammoth than by a rabbit. Obviously the risks involved must be outweighed by the potential benefits, or else people would never choose to pursue potentially dangerous prey animals.

While the vast majority of anthropologists would shudder at the thought of approaching 1,000+ kg animals armed only with hand propelled weaponry, our hesitance cannot and should not be imposed onto others. Recent forager-level hunters of elephant (e.g., Marks, 1976; Steinhart, 2000) and whale (e.g., Boeri, 1983; Alvard and Nolin, 2002) are well aware of the potential risks involved in their predatory endeavors. Although mediated to some extent through careful planning, labor organization, and technology, the risks stay quite high, yet hunters remain undeterred. In some ethnographic cases the pursuit of large dangerous animals serves to increase the social prestige associated with hunting, while in others it may simply result from economic necessity. Either way, the many gruesome risks that we can imagine befalling a Pleistocene hunter of megafauna are plausible but not necessarily accurate reasons for not pursuing the available large game species of the Pleistocene.

Likewise, variance in hunting returns increases with prey size as the hunting of large prey is subject to a more pronounced boom and bust cycle (Hawkes et al., 1991; Waguespack, 2003). Unless large prey are spatially aggregated (which they may be due to behavioral or habitat attributes), their populations are generally thinly spread across the landscape. Compounding the distributional effect is the simple fact that large animals exist at lower numbers than small animals in any given environment. In general, hunted resources exhibit a great degree of variance among individual return rates. Put simply, associated hunters individually pursuing game are likely to experience differential success rates and such differences are only exacerbated when large prey is the focus of predation. Whether variance is measured as an individual's hunting returns over a period of time or among multiple individuals within a given time period, a high degree of variance in the subsistence economy can be a problem that must be mediated. Numerous mechanisms can alleviate intra-group variance. Practices such as food sharing (Cashdan, 1985; Hawkes, 1992; Waguespack, 2002), food storage (Testart, 1982; Hayden, 1994), and cooperative hunting (Smith, 1991; Alvard and Nolin, 2002) have been identified as successful means of redistributing resources among an aggregate of consumers by minimizing the variance in returns between procurers. However the extent to which these "pre-" and "post-kill" mechanisms were employed by Pleistocene foragers of the Americas is not known.

In addition to risk and variance, there are more general concerns that must be acknowledged concerning large game predation. All things being equal, the larger an animal is, the lower its population density. So it must be kept in mind that if the frequency with which prey are encountered is directly related to their density, it can be expected that a hunter would have lower encounter rates and fewer opportunities to capture large prey items. The frequency of encounter should have no impact on prey rank, but if large high ranked prey are rarely encountered it simply cannot provide a consistent or frequent source of calories. For instance, mid-1900s ethnographic reports concerning the Ju/'hoansi of Africa indicate that giraffe was considered a prized resource (providing a windfall of calories and prestige) (Lee, 1979:230–232), but the species was so rarely brought down that it played little role in the everyday subsistence economy. Thus, however highly ranked giraffe might be, characterizing the Ju/'hoansi as specialized giraffe hunters would be wholly inappropriate since the vast majority of their kills and caloric intake is derived from smaller game species.

One additional concern is the potential handling costs associated with extremely large game. It has been argued that the positive relationship between caloric benefits and game size has an upper limit – whereby at some point prey became so large that the handling costs (e.g., transport, butchery, and processing activities that render a kill edible) become so high that caloric return rates begin to drop. The time and energy required to handle fauna the size of mammoths has been modeled to be so costly that return rates make them no more profitable than far smaller prey (the size of deer or bison) (Byers and Ugan, 2005). It is easy to imagine why disarticulating and transporting a mammoth carcass would be exceedingly costly, as moving a single limb would require multiple handlers. Much like risk and variance, however, handling costs can be behaviorally altered. Estimated handling costs based on butchery and processing of African elephants by Efe and Lese hunters include the costs of a crew of butchers traveling to the kill site, establishing a temporary camp, dismembering and butchering, the building of drying racks, the stripping of meat for drying, and transporting the meat back to residential locations (Byers and Ugan, 2005). In this case, elephant kills are infrequent occurrences (Duffy, 1984; Fisher, 1993) and butchery appears to present a costly diversion of labor and resources. However, while recent ethnographic examples attest to the enormous handling costs associated with extremely large prey, two things must be kept in mind: (1) Efe and Lese peoples are still willing to accept these handling costs, and (2) if extremely large-bodied prey are killed comparatively more frequently, then mobility regimes and labor could be organized in ways to reduce handling costs. Ethnographic examples of other high handling costs endeavors, such as whale butchery and the processing of mass bison and caribou kills, suggest that coordination between hunters and butchers (e.g., organized logistical and residential mobility to ensure labor availability and cost effective transport of people and resources) effectively maintain high return rates for large prey by decreasing handling costs.

It should be clear that how foragers make decisions regarding which prey to pursue involves the relative costs, benefits, and attributes of the socio-ecological context in which hunting occurs. Large game in particular can have enormous benefits and exceedingly high costs. Theoretically, Pleistocene foragers were capable of megafauna predation and generally agreed upon attributes of the Early Paleoindian lifestyle such as low population density, frequent residential mobility, relatively high investment in hunting technology, and minimal game processing all support the plausibility that megafauna was consistently utilized. While assuming that low human population levels rendered megafauna a sufficiently abundant source of calories, that mobility served to enhance large game encounter rates, and that technological investment and minimal resource processing indicate some degree of economic "success" all seem plausible, the fact remains that regular predation of extremely large prey presents something of an anomaly among ethnographically known foragers. This may be reason enough for many archeologists to dismiss the potential role of megafauna hunting in Pleistocene societies. However, doing so runs the dangerous risk of limiting hunter-gatherer diversity of the past to the documented present. Similarly, assuming that Pleistocene hunters preferred megafauna as prey simply because they could, belies the unique socio-environmental attributes that make such a strategy possible (e.g., behavioral means of mediating risk, variance, and handling costs).

Identifying Early Paleoindian Diet Breadth

If the extinction of >30 genera of mammalian megafauna in North America was caused directly or indirectly by human hunting, there should be archeological evidence indicating that humans not only hunted, but preferentially hunted, now extinct fauna. As outlined above, there is compelling, albeit purely theoretical, reasons why this may have been the case. Furthermore, because animal populations are capable of sustaining some level of predation depending upon their species-specific demographic attributes (e.g., Brook and Bowman, 2005; Mithen, 1993; Stiner et al., 1999, 2000; Alroy, 2001), the degree to which Early Paleoindians focused their subsistence efforts upon large mammals relates directly to the plausibility of the Overkill hypothesis (Alroy, 2001). If humans only rarely killed large mammals, then human hunting alone may not be sufficient to explain Pleistocene extinctions. If large mammals were regularly and preferentially targeted, Overkill is plausible. Thus, determining whether and to what extent Early Paleoindians hunted extinct fauna is critical to the Overkill debate (Martin, 1973, 1984; Martin and Steadman, 1999; Grayson, 2001; Grayson and Meltzer, 2002, 2003; Haynes, 2002a, b).

A Brief Review of Recent Research

A number of recent studies have examined the Early Paleoindian faunal record to address the question of what the first people in North America hunted and/or should have hunted (Grayson and Meltzer, 2002; Haynes, 2002a, b; Waguespack and Surovell, 2003; Barton et al., 2004; Cannon and Meltzer, 2004; Byers and Ugan, 2005). Among these studies, two camps have emerged: (1) Those who believe that Early Paleoindians were "large game specialists" and were characterized by a "narrow diet breadth" (Haynes 2002a,b: Waguespack and Surovell, 2003; Barton et al., 2004), and (2) Those who believe that Early Paleoindians were "generalists" and were characterized by a "broad diet breadth" (Grayson and Meltzer, 2002; Cannon and Meltzer, 2004; Byers and Ugan, 2005).

For the sake of brevity, we focus on three studies which have attempted to systematically examine the question. In 2003, we constructed a dataset of faunal remains from 33 late Pleistocene archeological sites in North America (Waguespack and Surovell, 2003). Examining the relationship between body size and archeological abundance, we found that large mammals were the most abundant and regularly occurring animals in Early Paleoindian faunal assemblages. Because there is a strong inverse correlation between body size and population density among animals, we argued that this faunal record could only have been produced by hunter-gatherers who were bypassing opportunities to take small game in favor of concentrating on larger prey. Furthermore, we argued that large game specialization is expected for the first inhabitants of North America since small numbers of highly mobile foragers could maintain fairly regular access to large-bodied animals. We considered the Pleistocene landscapes of the Americas to present a relatively unique demographic and ecological context for a hunter-gatherer subsistence strategy, comparatively rare in the modern ethnographic record, to have emerged.

Examining the same record, Cannon and Meltzer (2004) arrived at a very different conclusion. They concluded that "the faunal record provides little support for the idea that all, or even any, Early Paleoindian foragers were megafaunal specialists" (Cannon and Meltzer, 2004:1955). There are two primary reasons they invoke to explain this difference of opinion. The first is methodological. While we included all fauna recovered from Clovis components, whether strong evidence for subsistence use existed or not, Cannon and Meltzer (2004) only included animals for which "secure" subsistence associations were present. The second is bias. Cannon and Meltzer (2004) performed a number of statistical tests designed to demonstrate that the current sample of late Pleistocene archeological sites is a poor reflection of Paleoindian diets because it is highly biased in favor of large game. They argue that greater probabilities of discovery and research attention are afforded to mammoth-bearing archeological deposits resulting in an over-representation of large fauna sites in the record. Finally, Cannon and Meltzer (2004) suggested that Early Paleoindian diets would have varied considerably across the continent in response to environmental variability and therefore prey species availability.

Byers and Ugan (2005) took a slightly different approach, asking the question: "Should we expect large game specialization in the late Pleistocene?" Using known relationships between body size, return rates, handling costs, and population density, they built a formal prey-choice model based on estimated encounter rates, handling costs, and return rates, which allowed them to predict which species should have been included in Early Paleoindian diets given a foraging goal of maximizing post-encounter returns. Using various estimates of encounter rates for large mammals, they concluded that Paleoindians would have regularly killed animals at least down to the size of hares upon encounter, or possibly even smaller mammals such as ground squirrels. Their model causes them "to question whether Paleoindian subsistence would ever have been narrow and specialized..." (Byers and Ugan, 2005:1633) given the wide array of prey species available to them.

Despite these disagreements, it is our contention that there is actually more common ground among these studies than is immediately apparent. For example, we all agree what the Early Paleoindian faunal evidence looks like in terms of species representation, although there are clear differences about how datasets should be constructed and interpreted. From our perspective, there are two major points of disagreement. First is the extent the record is biased and how accurately it reflects Paleoindian prey choice decisions; second is how the terms "large game specialist" and "generalist" should be defined. We begin with the latter question.

Measuring Diet Breadth

Strangely, though much has been written about "diet breadth" and whether Early Paleoindians were "large game specialists" or "generalized foragers," rarely are these phrases defined in the Paleoindian literature. We believe this simple fact explains much of the disagreement about Early Paleoindian subsistence. As we demonstrate below, despite claims to the contrary, some aspects of the argument do not concern the prey species targeted by Paleoindians nor the composition of the faunal record. Instead, they are about what we should label Early Paleoindian subsistence practices.

Two basic approaches to the problem can be identified. One approach is to define the term large game specialist so narrowly (usually implicitly) that it must be false. It is essentially argued that if late Pleistocene foragers used any resources that are not large bodied mammals, whether plant or animal, then Early Paleoindians were not large game specialists (Byers and Ugan, 2005:1637). By this definition, there is no disagreement whatsoever about whether Clovis peoples were large game specialists. They were not. There is archeological evidence of the use of small mammals, reptiles, and possibly birds and fish in the late Pleistocene archeological record of North America (e.g., Haynes and Haury, 1982; Dent and Kauffman, 1985; Johnson, 1987; Storck and Spiess, 1994; Yates and Lundelius, 2001). In fact, by this definition, large game specialization is a strategy that has *never* been employed by humans. Even recent caribou hunting societies of the Arctic and bison hunting societies of the Great Plains would not meet this definition since they are and were known to exploit small bodied prey. If anthropologists are fundamentally interested in studying variation in human behavior, then it should be clear that such a strict definition of *large game specialist* is unworkable since it recognizes no variation whatsoever. While cloaked in the terminology of optimal foraging theory, diet breadth defined in this manner provides no insight into hunter-gatherer economics.

All hunter-gatherers have a diverse set of edible resources available to them, but not all of those resources are regularly exploited. Diet breadth in this sense refers to the actual subsistence choices made in light of the total possible range of choices that could be made based on selectivity. Having the broadest possible diet means regularly exploiting any edible resource upon encounter. Having a narrow diet breadth means often passing up opportunities to exploit certain resources in order to invest greater time and energy into searching for more profitable ones. In this sense, the term "large game specialist" refers to hunter-gatherers who frequently ignore opportunities to go after small prey in favor of searching for larger quarry (Waguespack and Surovell, 2003).[1] The distinction between specialist and generalist strategies relates to the extent to which small bodied prey are not exploited. A generalist regularly attacks small prey upon encounter, and a specialist often ignores small prey upon encounter. This definition is preferable because it recognizes the broad diversity of subsistence strategies known to have been used by hunter-gatherers, and addresses the decision making component of resource procurement. In this framework, the issue is not whether small game was taken by Early Paleoindians but the extent to which it was taken *relative to large game*.

If large game specialization is defined as regularly ignoring opportunities to take small prey upon encounter, how can it be measured archeologically? After all, the species which humans choose to utilize should produce clear archeological residues, while those not taken should remain absent from the record. There is a fairly straightforward solution to this problem. Strong global relationships exist between body size and population density (Peters, 1986) that can be used to estimate relative encounter frequencies for prey species in any ecosystem, present or past (Fig. 5.1a). Generally speaking, there is a negative power relationship between body size and population density, meaning that, all things being equal, small bodied animals are more common than large ones. Thus, the faunal record produced by a generalist who regularly takes any prey upon encounter should reflect ecological population densities and be dominated by small animals. The faunal record of a large game specialist will be dominated by large bodied mammals, although limited use of small prey may be evident as well.

The degree to which high and low ranked taxa are used is the critical distinction between a hunter's selectivity when choosing prey. Relating prey selectivity to the natural abundance of available prey species creates a simple framework for examining the types of prey utilized by hunter-gatherers of the past and present. The two common strategies, *specialized* and *generalized*, developed here do not make explicit which particular species of prey should be used (Fig. 5.1b). Only general categories of prey based on body size and the relative frequency of their encounter are predicted to be used in different proportions by each strategy. Although the terms generalized and specialized could imply two pure strategies, that is not their intended meaning in this discussion. As used here they refer to ends of an idealized continuum and concern decisions regarding which prey, among those available, a forager will target for predation.

Because encounter rate is primarily a function of prey population density, estimated population densities based on body size provide a reasonable approximation of prey encounter rates for specific environments. Actual encounter rates are difficult to establish in real world settings and deriving comparable values in prehistoric contexts is exceedingly difficult. Population density must then be considered only a proxy measure of encounter rate. A comparison of actual encounter rates and population densities of eight prey species used by the Ache of Paraguay in the Mbaracayu Reserve shows the disparity between these two measures. In this case, encounter rates equal the number of animals seen in over 2,000 km of transects walked by Ache informants and anthropologists (Hill and Padwe, 2000). Plots of mass and population density and corresponding encounter rates are somewhat different. Importantly, however, the overall trend between body size (Fig. 5.2a) and encounter rate (Fig. 5.2b) is approximated by population density. So although density does not provide an exact measure of encounter rate it does reasonably approximate prey encounters.

Prey Utilized by Recent Subsistence Hunters

If prey selection strategies can be characterized as either generalized or specialized based on the size of prey as an indicator of their population density, then clear trends should be apparent in the accumulated prey assemblages of hunting peoples of the past and present. Comprehensive prey species data derived from ethnographic observation were compiled for 12 societies of subsistence hunters from published literature (Table 5.1) (Waguespack, 2003). The data include a range of observation periods from approximately 1 month (Ju/'hoansi [or !Kung San]) to upwards of 2 years (Ache) and includes observations of societies in a diverse array of ecological settings. The sample for each group includes the total number and average weight of each prey species captured during the observation period. We examine the issue with ethnographic data in order to establish a degree of confidence in our application of the modeled relationship between prey size and prey choice decisions to Paleoindian contexts.

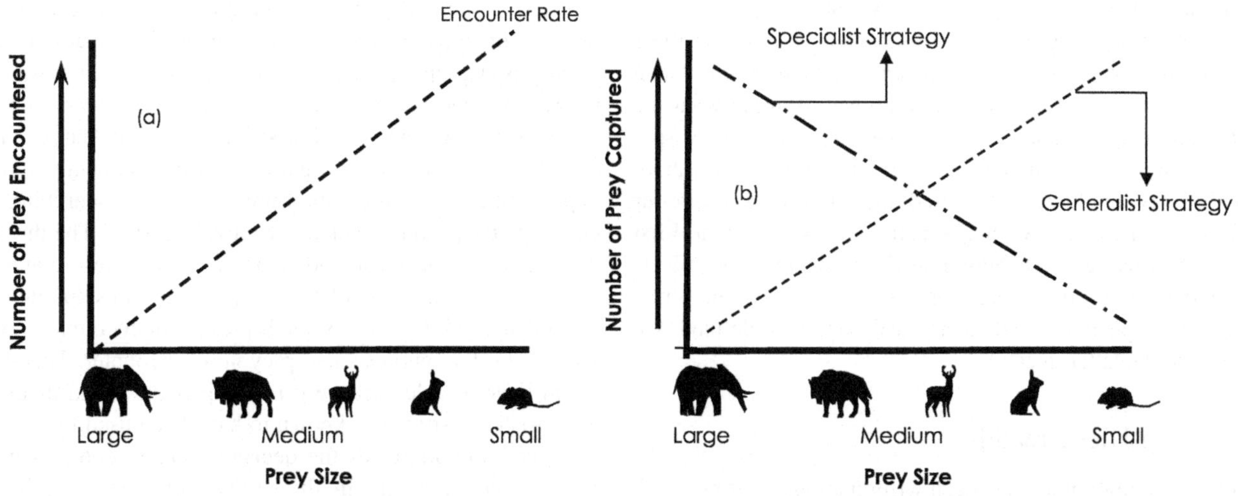

FIGURE 5.1a. Generalized relationship between prey size and encounter rate. b. Generalized relationship between body size and number of prey captured for the "large game specialist" and "generalist" prey choice strategies.

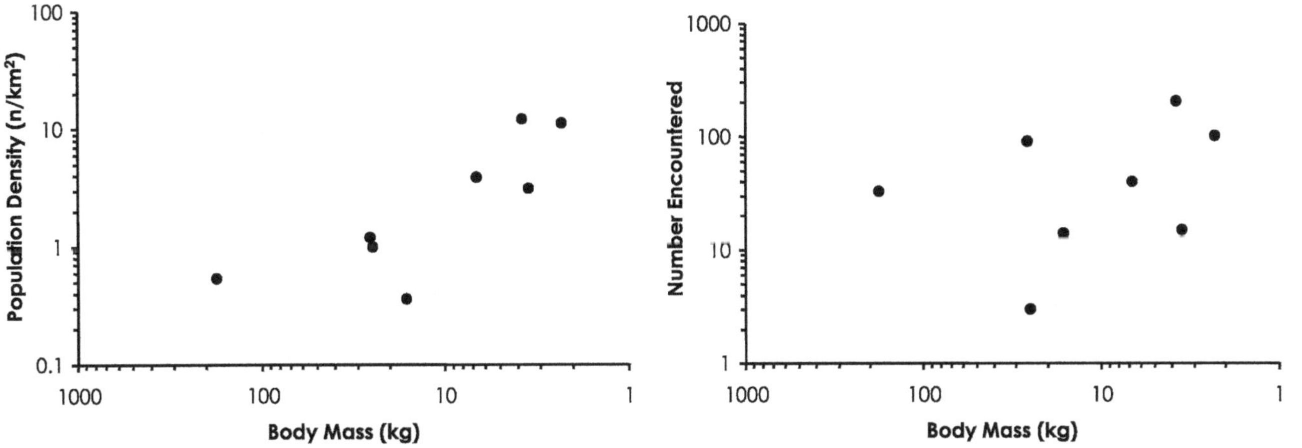

FIGURE 5.2a. Body size versus population density for eight prey species commonly taken by Ache hunter-gatherers. b. Body size versus number of prey encountered for Ache hunter-gatherers (Data from Hill and Padwe, 2000.) X- and y-axes are log-scaled, and x-axis is reversed for both graphs.

TABLE 5.1. Summary of prey choice data for recent subsistence hunters.

Group	Location	Sample duration	n people	N kills	Reference
Gidra	New Guinea	80 Days	20+	207	Ohtsuki and Suzuki 1990:42, Table 1
Pume	Venezuela	60+ Days	10+	155	Greaves, 1997:293, Table 2
Yanomamo	Venezuela	30+ Days	10+	171	Hames, 1979:234, Table II
Ye-kwana	Venezuela	30+ Days	10+	716	Hames, 1979:234, Table II
Inujjuamiut	Quebec	1 Year	?	317	Smith, 1991:1997–227
Siriono	Bolivia	1 Year	?	2300	Townsend, 2000:272, Table 13–1
Ju/'hoansi	Botswana	26 Days	5	18	Lee, 1979:266, Table 9.6
Gwi	Botswana	1 Year	50	171	Tanaka, 1980:68, Table 9.6
Yiwara	W. Australia	90 Days	10	468+	Gould, 1980:65, Table 5
	W. Australia	51 Days	40	337+	Gould, 1980:65, Table 6
Bisa	Zambia	1 Year	8	101	Marks, 1976:206, Table 38
Ache	Paraguay	Multiple years	?	3503	Hill and Padwe, 2000:95, Table 5.2
Nunamiut	Alaska	1 Year	20+	745	Binford, 1991:107, Table 39

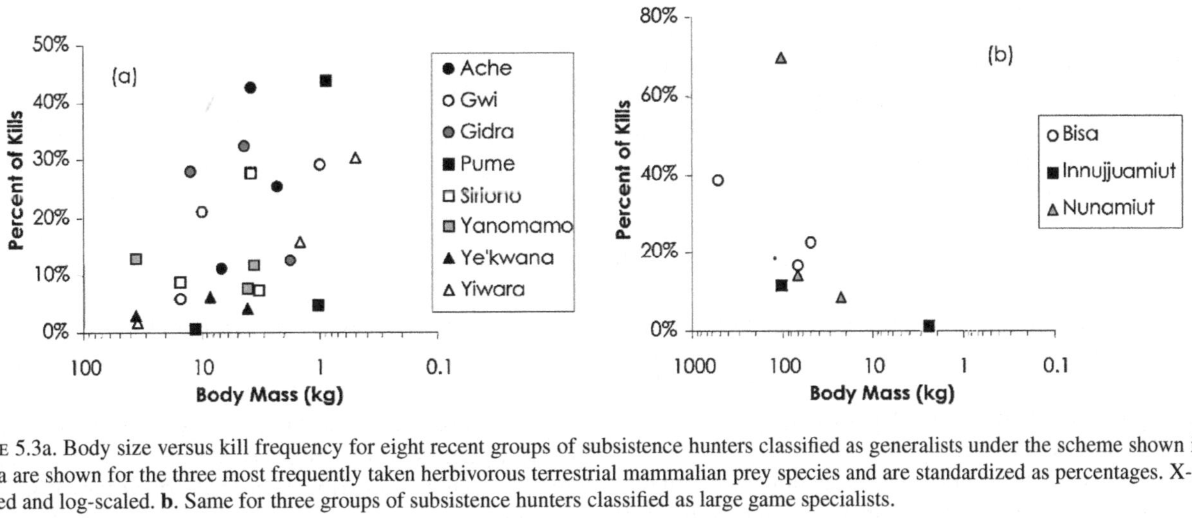

FIGURE 5.3a. Body size versus kill frequency for eight recent groups of subsistence hunters classified as generalists under the scheme shown in Fig. 1. Data are shown for the three most frequently taken herbivorous terrestrial mammalian prey species and are standardized as percentages. X-axis is reversed and log-scaled. **b**. Same for three groups of subsistence hunters classified as large game specialists.

Based only on the three most frequently utilized terrestrial herbivorous mammalian species, all groups, with three notable exceptions, exhibit a predominately negative trend in the percentage of animals killed relative to animal body size (Fig. 5.3). In the majority of groups, the largest animal hunted contributes the least in terms of the total number of individuals captured (Fig. 5.3a). For these societies, prey assemblages generally conform to the predicted distributions of an encounter-based or generalized prey selection strategy, an interpretation well supported in the ethnographic literature. Three groups, the Nunamiut, Inujjuamiut, and Bisa appear to utilize large game species more frequently than smaller prey (Fig. 5.3b). Both the Nunamiut and Inujjuamiut occupy arctic environments and their prey frequencies deviate primarily due to the inordinate proportion of caribou in their assemblages. High latitude environments are associated with lower levels of primary productivity (Begon et al., 1996), generally support a lower density and diversity of mammalian species (Andrewartha and Birch, 1961), and often maintain relatively large populations of large-bodied animals than more temperate or tropical ecosystems. While arctic environments may be characterized by a more limited diversity of potential prey, there are small prey species available. Both of the arctic hunting groups are well known ethnographically to purposefully elevate their encounter frequencies with large game, particularly caribou, by maintaining high levels of mobility and concentrating their hunting efforts along ungulate migration routes (Gubser, 1965; Binford, 1978; Smith, 1991). The Bisa are sedentary horticulturalists occupying a tropical dry savannah environment. Hunting is explicitly focused on the procurement of ungulates such as African buffalo, impala, and warthog, and large species are taken more frequently than smaller game. The area is renowned for its density of large mammals which frequent the numerous permanent water sources in the area (Marks, 1976).

While it apparent that within each strategy the relative contribution of the most commonly hunted mammalian species to the total inventory of hunted fauna varies, the distinction between generalists and specialists remains clear. Importantly, specialized large-game predation as defined here does not necessarily imply exclusive hunting of a single species of large prey or only large prey. In zooarcheological contexts assemblage dominance by a single prey species, particularly caribou (e.g., Enloe, 1999; Grayson and Delpech, 2001) and bison (e.g., Todd, 1987; Brugal et al., 1999), is often presented as the only defining characteristic of a specialized hunting economy. While dominance may accurately identify large-game hunting in particular contexts, as it would for the Nunamiut where nearly 70% of the animals taken are caribou, it may not be appropriate for hunting societies which utilize a suite of large species more frequently than smaller taxa. Neither the Bisa's nor Inujjuamiut's total faunal inventories are "dominated" (i.e., show high MNI percent values) by a single large taxon (Marks, 1976; Smith, 1991), but both utilize large game relatively more frequently than would be predicted by a generalist strategy.

As specialization is defined here, it assumes that smaller prey species are available but are relatively underutilized. It is difficult to establish whether the three groups identified as specialists are regularly passing up the opportunity to capture smaller animals. But there is evidence to suggest the Nunamiut and Inujjuamiut do deliberately forego small game:

Nunamiut

"The Nunamiut had little to say about red squirrels. They know the animal occurs in the timber, but apparently have never utilized them" (Gubser, 1965:263).

"The snowshoe hare occurs rarely in the Brooks Range…The arctic hare is perhaps a bit more common. Nunamiut have reported the presence of both hares…in occasional years…Neither hare is significant as a food source" (Gubser, 1965:276).

"The marmot occurs throughout the Brooks Range, usually in small, localized colonies…Summer may be a hungry time, and the fat meat of marmots is a welcome supplement…They are easy to catch as far as skill is concerned but difficult in the amount of labor required…" (Gubser, 1965:279)

Inujjuamiut

"Although the expressed purpose of the hunts was to locate and capture caribou…While hunting inland potential prey included fox, ptarmigan, and lake trout; the fish were never pursued, but the first two prey types were taken on occasion…In the case of ptarmigan, it appears they were sometimes ignored when encountered…The matter of fox trapping is more complicated. For one thing, foxes are harvested primarily for their pelts, which have exchange-value (as trade items) and use-value (as parka ruffs); food consumption is incidental to this…It is probably enough to say that the small loss in foraging efficiency entailed by trapping effort on these hunts was more than offset by the monetary return…" (Smith, 1991:220–2).

With regard to the Nunamiut, Gubser's anecdotal observations suggest that some small prey species are ignored in favor of caribou. For the Inujjuamiut, it seems clear that ptarmigan are overlooked in favor of larger mammals. And the occasional capture of foxes is, at least in part, a direct result of the hunting in exchange for money rather than for food. In both cases there is at least limited evidence to suggest that smaller prey are encountered more frequently then they are pursued and/or captured.

Specialization Past and Present

Outside of arctic environments and the unique case of Bisa farmers, none of the societies sampled express a prey selectivity strategy biased towards exceptionally large game. Why not? Assuming that a specialized strategy is viable only when large game are available in sufficient quantities, the environmental conditions present in the majority of ethnographic contexts examined may not be capable of supporting a selective strategy. Subsistence hunters documented in the modern era occupy ecosystems that have been inhabited by humans for thousands of years, and human populations likely exist at relatively high density levels. Also, most recent subsistence hunters occupy primarily "marginal" environments; consequently their subsistence options are likely more constrained than those of hunter-gatherers of the past. In some non-arctic contexts that are no longer represented in the ethnographic record large mammals may have been much more predictable and/or relatively easy to procure. For instance, the risks associated with hunting could be minimal when large prey are abundant relative to human population sizes and/or are naïve to human predation. These prey characteristics are not present in the majority of modern foraging environments, but may have contributed to prey selection criteria among prehistoric foragers. Importantly, comparison of prey use frequency relative to body size and the predicted relationships outlined here provides a method for evaluating prehistoric prey use strategies. Analysis of ethnographically documented subsistence hunters establishes the validity of the more general expectations about specialized and generalized hunting strategies and their faunal assemblage attributes.

Building a Paleoindian Dataset

In an ideal world, our archeological sample of Early Paleoindian subsistence choices would be large, unbiased, and well-studied. In the real world, it is not large, it is likely biased, and it is not well-studied. There are less than 50 sites from the entirety of North America that are likely to tell us something about Early Paleoindian subsistence choices (Waguespack and Surovell, 2003; Cannon and Meltzer, 2004) (Fig. 5.4, Table 5.2). Potential biases related to discovery, excavation, and research may be skewing our view of Early Paleoindian behavior (Grayson, 1988; Meltzer, 1989, 1993; Grayson and Meltzer, 2002; Waguespack and Surovell, 2003; Cannon and Meltzer, 2004; Byers and Ugan, 2005). Finally, standards of excavation, analysis, and reporting vary widely (Waguespack and Surovell, 2003; Cannon and Meltzer, 2004). In combination, these issues not only make it difficult to determine what people were killing and eating in the late Pleistocene of North America, but also they serve as fodder for debate.

One of the fundamental issues that must be dealt with is how one should go about building a database of Paleoindian subsistence choices from a series of faunal remains reported in archeological monographs and papers. Taphonomy must be considered, but performing a taphonomic analysis on actual specimens is not the same as doing what we call "textual taphonomy," or the evaluation of taphonomic factors on the basis of published text and images. Textual taphonomy has a long tradition in Paleoindian archeology (e.g., Grayson, 1984; Haynes and Stanford, 1984) but remains problematic in that it can be extremely difficult to evaluate evidence for subsistence use on the basis of published accounts which vary widely in quality.

Haynes and Stanford (1984:217–222) provide a useful framework for evaluating subsistence use in terms of levels of association. The weakest evidence for subsistence use they term "contemporaneity," meaning evidence which simply demonstrates spatio-temporal association. Contemporaneity is established solely on the basis of stratigraphic association

FIGURE 5.4. Map of sites included the Surovell and Waguespack and Cannon and Meltzer (2004) Early Paleoindian faunal datasets: 1. Manis, 2. Charlie Lake Cave, 3. Wally's Beach, 4. Colby, 5. Murray Springs, Lehner, Naco, Escapule, and Leikem, 6. Sheaman, 7. Dent, 8. Lange-Ferguson, 9. Jake Bluff, 10. Domebo, 11. Miami, 12. Blackwater Draw, 13. Lubbock Lake, 14. McLean, 15. Kincaid, 16. Gault, 17. Aubrey and Lewisville, 18. Wacissa River, 19. Little Salt Spring, 20. Kimmswick, 21. Boaz, 22. Schaefer and Hebior, 23. Martin's Creek, 24. Shawnee-Minisink, 25. Hiscock, 26. Udora, 27. Whipple, 28. Bull Brook, 29. Guest, 30. Holcombe Beach.

TABLE 5.2. Sites used in this study for analyses of Clovis diet breadth.

Site (state/prov.)	Region	Hearths?[a]	Kill site?	References
Aubrey (TX)	Central	Y	N	Ferring, 2001; Yates and Lundelius, 2001
Blackwater Draw (NM)	Central	Y	Y	Lundelius, 1972
Charlie Lake Cave (BC)	Central	–	N	Driver, 1999; Fladmark et al., 1988
Boaz Mastodon (WI)	East	N	Y	Palmer and Stoltman, 1975
Bull Brook (MA)	East	Y	N	Byers, 1955; Spiess et al., 1985
Colby (WY)	Central	N	Y	Frison and Todd, 1986; Walker and Frison, 1980
Dent (CO)	Central	N	Y	Figgins, 1933; Brunswig and Fisher, 1993
Domebo (OK)	Central	N	Y	Leonhardy, 1966; Slaughter, 1966; Leonhardy and Anderson, 1966
Escapule (AZ)	West	N	Y	Hemmings and Haynes, 1969; Saunders, n.d.
Gault (TX)	Central	N	N	Collins, 1999
Guest (FL)	East	N	Y	Hoffman, 1983; Rayl, 1974
Hebior (WI)	East	N	Y	Overstreet, 1996; Overstreet et al., 1995; Overstreet and Stafford, 1997
Hiscock (NY)	East	N	Y?	Laub et al., 1988; Tankersley et al., 1998; Steadman, 1988
Holcombe Beach (MI)	East	-	N	Cleland, 1965; Fitting et al., 1966; Spiess et al., 1985
Jake Bluff (OK)	Central	N	N?	Bement and Carter, 2003 as cited by Cannon and Meltzer, 2004
Kimmswick (MO)	Central	N	Y	Graham et al., 1981; Graham and Kay, 1988
Kincaid Shelter (TX)	Central	N	N	Collins et al., 1989
Lange-Ferguson (SD)	Central	N	Y	Hannus, 1989, 1990; J. Martin, 1984
Lehner (AZ)	West	Y	Y	Haury et al., 1959; Haynes and Haury, 1982; Lance, 1959; Saunders, n.d.
Leikem (AZ)	West	N	Y	Saunders, 1980; Saunders, n.d
Lewisville (TX)	Central	Y	N	Crook and Harris, 1957, 1958
Little Salt Spring (FL)	East	N	N	Clausen et al., 1979
Lubbock Lake (TX)	Central	N	Y?	Johnson, 1987
Manis (WA)	West	N	Y	Gustafson et al., 1979
Martin's Creek (OH)	East	N	Y	Brush and Smith, 1994; Brush et al., 1994
McLean (TX)	Central	N	Y	Ray, 1930, 1942; Ray and Bryan, 1938
Miami (TX)	Central	N	Y	Sellards, 1952; Holliday et al., 1994
Murray Springs (AZ)	West	Y	Y	Haynes, 1993; Saunders, 1980; Saunders, n.d.
Naco (AZ)	West	N	Y	Haury, 1953; Lance, 1959; Saunders, n.d.
Schaefer (WI)	East	N	Y	Overstreet, 1996; Overstreet et al., 1995; Overstreet and Stafford, 1997
Shawnee-Minnisink (PN)	East	Y	N	Eisenberg, 1978; Dent and Kauffman, 1985
Sheaman (WY)	Central	N	N	Frison, 1982
Udora (ON)	East	Y	N	Storck and Spiess, 1994
Wacissa River (FL)	East	N	Y	Webb et al., 1984
Wally's Beach (AB)	Central	–	Y?	Kooyman et al., 2001, 2006
Whipple (NH)	East	Y	N	Spiess et al., 1985

[a] Data taken directly from Cannon and Meltzer (2004:Table 6). Cell values left blank represent those sites not considered by Cannon and Meltzer (2004)

of archeological and faunal specimens. Of greater reliability is "association" referring to spatio-temporal contemporaneity of artifacts and animal remains demonstrably discarded by humans in a single event, such as during occupation of a campsite. Finally, "utilization" shows not only spatio-temporal contemporaneity but also clear evidence of subsistence use, whether derived through hunting or scavenging. Utilization can be established by tight spatial association between artifacts and faunal remains that display human modifications, usually cut marks and/or impact fractures. The most conservative approach to building a database of subsistence choices would be to rely on utilization evidence only, but as Haynes and Stanford (1984:222) caution, by doing so, "we run the risk of ignoring true archaeological specimens." It is unrealistic to expect that all utilized fauna will present and retain evidence of butchery. At the other extreme, one could use all evidence established through contemporaneity but risk creating "fictive data" (Haynes, 2002a:182) through the inclusion of specimens that are not archeological. Both approaches have been used and both likely present skewed views of the record. The question is how skewed?

In 2003, we compiled a dataset including 33 late Pleistocene archeological sites from North America (Waguespack and Surovell, 2003) based solely on "contemporaneity" evidence in the Haynes and Stanford (1984) scheme, essentially assuming that that if a species is present, it was used. This assumption is obviously incorrect, and undoubtedly produced a biased sample, a sample which we felt was probably "skewed in favor of small game" (Waguespack and Surovell, 2003: 339). Thus, we knowingly included species which had a very low probability of use, but no preference was given to species with regard to body size. For example, we included bison from the Naco site on the basis of the presence of scattered tooth fragments (Lance, 1959:37), which we were fairly confident had no association with the eight Clovis points scattered throughout the body of the nearby mammoth. Similarly, from the Lange Ferguson mammoth kill in South Dakota (Hannus, 1989, 1990), we included bison, cervids, rabbits, fish, two species of birds, two species of amphibians, ten species of rodents, three species of insectivores, and two species of snakes (J. Martin, 1984).

In contrast, Grayson and Meltzer (2002) and Cannon and Meltzer (2004) attempted to cleanse the data of species which show no evidence of subsistence use. Cannon and Meltzer explicitly criticized us (Waguespack and Surovell, 2003) and Haynes (2002a, b) for the very reason outlined above:

> These publications are welcome contributions to the literature, but they unfortunately devote little attention to taphonomic issues that must be addressed before animal remains can be treated as evidence of human subsistence practices (Cannon and Meltzer, 2004:1956).

This is a fair criticism, at least in our case, although it would have been judicious to note that we pointed out this weakness ourselves. It is worth questioning whether Cannon and Meltzer (2004) truly treat all faunal remains equally. It is our contention that they do not, and that by taphonomically vetting the record, they introduce biases as well.

Two sites that figure prominently in the Clovis subsistence debate are Aubrey and Lewisville located in the upper Trinity River basin of North Texas (Crook and Harris, 1957, 1958; Ferring, 1995, 2001; Yates and Lundelius, 2001). Cannon and Meltzer lean heavily on these sites; of the 14 strong cases for use of small game (lagomorphs, rodents, birds, turtles, other reptiles, and fish) that they identify, Aubrey and Lewisville account for ten (Table 5.3). In other words, these sites are critical to making the argument that Early Paleoindians were not large game specialists. A wide array of species have been recovered from the Clovis occupation at Aubrey including mammoth, bison, ground sloth, deer, rodents, turtles and tortoises, snakes, birds, amphibians, and fish.

Regarding the small game from Aubrey, Cannon and Meltzer (2004: Table 5), accept rabbits, rodents, birds, turtles, snakes, and fish as having been exploited by Clovis peoples, based on the association of burned bone with hearths, spatial analysis, and the unlikelihood of natural burning producing observed spatial patterns (Cannon and Meltzer, 2004:1969). Regarding the hearths at the site, Ferring (2001:124) noted: "Despite extremely slow troweling in areas yielding burned material, no hearth or pit outlines of any kind could be found." Instead, hearths were identified on the basis of clusters of burned bone and charcoal (Ferring, 2001:124–125). This is a common practice in Early Paleoindian archeology, and we have done the same thing ourselves (Surovell and Waguespack, 2007). Unfortunately, there is a bit of circular reasoning here. Hearths are presumed to be cultural on the basis of clustering in burned bone, and burned bone is assumed to be cultural on the basis of association with hearths. Thus, in one fell swoop, we could eliminate all of the evidence for use of small game from Aubrey since it is in large part based on clustering in association with hearths. Furthermore, burned bone occurs at Aubrey in sediments pre- and post-dating the Clovis occupation (Yates and Lundelius, 2001:115). We point this out not to make the argument that Clovis people at Aubrey did not use small game. They may have, and we included a greater array of small game from Aubrey in our study (Waguespack and Surovell, 2003: Table 2) than did Cannon and Meltzer (2004). Instead, we do this to show how easy it is to eliminate any evidence for subsistence use by careful reading of site reports and the use of clever arguments to exclude things that do not fit our preconceptions.

Turning to the Lewisville site, Cannon and Meltzer (2004:1969–1970) accept evidence for use of horse, deer, carnivores, rabbits, rodents, birds, turtles, and snakes. The Lewisville site was excavated between 1949 and 1951 in conjunction with construction of the Lewisville dam on the Trinity River. A diverse Pleistocene fauna was recovered in association with 21 burned features interpreted to be hearths (Crook and Harris, 1957, 1958). Cannon and Meltzer accept the evidence from Lewisville on the following basis:

> [B]ecause the findings at Lewisville have essentially been replicated at nearby Aubrey, a reasonable case can be made for subsistence use of at least those taxa, listed above, for which specimens described as burned have been recovered from hearth contexts (Cannon and Meltzer, 2004:1970–1971).

There are two significant differences between the Aubrey and Lewisville sites. Aubrey has produced just under 10,000 artifacts from the Clovis level (Ferring, 2001: 130). Lewisville has produced only one artifact from buried deposits, a Clovis point, which some felt was intrusive or possibly planted at the site (Sellards, 1960; Krieger, 1962). It is difficult to attribute the lack of artifacts at Lewisville to excavation practices since a wide variety of small animals were recovered (Crook and Harris, 1957).[2] Also, a radiocarbon date on charcoal (some think lignite) from one of the "hearths" produced an infinite age (Crook and Harris, 1958). In other words, there is significant reason to doubt whether the Lewisville site is archeological at all, and yet just like Aubrey it shows clustering in charcoal, burned sediments, and faunal remains interpreted to be "hearths."

Not only does this place further doubt on the cultural nature of hearths and the use of small game at Aubrey, but it also shows how difficult it is to equally apply textual taphonomic criteria to all sites. For example, if we apply the criterion of whether "findings have essentially been replicated" elsewhere to a serious of questionable mammoth and mastodon kills which have been rejected by some (Grayson and Meltzer, 2002; Cannon and Meltzer, 2004), many additional sites might be included in a database of "secure" subsistence associations. These would include the Boaz mastodon (Palmer and Stoltman, 1975), the Manis mastodon (Gustafson et al., 1979),

TABLE 5.3. Presence/absence of vertebrate taxa by site for Surovell and Waguespack Early Paleoindian subsistence dataset.

Taxonomic group	Body mass (kg)	Body Size Class	Aubrey (TX)	Blackwater Draw (NM)	Charlie Lake Cave (BC)	Boaz Mastodon (WI)	Bull Brook (MA)	Colby (WY)	Dent (CO)	Domebo (OK)	Escapule (AZ)	Gault (TX)	Guest (FL)	Hebior (WI)	Hiscock (NY)	Holcombe Beach (MI)	Jake Bluff (OK)	Kimmswick (MO)	Kincaid Shelter (TX)
Herbivorous mammals																			
Sorex	0.006	1	–	–	–	–	–	–	–	–	–	–	–	–	–	–	–	–	–
Perognathus	0.01	1	+	–	–	–	–	–	–	+	–	–	–	–	–	–	–	–	+
Reithrodontomys	0.01	1	+	–	–	–	–	–	–	–	–	–	–	–	–	–	–	–	–
Family Soricidae	0.01	1	+	–	–	–	–	–	–	–	–	–	–	–	–	–	–	–	–
Clethrionomys	0.02	1	–	–	–	–	–	–	–	–	–	–	–	–	–	–	–	–	–
Zapus	0.025	1	–	–	–	–	–	–	–	–	–	–	–	–	–	–	–	–	–
Onychomys	0.03	1	–	–	–	–	–	–	–	–	–	–	–	–	–	–	–	–	–
Synaptomys	0.03	1	+	–	–	–	–	–	–	+	–	–	–	–	–	–	–	+	–
Microtus	0.05	1	+	+	+	–	–	–	–	+	–	–	–	–	–	–	–	+	–
Peromyscus	0.05	1	+	–	+	–	–	–	–	–	–	–	–	–	–	–	–	+	–
Oryzomys	0.06	1	+	–	–	–	–	–	–	–	–	–	–	–	–	–	–	–	–
Eutamias	0.08	1	–	–	–	–	–	–	–	–	–	–	–	–	–	–	–	–	–
Scalopus	0.1	1	+	–	–	–	–	–	–	–	–	–	–	–	–	–	–	–	–
Sigmodon	0.15	1	+	–	–	–	–	–	–	+	–	–	–	–	–	–	–	–	–
Thomomys	0.15	1	–	–	–	–	–	–	–	–	–	–	–	–	–	–	–	–	–
Blarina	0.19	1	–	–	–	–	–	–	–	–	–	–	–	–	–	–	–	+	–
Neotoma	0.2	1	+	–	–	–	–	–	–	–	–	–	–	–	–	–	–	+	–
Geomys	0.24	1	+	–	–	–	–	–	–	+	–	–	–	–	–	–	–	+	–
Spermophilus	0.3	1	+	–	+	–	–	–	–	–	–	–	–	–	–	–	–	+	–
Sciurus	0.52	1	–	–	+	–	–	–	–	–	–	–	–	–	–	–	–	+	–
Cynomys	1	2	–	+	–	–	–	–	–	–	–	–	–	–	–	–	–	–	–
Ondatra	1.15	2	+	+	+	–	–	–	–	+	–	–	–	–	–	–	–	+	–
Sylvilagus	1.42	2	+	–	–	–	–	–	–	–	–	–	–	–	–	–	–	+	–
Lepus	3	2	+	+	+	–	–	+	–	–	–	–	–	–	–	–	–	–	–
Dasypus	4.2	2	–	–	–	–	–	–	–	–	–	–	–	–	–	–	–	–	+
Marmota	5.4	2	–	–	–	–	–	–	–	–	–	–	–	–	–	–	–	+	–
Capromeryx	12.5	2	–	–	–	–	–	–	–	–	–	–	–	–	–	–	–	–	–
Castor	22	2	–	–	–	–	+	–	–	–	–	–	–	–	–	–	–	–	–
Antilocapra	46	3	–	+	–	–	–	+	–	–	–	–	–	–	–	–	–	–	–
Odocoileus	55	3	+	+	–	–	–	–	–	–	–	–	+	–	–	–	–	+	–
Mylohyus	75	3	–	–	–	–	–	–	–	–	–	–	–	–	–	–	–	+	–
Palaeolama	80	3	–	–	–	–	–	–	–	–	–	–	–	–	–	–	–	–	–
Rangifer	86	3	–	–	–	–	+	–	–	–	–	–	–	–	+	+	–	–	–
Hemiauchenia	110	3	–	+	–	–	–	–	–	–	–	–	–	–	–	–	–	–	–
Platygonus	110	3	–	+	–	–	–	–	–	–	–	–	–	–	–	–	–	–	–
Family Cervidae	120	3	–	–	–	–	–	–	–	–	–	–	–	–	–	–	–	–	–
Cervus	218	4	–	+	–	–	–	–	–	–	–	–	–	–	+	–	–	–	–
Holmesina	250	4	–	–	–	–	–	–	–	–	–	–	–	–	–	–	–	–	–
Tapirus	323	4	–	–	–	–	–	–	–	–	–	–	–	–	–	–	–	–	–
Equus	400	4	+	+	–	–	–	+	–	–	+	+	–	–	–	–	–	–	+
Megalonyx	600	4	–	–	–	–	–	–	–	–	–	–	–	–	–	–	–	–	–
Bison	900	4	+	+	+	–	–	+	–	+	–	+	+	–	–	–	+	–	–
Camelops	900	4	–	+	–	–	–	+	–	–	–	–	–	–	–	–	–	–	–
Glyptotherium	1100	4	–	–	–	–	–	–	–	–	–	–	–	–	–	–	–	–	–
Glossotherium	1587	4	+	–	–	–	–	–	–	–	–	–	–	–	–	–	–	+	–
Mammut	4500	5	–	–	–	+	–	–	–	–	–	–	–	–	+	–	–	+	–
Order Proboscidea	5750	5	+	–	–	–	–	–	–	–	–	–	–	–	–	–	–	–	–
Mammuthus	7000	5	–	+	–	–	–	+	+	+	+	+	+	+	–	–	–	–	+

Carnivorous mammals																			
Procyon	4	–	–	–	–	–	–	–	–	–	–	–	–	–	–	–	–	–	+
Ursus	100	–	–	–	–	–	–	–	–	–	–	–	–	–	–	–	–	–	–
Vulpes	3.2	–	–	+	–	–	–	–	–	–	–	–	–	–	–	–	–	–	–
Alopex	3.45	–	–	–	–	–	–	–	–	–	–	–	–	–	–	–	–	–	–
Arctodus	720	–	–	–	–	–	–	–	–	–	–	–	–	–	–	–	–	–	–
Canis	32	–	–	+	–	–	–	–	–	–	–	–	–	–	–	–	–	–	–
Mephitis	2	–	–	–	–	–	–	–	–	–	–	–	–	–	–	–	–	–	–
Mustela	0.42	–	–	–	–	–	–	–	–	–	–	–	–	–	–	–	–	+	–
Smilodon	400	–	–	+	–	–	–	–	–	–	–	–	–	–	–	–	–	–	–
Taxidea	7.1	–	–	–	–	–	–	–	–	–	–	–	–	–	–	–	–	–	+
Order Carnivora	–	–	+	–	–	–	–	–	–	–	–	–	–	–	–	–	–	–	–
Nonmammals																			
Class Reptilia	–	–	+	+	–	–	–	–	–	+	–	–	–	–	–	–	–	+	+
Class Aves	–	–	+	–	+	–	–	–	–	–	–	–	–	–	+	–	–	+	–
Class Osteichthyes	–	–	+	–	+	–	–	–	–	–	–	–	–	–	+	–	–	+	–
Class Amphibia	–	–	–	–	–	–	–	–	–	+	–	–	–	–	–	–	–	+	–

(continued)

TABLE 5.3. (continued)

Taxonomic group	Body mass (kg)	Body Size Class	Lange-Ferguson (SD)	Lehner (AZ)	Leikem (AZ)	Lewisville (TX)	Little Salt Spring (FL)	Lubbock Lake (TX)	Manis (WA)	Martins Creek (OH)	McLean (TX)	Miami (TX)	Murray Springs (AZ)	Naco (AZ)	Schaefer (WI)	Shawnee-Minnisink (PA)	Sheaman (WY)	Udora (ON)	Wacissa River (FL)	Wally's Beach (AB)	Whipple (NH)
Herbivorous mammals																					
Sorex	0.006	1	+	–	–	–	–	–	–	–	–	–	–	–	–	–	–	–	–	–	–
Perognathus	0.01	1	–	+	–	–	–	+	–	–	–	–	–	–	–	–	–	–	–	–	–
Reithrodontomys	0.01	1	+	–	–	–	–	+	–	–	–	–	–	–	–	–	–	–	–	–	–
Family Soricidae	0.01	1	–	–	–	–	–	–	–	–	–	–	–	–	–	–	–	–	–	–	–
Clethrionomys	0.02	1	+	–	–	–	–	–	–	–	–	–	–	–	–	–	–	–	–	–	–
Zapus	0.025	1	+	–	–	–	–	–	–	–	–	–	–	–	–	–	–	–	–	–	–
Onychomys	0.03	1	+	–	–	–	–	+	–	–	–	–	–	–	–	–	–	–	–	–	–
Synaptomys	0.03	1	–	–	–	–	–	–	–	–	–	–	–	–	–	–	–	–	–	–	–
Microtus	0.05	1	+	+	–	+	–	+	–	–	–	–	+	–	–	–	–	–	–	–	–
Peromyscus	0.05	1	+	–	–	+	–	+	–	–	–	–	–	–	–	–	–	–	–	–	–
Oryzomys	0.06	1	–	–	–	–	–	–	–	–	–	–	–	–	–	–	–	–	–	–	–
Eutamias	0.08	1	+	–	–	–	–	–	–	–	–	–	–	–	–	–	–	–	–	–	–
Scalopus	0.1	1	–	–	–	+	–	–	–	–	–	–	–	–	–	–	–	–	–	–	–
Sigmodon	0.15	1	–	–	–	–	–	–	–	–	–	–	–	–	–	–	–	–	–	–	–
Thomomys	0.15	1	+	+	–	–	–	+	–	–	–	–	–	–	–	–	–	–	–	–	–
Blarina	0.19	1	+	–	–	–	–	+	–	–	–	–	–	–	–	–	–	–	–	–	–
Neotoma	0.2	1	–	–	–	+	–	+	–	–	–	–	+	–	–	–	–	–	–	–	–
Geomys	0.24	1	–	–	–	–	–	+	–	–	–	–	–	–	–	–	–	–	–	–	–
Spermophilus	0.3	1	+	–	–	+	–	+	–	–	–	–	–	–	–	–	–	–	–	–	–
Sciurus	0.52	1	–	–	–	+	–		–	–	–	–	–	–	–	–	–	–	–	–	–
Cynomys	1	2	–	–	–	+	–	+	–	–	–	–	–	–	–	–	–	–	–	–	–
Ondatra	1.15	2	+	+	–	–	–	+	+	–	–	–	–	–	–	–	–	–	–	–	–
Sylvilagus	1.42	2	–	+	–	+	+	–	–	–	–	–	–	–	–	–	–	–	–	–	–
Lepus	3	2	+	+	–	+	–	+	–	–	–	–	–	–	–	–	–	+	–	–	–
Dasypus	4.2	2	–	–	–	–	–	–	–	–	–	–	–	–	–	–	–	–	–	–	–
Marmota	5.4	2	–	–	–	–	–	–	–	–	–	–	–	–	–	–	–	–	–	–	–
Capromeryx	12.5	2	–	–	–	–	–	+	–	–	–	–	–	–	–	–	–	–	–	–	–
Castor	22	2	–	–	–	–	–	–	–	–	–	–	–	–	–	–	–	–	–	–	–
Antilocarpa	46	3	–	–	–	–	–	–	–	–	–	–	–	–	–	–	+	–	–	–	–
Odocoileus	55	3	–	–	–	+	+	–	–	+	–	–	–	–	–	–	–	–	+	–	–
Mylohyus	75	3	–	–	–	–	–	–	–	–	–	–	–	–	–	–	–	–	–	–	–
Palaeolama	80	3	–	–	–	–	–	–	–	–	–	–	–	–	–	–	–	–	+	–	–
Rangifer	86	3	–	–	–	–	–	–	–	–	–	–	–	–	–	–	–	+	–	–	+
Hemiauchenia	110	3	–	–	–	–	–	+	–	–	–	–	–	–	–	–	–	–	+	–	–
Platygonus	110	3	–	–	–	+	–	+	–	–	–	–	+	–	–	–	–	–	–	–	–
Family Cervidae	120	3	+	–	–	–	–	–	–	–	–	–	–	–	–	–	–	–	–	–	–
Cervus	218	4	–	–	–	–	–	–	–	–	–	–	–	–	–	–	–	–	–	–	–
Holmesina	250	4	–	–	–	–	–	+	–	–	–	–	–	–	–	–	–	–	–	–	–
Tapirus	323	4	–	+	–	–	–	–	–	–	–	–	–	–	–	–	–	–	–	–	–
Equus	400	4	–	+	–	+	–	+	–	–	–	–	+	–	–	–	–	–	+	+	–
Megalonyx	600	4	–	–	–	–	+	–	–	–	–	–	–	–	–	–	–	–	–	–	–
Bison	900	4	+	+	–	+	+	+	+	–	–	–	+	+	–	–	+	–	+	–	–
Camelops	900	4	–	+	–	+	–	+	–	–	–	–	+	–	–	–	–	–	–	–	–
Glyptotherium	1100	4	–	–	–	+	–	–	–	–	–	–	–	–	–	–	–	–	–	–	–

Glossotherium	1587	4	–	–	–	–	–	–	–	–	–	–	–	–	–	–	–	–	–	–	–
Mammut	4500	5	–	–	–	–	–	–	+	+	–	–	–	–	–	–	–	–	–	–	–
Order Proboscidea	5750	5	–	–	–	–	+	–	–	–	–	–	–	–	–	–	–	–	–	–	–
Mammuthus	7000	5	+	+	+	+	–	+	–	–	+	+	+	+	+	–	–	–	–	–	–
Carnivorous mammals																					
Procyon	4	–	–	–	–	+	–	–	–	–	–	–	–	–	–	–	–	–	–	–	–
Ursus	100	–	–	+	–	+	–	–	–	–	–	–	–	–	–	–	–	–	–	–	–
Vulpes	3.2	–	–	–	–	–	–	+	–	–	–	–	–	–	–	–	–	–	–	–	–
Alopex	3.45	–	–	–	–	–	–	–	–	–	–	–	–	–	–	–	–	+	–	–	–
Arctodus	720	–	–		–	–	–	+	–	–	–	–	–	–	–	–	–	–	–	–	–
Canis	32	–	–	+	–	+	–	+	–	–	–	–	+	–	–	–	–		–	–	–
Mephitis	2	–	–	–	–	+	–		–	–	–	–	–	–	–	–	–	–	–	–	–
Mustela	0.42	–	–	–	–	–	–	–	–	–	–	–	–	–	–	–	–	–	–	–	–
Smilodon	400	–	–	–	–	–	–	–	–	–	–	–	–	–	–	–	–	–	–	–	–
Taxidea	7.1	–	–	–	–	–	–	–	–	–	–	–	–	–	–	–	–	–	–	–	–
Order Carnivora		–	–	–	–	–	–	–	–	–	–	–	–	–	–	–	–	–	–	–	–
Non–mammalian taxa																					
Class Reptilia		–	+	+	–	+	+	+	–	–	–	–	–	–	–	–	–	–	+	–	–
Class Aves		–	+	+	–	+	+	+	–	–	–	–	–	–	–	–	–	–	–	–	–
Class Osteichthyes		–	+	–	–	+	–	+	–	–	–	–	–	–	–	+	–	–	–	–	–
Class Amphibia		–	+	+	–	+	–	+	–	–	–	–	–	–	–	–	–	–	–	–	–

the Leikem mammoth (Saunders, 1980, n.d.), the McLean mammoth (Ray, 1930, 1942; Ray and Bryan, 1938), and the Martin's Creek mastodon (Brush and Smith, 1994; Brush et al., 1994). Though questions of association remain unanswered, at all of these sites artifacts have reportedly been recovered in spatial association with proboscidean remains, a pattern which has been repeated at numerous sites, including Aubrey itself.

The two approaches discussed above each offer varying degrees of "security" and reliability in the Early Paleoindian faunal record. We included all species on the basis of spatial and stratigraphic association, whether there was strong evidence for subsistence use or not (Waguespack and Surovell, 2003). Using this approach, we can be confident that our dataset includes all species recovered that were used by Early Paleoindians, but it does so by the inclusion of many species that were likely not used. In contrast, Cannon and Meltzer (2004) included only those species which they felt could be securely linked to human subsistence behavior. This approach produces a much smaller sample, but one that might be a better reflection of past subsistence choices, although there is a much greater likelihood of eliminating taxa that were actually used. The downside of this approach, we argue, is that it is difficult to equally apply such criteria to all sites, and as such, one is likely to introduce additional research bias into the data.

What Does the Record Tell Us?

In Tables 5.2 and 5.3, we present our original dataset (Waguespack and Surovell, 2003) with the addition of three new sites: Udora (Storck and Spiess, 1994), Jake Bluff (Bement and Carter, 2003 as cited by Cannon and Meltzer, 2004), and Wally's Beach (Kooyman et al., 2001, 2006) (Fig. 5.4). The addition of these sites does not significantly change any patterns we previously identified. We have reformatted the data to facilitate comparison with the Cannon and Meltzer dataset. Mammalian data are presented as presence/absence values by genus, family, or order. Data for non-mammals are presented as present/absent by class. In Table 5.4, we present the Cannon and Meltzer (2004) dataset. The largest difference between the two datasets is sample size. Our dataset includes a minimum of 241 occurrences of animals (Table 5.3), and the Cannon and Meltzer dataset includes a minimum of 46 occurrences (Table 5.4). The total number of associations, however, is not as critical to the diet breadth debate as relative species representation. Does species representation differ significantly between the two datasets?

To perform this analysis, we grouped herbivorous mammalian genera into five body-size classes, as shown in Table 5.3. From Class 5 to 1 respectively, these are roughly mammoth-, bison-, deer-, rabbit-, and small rodent-sized animals. Body size estimates for extinct and extant species were derived from Smith et al. (2003). Body size classes were further compressed for the purpose of performing chi-square tests, which are sensitive to small cell values. Two chi-square tests were performed, one based on the number of sites showing the presence or absence of each taxonomic group (Table 5.5), and one based on the total number of occurrences of each taxonomic group in each dataset (Table 5.6). Looking at the relative frequencies of sites showing each taxonomic group (Table 5.5, Fig. 5.5a), there are no significant differences between the two datasets ($X^2 = 1.46$, df = 4, p = 0.83).

When viewed this way, the same result in terms of taxonomic representation is obtained whether a taphonomic filter is applied to the data or not. The total number of occurrences of each taxonomic group (Table 5.6, Fig. 5.5b) does, however, differ significantly ($X^2 = 17.9$, df = 4, p = 0.001). Large mammals (Body Size Classes 3–5) are overrepresented in the Cannon and Meltzer dataset, and small mammals (Body Size Classes 1 & 2) are overrepresented in our dataset. This finding supports our hypothesis that by including all species present within an assemblage, whether there is clear evidence for subsistence use or not, we disproportionately included larger frequencies of small fauna.

What the preceding analysis demonstrates is quite clear. Although arguments have been made to the contrary, the debate is not about relative species representation. If this was the crux of the debate, the conclusions of these two studies should have been reversed or perhaps identical (Fig. 5.5). Instead, the argument is in large part about what we should label the subsistence strategy that produced this faunal record and the extent to which the current faunal record reflects Early Paleoindian subsistence choices.

The issue of semantics aside, regarding prey choice and its relationship to Pleistocene extinctions, the more important question is whether either or both datasets show evidence of selective hunting of large-bodied animals. Because the ecological relationship between body size and population density is dependent upon metabolism and trophic level (Peters, 1986), we limit this analysis to herbivorous mammals, but increasing the taxonomic scope would not change the outcome. To control for variation in taxonomic diversity within each body size class, we divide the total number of occurrences by the number of genera represented within each body size class (Table 5.7). For both datasets, there is a significant positive correlation between body size and archeological abundance (Cannon and Meltzer, Spearman's $\rho = 0.975$, $p = 0.005$; Surovell and Waguespack, Spearman's $\rho = 1.00$, $p < 0.001$) (Fig. 5.6).

In other words, whether the data are taphonomically vetted or not, the largest and rarest mammalian herbivores in late Pleistocene landscapes are the most frequent species to occur in Early Paleoindian faunal assemblages. This pattern is robust and occurs even within our dataset which includes 93 occurrences of small mammal fauna. In fact, large game is apparently more abundant when taphonomic considerations are used in dataset construction. This finding provides strong support for the large game specialist hypothesis. If Early Paleoindians regularly took small game upon encounter, small animals such as hares should vastly outnumber mammoths and bison in the archeological record. They do not (Fig. 5.6).

TABLE 5.4. Presence/absence of vertebrate taxa with "secure evidence" of subsistence use by site for the Cannon and Meltzer (2004) Early Paleoindian subsistence dataset.

Taxonomic group	Body Size Class	Aubrey (TX)	Blackwater Draw (NM)	Bull Brook (MA)	Colby (WY)	Dent (CO)	Domebo (OK)	Escapule (AZ)	Hebior (WI)	Jake Bluff (OK)	Kimmswick (MO)	Lange-Ferguson (SD)	Lehner (AZ)	Lewisville (TX)	Lubbock Lake (TX)	Miami (OK)	Murray Springs (AZ)	Naco (AZ)	Pleasant Lake (MI)	Shawnee-Minisink (PA)	Wacissa River (FL)	Udora (ON)	Whipple (NH)
Herbivorous mammals																							
Order Rodentia	1&2	+	−	+	−	−	−	−	−	−	−	−	−	+	−	−	−	−	−	−	−	−	−
Family Leporidae	2	+	−	−	−	−	−	−	−	−	−	−	−	+	−	−	−	−	−	−	−	+	−
Family Cervidae	3	+	−	−	−	−	−	−	−	−	+	−	−	+	−	−	−	−	−	−	−	−	−
Rangifer	3	−	−	+	−	−	−	−	−	−	−	−	−	−	−	−	−	−	−	−	−	+	+
Equus	4	−	+	−	−	−	−	−	−	−	−	−	−	+	−	−	−	−	−	−	−	−	−
Camelops	4	−	−	−	−	−	−	−	−	−	−	−	+	−	−	−	−	−	−	−	−	−	−
Bison	4	+	+	−	−	−	−	−	−	+	−	−	+	−	−	−	+	−	−	−	+	−	−
Mammut	5	−	−	−	−	−	−	−	−	−	+	−	−	−	−	−	−	−	+	−	−	−	−
Mammuthus	5	−	+	−	+	+	+	+	+	−	−	+	+	−	+	+	+	+	−	−	−	−	−
Carnivorous Mammals																							
Order Carnivora	−	−	−	−	−	−	−	−	−	−	−	−	+	+	−	−	−	−	−	−	−	+	−
Non-mammalian taxa																							
Class Aves	−	+	−	−	−	−	−	−	−	−	−	−	−	+	−	−	−	−	−	−	−	−	−
Class Reptilia	−	+	+	−	−	−	−	−	−	−	+	−	−	+	−	−	−	−	−	−	−	−	−
Class Osteichthyes	−	+	−	−	−	−	−	−	−	−	−	−	−	−	−	−	−	−	−	+	−	−	−

TABLE 5.5. Chi-square test comparing the number of sites showing the presence or use of all taxonomic groups present in the Surovell and Waguespack and Cannon and Meltzer datasets.

	Number of sites	
Body Size Class	S&W	C&M
HMBS 5	26 (28.5)	**14 (11.5)**
HMBS 3&4	**28 (27.8)**	11 (11.2)
HMBS 1&2	**15 (13.6)**	4 (5.4)
Carnivores	**9 (8.6)**	3 (3.4)
Nonmammals	**14 (13.6)**	5 (5.4)

$X^2 = 1.46$, $df = 4$, $p = 0.834$

Expected values shown in parentheses. Cells for which observed values exceed expected values are shown in bold.

TABLE 5.6. Chi-square test comparing the number of occurrences of all taxonomic groups present in the Surovell and Waguespack and Cannon and Meltzer datasets.

	Number of occurrences	
Body Size Class	S&W	C&M
HMBS 5	26 (33.5)	**14 (6.5)**
HMBS 3&4	70 (71.2)	**15 (13.8)**
HMBS 1&2	**90 (80.5)**	6 (15.5)
Carnivores	**18 (17.6)**	3 (3.4)
Nonmammals	34 (35.2)	**6 (6.8)**

*$X^2 = 17.9$, $df = 4$, $p = 0.001$*Expected values shown in parentheses. Cells for which observed values exceed expected values are shown in bold.

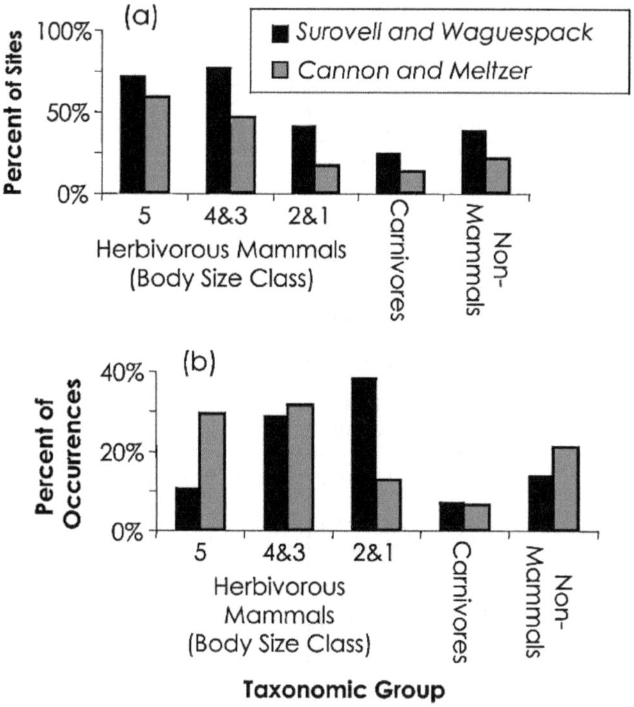

FIGURE 5.5a. Percent of sites showing the presence of prey taxa for the Surovell and Waguespack and Cannon and Meltzer (2004) datasets. **b.** Percent of occurrences of mammalian prey taxa for both datasets.

TABLE 5.7. Number of occurrences of herbivorous mammals standardized to taxonomic diversity by body class for the Surovell and Waguespack and Cannon and Meltzer datasets.

	Cannon and Meltzer			Waguespack and Surovell		
Body Size Class	Occurrences	Genera	Occurrences per genus	Occurrences	Genera	Occurrences per genus
5	14	2	7.00	26	2	13.00
4	9	3	3.00	44	9	4.89
3	6	2	3.00	26	7	3.71
2	4	3[a]	1.33	30	8	3.75
1	2	2[b]	1.00	63	19	3.32

[a] Cannon and Meltzer (2004) do not specify which genera are included in this body size class. We can infer at least two genera from their data: *Castor* and at least one genus of lagomorph. We assume three genera to be represented: *Castor*, *Lepus*, and *Sylvilagus*. If only two genera are assumed, it does not change the outcome of the analysis.

[b] Two rodent genera are assumed. Cannon and Meltzer identify three reliable associations with rodents, one of which is beaver at Bull Brook, which falls within our Body Size Class 2. We assume the other two associations to be small rodents (Body Size Class 1), and that two genera are represented.

The data also provide an opportunity to directly test the Byers and Ugan (2005) Early Paleoindian diet breadth model. They argued that Early Paleoindian diets should have consistently included animals down to the size of hares, and possibly even smaller mammals depending upon estimated encounter rates (Byers and Ugan, 2005:1633). In other words, they predicted all animals roughly 2 kg or larger would have always been taken upon encounter. Therefore, one would expect prey frequencies in Paleoindian faunal assemblages to be directly proportional to their estimated relative encounter rates (Byers and Ugan, 2005: Table 5.7)[3]. In Table 5.8, we show the predicted number of occurrences of each body size class stand-

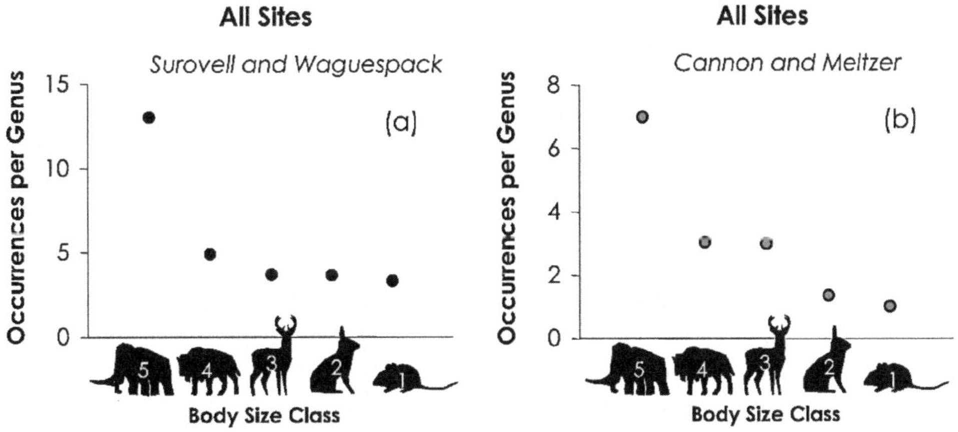

FIGURE 5.6a. Body size class versus archeological abundance standardized to taxonomic diversity for all sites in the Surovell and Waguespack database. **b**. Same for Cannon and Meltzer (2004) dataset.

TABLE 5.8. Predicted relative archeological frequency by body size class for herbivorous mammals estimated using the Byers and Ugan (2005) model.

Body Size Class	Predicted Relative Archeological Frequency[a]
5	1.0
4	7.6
3	25.7
2	630.6
1	15737.4

aPredicted relative archeological frequency was calculated using the equation relating body mass (kg) to population density (indiv.*km^{-2}) in Byers and Ugan (2005:Table 7): density = 4.33 – 0.75*log(mass). Population density for each Body Size Class was based on a weighted average of density by body size class for genera listed in Table 5.3. To calculate predicted archeological relative frequency, population densities for each body size class were standardized to that of Body Size Class 5 (proboscideans) assuming that encounter and capture rate are directly proportional to population density following Byers and Ugan (2005). These values assume constant taxonomic diversity for each Body Size Class.

ardized to proboscideans, which according to their model should be least frequently encountered and therefore least abundant in Early Paleoindian faunal assemblages. In short, the Byers and Ugan model predicts that for every mammoth or mastodon present in Early Paleoindian faunal assemblages, there should be approximately 7 bison-, 26 deer-, and 630 hare-sized mammals. If small rodents are included in the diet, there should be approximately 15,737 of these animals per mammoth or mastodon. Obviously, these predictions are easily falsified, but it should also be noted that the predicted encounter rates shown in Table 5.8 assume constant taxonomic diversity for each body size class. Because taxonomic diversity is generally inversely correlated with body size, these numbers are dramatic underestimates. Furthermore, because there are at least 14 known secure associations with mammoth and/or mastodon from Clovis contexts (Grayson and Meltzer, 2002; Cannon and Meltzer, 2004), the actual number of associations with non-proboscidean fauna predicted by the Byers and Ugan (2005) for our current sample of sites is much greater (e.g., 106 bison, 360 deer, 15,120 hares, 15,120 rabbits, etc).

Thus, the Byers and Ugan model does not appear to be a very good predictor of Paleoindian subsistence behavior. In contrast to the strong negative correlation between body size and archeological abundance predicted by their model, a strong positive correlation is seen. In order to stress this point, we repeat it: *The largest and the least common animals on Pleistocene landscapes are the most abundant and the most regular constituents of Paleoindian faunal assemblages.* Why does the Byers and Ugan (2005) model fail? There are three possible reasons. First, it may be built upon faulty assumptions such that estimated encounter and/or return rates are highly inaccurate. Second, the model may not be a good reflection of Paleoindian subsistence decisions. For example, post-encounter return rates may not have been the sole currency upon which the decision to kill or not to kill was made. Finally, it is possible that their model is a good representation of Paleoindian behavior, but that the archeological record of Paleoindian prey choice is extremely biased.

Is the Record Biased?

The simplest answer to this question is: probably, but it is difficult to know with certainty. Although numerous studies have discussed the possibility of bias in the Early Paleoindian faunal record (Grayson, 1988; Meltzer, 1989; Waguespack and Surovell, 2003; Cannon and Meltzer, 2004; Byers and Ugan, 2005), it is extremely difficult to directly test hypotheses about sample bias. Bias occurs when certain portions of populations have a greater or lesser likelihood of being sampled. Without knowing the distribution of the sampled population, the only way to directly test a hypothesis about sample bias is to have a theoretical or empirical model of what the underlying population should look like.

In the case of the Early Paleoindian faunal record, we have no clear basis for developing such null models. Here is an example. Hypothesis: *Large game kill sites are overrepresented in the Early Paleoindian archeological record because they are more easily discovered than campsites* (Grayson, 1988; Meltzer, 1989; Cannon and Meltzer, 2004). *Thus, large game are overrepresented in the Paleoindian archeological record.* To directly test this hypothesis, one would need some way to determine the expected relative frequencies of kill and non-kill sites in the case of no bias (the relative frequencies in the population). A theoretical or empirical null model, in theory, could be used. For example, one could examine hunter-gatherer ethnography to determine approximately how many large game kill sites are expected to occur per campsite and compare this to the archeological record, but this approach is problematic because it rests upon a tenuous uniformitarian assumption that the selected ethnographic case or cases are suitable analogies for the Paleoindian case. In other words, if the Paleoindian dataset is found to differ significantly from the null model, is it because the sample is biased or because the null model is inappropriate?

Despite these considerations, Cannon and Meltzer (2004:1974–1978) claim to have tested this and other hypotheses concerning bias. For example, they compare taxonomic richness and species representation between sites which were discovered on the basis of large mammal bones and those that were not. Not surprisingly, they find that sites which were discovered due to the presence of large mammal bones tend to contain greater proportions of large mammals in their assemblages (Cannon and Meltzer, 2004:1978). It is not difficult to show that this does not tell us one way or another whether the record is biased. To do so would require demonstrating that our method for sampling the archeological record produces a faunal record that actually differs from Paleoindian subsistence choices. They conduct a similar test with regard to kill sites versus campsites showing that large game kill sites tend to have lower taxonomic diversity and greater relative frequencies of large game (Cannon and Meltzer, 2004:1980–1981). From this analysis, one could argue that if campsites are overrepresented in our current sample of Paleoindian subsistence, then small game would be overrepresented as well, but it is unclear if our sample of kill and campsites differs significantly from the underlying population.

Nonetheless, we suspect that the record is biased and that large game are overrepresented (Waguespack and Surovell, 2003; Surovell and Waguespack, 2008). It seems very unlikely to us that more than one-half of the bone-bearing sites produced by Early Paleoindians were mammoth or mastodon kills, something which the record at face value would suggest (Table 5.2). This contention admittedly is not based on a theoretical or empirical null model per se but is instead a hunch. More importantly, it is worth asking whether excavating more campsites would really change our conclusion that Early Paleoindians preferentially targeted large mammals. Specifically, if we were to limit our analysis to sites that are not kills, would we come to a different conclusion about Paleoindian prey choice?

Again, we perform this analysis for both datasets. Cannon and Meltzer (2004) used the presence or absence of hearths to make the distinction between camp and kill sites. In contrast, for our dataset, to identify kill sites we use the criterion of whether artifacts are found in direct association with the carcasses of individual animals, or in the case of the Wacissa River site, a projectile point embedded in bone. Thus, certain sites which Cannon and Meltzer (2004) considered campsites due to the presence of hearths, such as Murray Springs, Lehner, and Blackwater Draw, we do not include in our non-kill site sample, and Jake Bluff, which Cannon and Meltzer consider to be a kill site, we include in our non-kill sample. This reduces our sample to 14 sites, and the Cannon and Meltzer sample to nine sites. It may be surprising to discover that when the sample is limited to non-kill sites, the same pattern of large game specialization persists (Fig. 5.7, Table 5.9). For both datasets, there is a significant positive

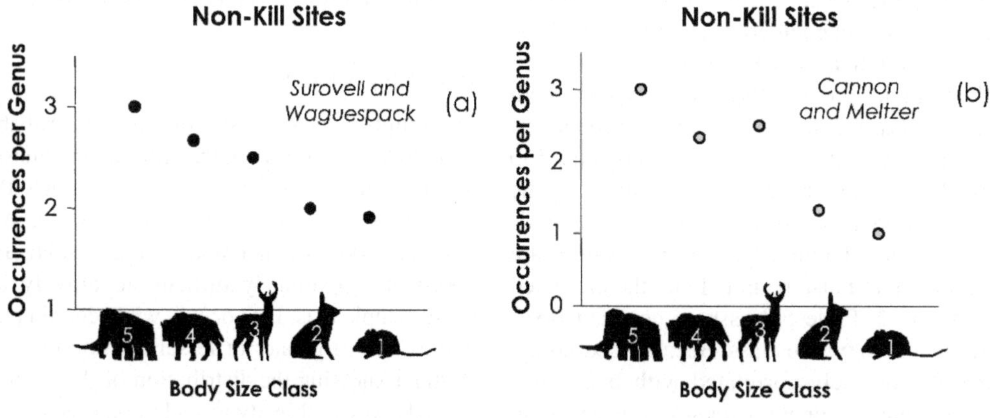

FIGURE 5.7a. Body size class versus archeological abundance standardized to taxonomic diversity for non-kill sites in the Surovell and Waguespack database. **b** Same for Cannon and Meltzer (2004) dataset.

TABLE 5.9. Number of occurrences of herbivorous mammals standardized to taxonomic diversity by body class for non-kill sites (campsites and processing sites) in the Surovell and Waguespack and Cannon and Meltzer (2004) datasets.

	Cannon and Meltzer			Waguespack and Surovell		
Body Size Class	Occurrences	Genera	Occurrences per genus	Occurrences	Genera	Occurrences per genus
5	3	1	3.00	6	2‡	3.00
4	7	3	2.33	16	6	2.67
3	5	2	2.50	10	4	2.50
2	4	3[a]	1.33	12	6	2.00
1	2	2[b]	1.00	23	12	1.92

For the Cannon and Meltzer (2004) dataset, sites with hearths were included in this analysis (see Table 5.2). For our dataset, we included those sites labeled "N" in the Table 5.2 "Kill Site" column.

[a,b]See notes in Table 6

[b]Only mammoth has been identified from these sites, but at Little Salt Springs and Aubrey, proboscideans were only identified to Order. We assume two genera here, but assuming only one genus may be more realistic.

correlation between body size and archeological abundance (C&M, Spearman's $\rho = 0.90$, $p = 0.037$; S&W, Spearman's $\rho = 1.00$, $p < 0.001$). Therefore, using only the campsite sample, again the largest and rarest species are the most common in Paleoindian faunal assemblages. This analysis suggests that if our current sample had a greater proportion of campsites, Early Paleoindians subsistence would still appear to have been focused upon large game.

Finally, we address what we call "geographic bias." To this point, we have only examined the record at a continental scale. Some have argued that the bulk of sites informing us about Paleoindian subsistence occur in the mid-continent, particularly the Western Great Plains and Southwest, and that we cannot and should not extrapolate subsistence patterns from these areas to far eastern or western North America (Meltzer and Smith, 1986; Grayson, 1988; Meltzer, 1989, 1993; Cannon and Meltzer, 2004). Thus, the appearance of large game specialization may be due to a record biased toward regions where large game hunting was more prevalent. It is true that there are large swaths of the continent where we know little or nothing about Early Paleoindian subsistence (Fig. 5.4), the Great Basin being one obvious example. Any discussion of Early Paleoindian subsistence in the Great Basin, therefore, must by necessity be based on indirect evidence because there is little direct evidence to speak of (e.g., Heizer and Baumhoff, 1970; Grayson, 1993; Beck and Jones, 1997). There are two important points to be made here. First, in areas where we have no evidence of Early Paleoindian subsistence, we simply do not know what people were killing and eating. Second, arguments about geographic bias implicitly argue that if we had more sites in the far east or possibly far west, they would attest to a more generalized subsistence pattern. The available data allow us to begin to explore this idea.

One last time, we turn to both datasets. We divided each dataset into three regions: Eastern North America (east of the Mississippi River), Western North America (west of the Continental Divide), and Central North America (east of the Continental Divide and west of the Mississippi River) as shown in Table 5.2. For each dataset and region, we once again examined body size vs. archeological abundance among herbivorous mammals (Table 5.10). Central North America accounts for the majority of associations in both datasets representing 57% of occurrences in the Cannon and Meltzer sample, and 72% of our sample. The remaining occurrences are equally divided between eastern and western North America for both datasets, but these samples are both problematic. The sample for western North America is comprised exclusively of large game kill sites by our definition, and the sample for eastern North America contains a series of very poorly preserved faunal assemblages.

In our dataset, for all regions, there are positive though not statistically significant correlations between body size and archeological abundance (Fig. 5.8). In the Cannon and Meltzer dataset, Western and Central North America show similar correlations. For Eastern North America, however, there is a nonsignificant but negative correlation between body size and archeological abundance. This analysis again suggests that across the continent, with the possible exception of Eastern North America, late Pleistocene hunter-gatherers preferentially targeted large-bodied prey. Because Eastern North America stands out and because it has been argued that large game specialization might not be expected for this region (e.g., Meltzer and Smith, 1986; Meltzer, 1988), it is worth taking a closer look at this record. The Cannon and Meltzer dataset includes a total of eight occurrences of herbivorous mammalian fauna: one mammoth, one mastodon, one bison, three caribou, one hare, and one beaver. While this might suggest a generalized foraging pattern for eastern North America, of the eight secure associations identified by Cannon and Meltzer (2004), six are with megafauna, using the traditional definition of animals weighing more than 40–45 kg (Martin, 1984; Martin and Steadman, 1999; Stuart, 1999; Barnosky et al., 2004; Koch and Barnosky, 2006). Thus, large game occur at significantly higher frequencies than small game based upon relative encounter frequencies.

TABLE 5.10. Number of occurrences of herbivorous mammals standardized to taxonomic diversity by body class for by geographic region for the Surovell and Waguespack and Cannon and Meltzer (2004) datasets.

Body Size Class	Region: West			Region: Central			Region: East		
	Occurrences	Genera	Occ/gen	Occurrences	Genera	Occ/gen	Occurrences	Genera	Occ/gen
Cannon and Meltzer									
5	4	1	4.00	8	2	4.00	2	2	1.00
4	3	2	1.50	5	2	2.50	1	1	1.00
3	0	0	–	3	1	3.00	3	1	3.00
2	0	0	–	2	2	1.00	2	2	1.00
1	0	0	–	2	2	1.00	0	0	–
Surovell and Waguespack									
5	6	2	3.00	13	3	4.33	7	2	3.50
4	10	4	2.50	28	7	4.00	6	4	1.50
3	1	1	1.00	14	5	2.80	11	4	2.75
2	4	3	1.33	23	7	3.29	3	3	1.00
1	5	4	1.25	58	19	3.05	0	0	–

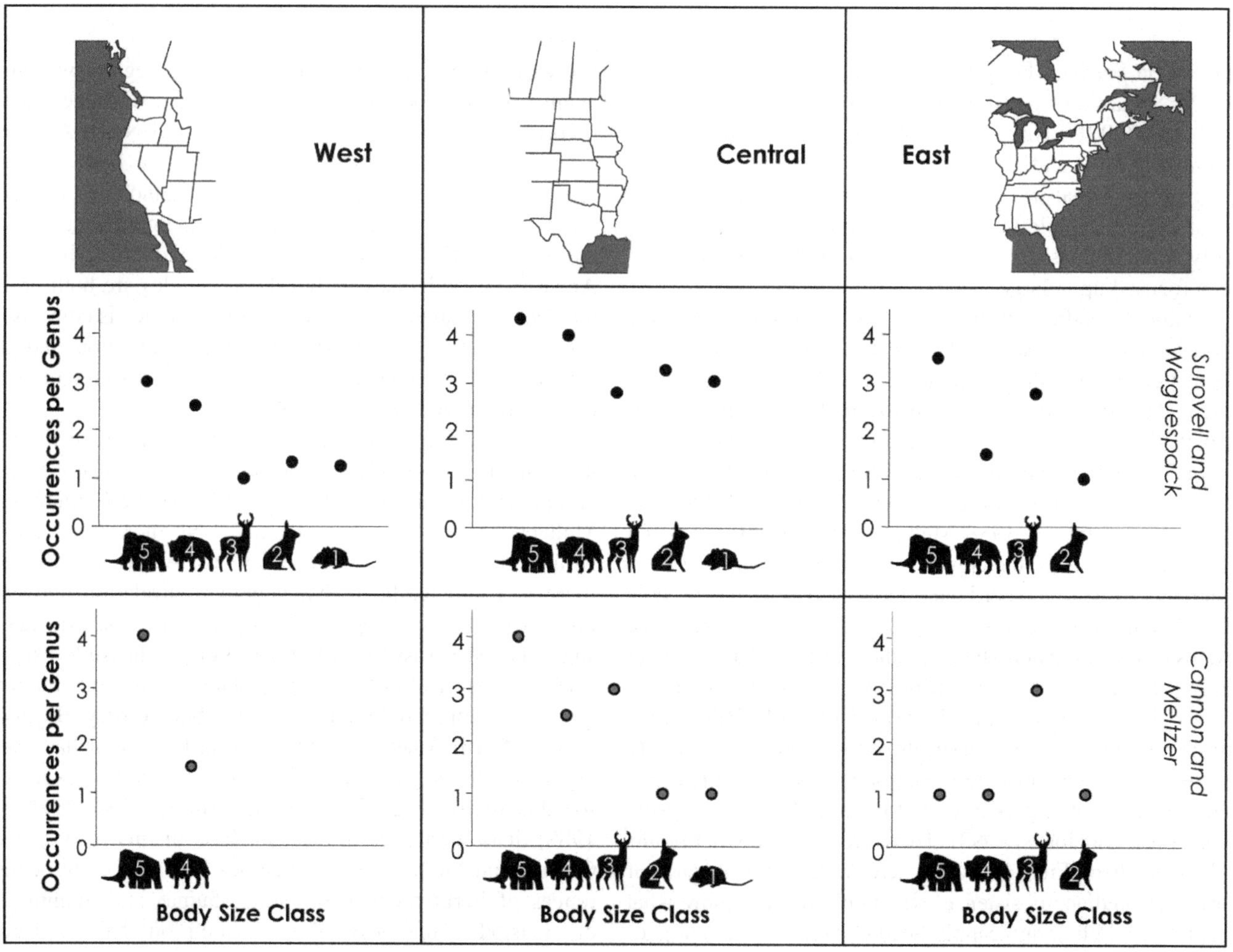

FIGURE 5.8. Body size class versus archeological abundance standardized to taxonomic diversity by geographic region for the Surovell and Waguespack and Cannon and Meltzer (2004) datasets.

Aside from sample size, there are other reasons why the record from eastern North America may differ. There is certainly no shortage of fluted point sites in eastern North America (Anderson and Faught, 2000), but there is a lack of bone from those sites, or in some cases a lack of clear associations (e.g, Dunbar, 1991; Fisher, 1984). The humid and acidic soils of the east dramatically reduce the probability of survival of late Pleistocene bone. Of Cannon and Meltzer's eight associations with herbivorous mammals, five are preserved due only to burning and calcination. These include the five smallest animals of the sample, all three caribou, the hare, and the beaver. In other words, the bulk of the record from eastern North America is based on an extremely limited sample of burned bones recovered from possible hearth contexts, which again raises the red flag of bias. While it is not a simple matter to test the hypothesis that the faunal record of eastern North America is biased, it is interesting to note that virtually all the small game associations that are accepted by Cannon and Meltzer for all regions are accepted on the basis of burning, while very few of the large game associations are based on this criterion. Nonetheless, even for the small and problematic sample known from eastern North America, megafauna regularly occur and dominate faunal assemblages, and when criteria for association are relaxed, the pattern for eastern North America mimics that of the remainder of the continent.

Paleoindian Prey Choice and North American Megafaunal Extinctions

Do the analyses above support the hypothesis that human hunting directly or indirectly caused the extinction of >30 genera of North American megafauna? In a general sense, we believe the answer is yes. Globally or locally extinct fauna regularly occur in Early Paleoindian faunal assemblages. In fact, extinct fauna occur in 83% to 97% of the sites used in this study (Table 5.3) depending upon how one defines "extinct". The high estimate includes *Bison antiquus* (ancestral to *B. bison* and locally extinct in parts of North America) and *Rangifer tarandus* (locally extinct in parts of North America) as extinct species; the low estimate does not. Including both species, only one site, Shawnee-Minisink, lacks evidence of hunting of extinct megafauna. Even if one only uses "secure" subsistence associations (Table 5.4), extinct megafauna still occur in 68% to 95% of Early Paleoindian faunal assemblages using the same criteria. Furthermore, as we have argued, there is clear evidence that Paleoindians not only hunted but preferentially hunted large-bodied prey. Because late Pleistocene extinctions in North America were similarly size-selective (Martin, 1984; Alroy, 1999; Lyons et al., 2004), large game specialization by Early Paleoindians provides circumstantial support for the Overkill hypothesis.

Selective targeting of the largest available animals would have meant that all potential prey would not have been subject to equal predation pressures. Generally speaking the ability of animal populations to sustain viable populations under predation pressure is negatively correlated with body size (Alroy, 2001), but there are exceptions to this pattern, such as turtles and tortoises (Stiner et al., 1999, 2000; Surovell, 1999). As a group, these animals are usually not large-bodied. Importantly however, they also suffered extinctions in the Pleistocene (Martin, 1984) and occur in a number of Early Paleoindian faunal assemblages (Waguespack and Surovell, 2003). Preferential human hunting of large, slowly reproducing species is thus one clear mechanism for producing the pattern of size bias in Pleistocene extinctions seen in North America and worldwide (Alroy, 1999, 2001; Lyons et al., 2004).

At best, this evidence must be seen as circumstantial because there are very few "secure" subsistence associations with the vast majority of extinct Pleistocene genera (Grayson and Meltzer, 2002, 2003; Cannon and Meltzer, 2004). Of globally extinct North American megafauna, according to Cannon and Meltzer (2004) only *Mammuthus*, *Mammut*, *Equus*, and *Camelops* can be shown to have been utilized by Early Paleoindians for subsistence purposes. Of these genera, *Mammuthus* accounts for the vast majority of "secure" associations (Table 5.4). For extinctions in the Order Proboscidea, we think a strong argument can be made for human causation for two reasons. First, over three continents and 800,000 years, evidence of human subsistence use of proboscideans is limited to the edges of human global colonization, suggesting that for Europe, Asia, North America, and South America local extinctions among the proboscidea coincided directly with human incursions into uninhabited regions (Surovell et al., 2005). Second, although 14 "secure" associations with mammoths and mastodons may not seem like a large number in the absolute sense (Dixon, 1999:216; Adovasio and Page, 2002:127; Grayson and Meltzer, 2002:248), it is a huge number of proboscidean kills when compared to similar evidence from other parts of the world (Haynes, 2002b:183; Surovell and Waguespack, 2008; Surovell et al., 2005). Strong evidence for hunting of mammoths and mastodons, however, is not easily translated into an argument that humans also hunted other extinct taxa for which there are few or no subsistence associations.

However, one must ask whether this absence of evidence is truly meaningful given our current sample of late Pleistocene archeological sites. At the start of 1926, the first year of excavations at the Folsom site in New Mexico, no one would have considered it odd that there were no recognized "secure" subsistence associations between Early Paleoindian artifacts and extinct fauna because so few sites relevant to the question had been investigated. Now, 80 years later, the sample has increased, but we still struggle to interpret what the numerous extinct fauna absent from the subsistence record means for Overkill. How large of an archeological sample is necessary to demonstrate that this absence of evidence is truly meaningful? Consider the "secure" subsistence associations identified by Cannon and Meltzer (2004) shown in Table 5.4. It is well known that there is a strong relationship between sample size and taxonomic diversity in faunal assemblages (Grayson and Delpech, 1998, 2001, 2002; Grayson, et al., 2001), and one could ask what is the probability of discovering >30 extinct genera represented in sufficient quantities to support Overkill from a total of 42 "secure" subsistence associations? Obviously, the probability is extremely low and it would require that virtually every association be taxonomically unique. In other words, this particular absence of evidence need not be evidence of absence.

From a brief examination of the set of sites which do tell us something about what Early Paleoindians hunted (Tables 5.3 and 5.4), it is not difficult to again make the argument that this absence may not be meaningful. First, one must ask in which sites we might expect to reliably see evidence of hunting of non-proboscidean extinct fauna? Of the 22 sites that Cannon and Meltzer (2004) consider to contain reliable evidence of Paleoindian subsistence behavior, 15 are in whole or in part mammoth, mastodon, or bison kills (Tables 5.2 and 5.4). While other utilized fauna do occasionally occur in these sites, particularly those with hearths such as Lehner and Blackwater Draw (Table 5.4), generally speaking we do not see taxonomically diverse subsistence associations at such sites (Cannon and Meltzer, 2004), and thus the absence of secure evidence for subsistence use of other extinct Pleistocene fauna at these large mammal kill sites may not be meaningful. Of the remaining six sites, four (Bull Brook, Shawnee-Minisink, Udora, and Whipple) have very small (NISP<20), very poorly preserved faunal assemblages, where the only skeletal elements preserved are those which happened to become calcined in hearths. Again, the absence of extinct Pleistocene fauna in these sites may not be meaningful.

This leaves us with three sites in the Cannon and Meltzer (2004) sample where one might reasonably expect to see evidence of the use of the remaining 29 genera of extinct Pleistocene fauna, for which we have little evidence of human hunting: Aubrey, Jake Bluff, and Lewisville. Cannon and Meltzer (2004) classify Jake Bluff as a kill site, and there are serious reasons to doubt whether Lewisville is an archeological site at all (see above). Thus, one could argue that the only site currently published in sufficient detail to evaluate taphonomically and likely to show evidence of use of the remaining extinct fauna is Aubrey. We do not consider the absence of evidence for hunting of 29 genera of extinct fauna at Aubrey to be meaningful.

Over 20 years ago, Donald Grayson made a similar argument and concluded that "the lack of human associations with certain extinct taxa may well be a function of the structure of the record as we happen to know it" (Grayson, 1984:220). Since that time Grayson seems to have firmed up his opinion on the matter and decided that the record is sufficient to address the Overkill question (Grayson, 2001; Grayson and Meltzer, 2002, 2003). But we feel the point is still valid. Given our prior arguments about bias, it should be clear we are not arguing that if we were to dig a number of well-preserved Early Paleoindian campsites from across North America, they would necessarily provide evidence for hunting of all extinct Pleistocene megafauna. It is tempting to invoke such negative evidence, but ultimately it is nothing more than speculation. Instead, we wish to make the point that although 80 years have passed since the Folsom discovery, our archeological sample of Early Paleoindian subsistence still may be inadequate for answering the question of whether this particular absence of evidence is truly problematic for the Overkill hypothesis. Clearly, we feel that the current sample is adequate for addressing general trends of prey-choice as it relates to body size, but it may not be adequate for addressing the detailed predation histories for most individual taxa that would be required for a true test of the Overkill hypothesis.

Conclusions

While much ink has been spilled over the issue of dataset construction and as easy as it may be to continue quibbling about the inclusion of various specimens from various sites, the archeological record presents a fairly consistent record of Paleoindian prey choice decisions. From an optimal foraging perspective, there is enormous potential economic and social benefit to hunting the largest prey available. So while mammoth predation may be odd in light of other carnivores and in comparison to the majority of ethnographically documented hunter-gatherer societies, it is a logical and arguably *predictable* human subsistence activity to occur when large prey is available in sufficient quantities. During the late Pleistocene in North America, and at varying times and places throughout the world, human population densities were undoubtedly extremely low. In such circumstances where large animals are available and encountered frequently enough to meet human subsistence needs, either because humans are few, prey are plentiful or some combination of the two, the option to exercise a specialized predation strategy exists (Waguespack and Surovell, 2003). Further, since specialized hunting economies can provide an efficient means of procuring animal resources, preferential predation of large game was likely far more common in past hunter-gatherer societies than represented in the recent ethnographic record. Importantly, the risks and variance associated with specialized hunting strategies due to lengthy search costs and/or unpredictable, infrequent successes have organizational implications for Paleoindian mobility (e.g., Kelly and Todd, 1988; Anderson and Gillam, 2000; Haynes, 2002a), demography (Anderson, 1995; Surovell, 2000; Marlowe, 2001; Meltzer, 2004), and labor (Waguespack, 2005).

While the case for Overkill is not, from our perspective, definitively settled, our investigation of Paleoindian subsistence provides the following relevant conclusions regarding Pleistocene hunting strategies and its potential impact on Pleistocene prey:

1. Specialized large-game predation strategies can provide economic and social benefits to hunter-gatherer populations. While rare ethnographically, when and where ecological conditions provide large-game in sufficient quantities and sociocultural mechanisms are in place to offset costs, a narrow diet-breadth selecting and utilizing prey based on their rank as opposed to encounter rate is the expected strategy.
2. Early Paleoindian hunters followed a specialized predation strategy, passing up some opportunities to procure

small prey in favor of larger animal resources. Medium and small sized animals were part of the diet, but were not used as frequently as one would expect based on relative encounter rates. Thus, Early Paleoindian diets included a wide variety of prey species, but large, rare prey remain the most frequently occurring and abundant animals in Early Paleoindian sites and assemblages.

3. The Early Paleoindian faunal record is likely biased. The combined effects of taphonomy, site discovery bias, research attention bias, and inconsistent site analysis and recording procedures, render imperfect the archeological record pertaining to Early Paleoindian hunting. Much like the archeological record in other times and regions, or for that matter all times and all places, it is difficult to address the extent of these biases. Our analysis suggests that the cumulative record, compiled from known sites and faunal inventories, presents a consistent pattern of size related hunting preferences that are difficult to dismiss on the basis of chance or deliberate bias.
4. The hunting strategy of Early Paleoindian foragers is compatible with the Overkill hypothesis of Pleistocene megafaunal extinction, but due to the lack of secure associations with most extinct genera, support for Overkill can be viewed only as circumstantial. By focusing their predation efforts on the largest available prey, Early Paleoindian foragers also deliberately hunted prey species that were the least able to sustain population growth or maintenance under hunting pressure. Quite simply: (a) Early foraging populations of America hunted really big prey, (b) Really large animals are highly susceptible to the deleterious impacts of predation, and (c) Primarily large animals went extinct. There is currently little evidence attesting to the regular subsistence use of *all* extinct Pleistocene fauna. Negative evidence is famously difficult to interpret, and as more sites are discovered we are left only to assume that they will either contain more extinct Pleistocene fauna or will contain the long lost plethora of rabbits, rodents, and other small game currently needed to change the archeological patterns identified here.

Notes

1 We consider the use of plants to be a separate issue. While Paleoindians certainly used plants for subsistence and other purposes, we are concerned solely with the degree to which hunting was selective with respect to prey body size.

2 Dennis Stanford conducted additional excavations at the Lewisville site from 1978–1980, but the results of this work have not been published. Microflakes (sand-sized?) were reportedly recovered during this work, possibly supporting the hypothesis that these deposits do represent a late Pleistocene archeological site. This evidence is difficult to evaluate (see also Grayson and Meltzer, 2002), but we do not find it convincing. First, it is unclear if the microflakes are truly artifacts. Also, it is difficult to understand how a group of hunter-gatherers killed, butchered, and cooked dozens of animals involving 21 hearths while leaving behind only one Clovis point and a handful of tiny flakes.

3 This assumes searching in a "fine-grained environment." This is a fundamental assumption of the prey choice model (Stephens and Krebs, 1986) and as such, it is an assumption also made by Byers and Ugan (2005).

References

Adovasio JM, Page J (2002) The first Americans: In pursuit of archaeology's greatest mystery. Random House, New York

Alroy J (1999) Putting North America's end-Pleistocene megafaunal extinctions in context: Large-scale analyses of spatial patterns, extinction rates, and size distributions. In: MacPhee R (ed) Extinctions in near time. Kluwer/Plenum, New York, pp 105–143

Alroy J (2001) A multispecies overkill simulation of the end-Pleistocene megafaunal mass extinction. Science 292:1893–1896

Alvard MS, Nolin DA (2002) Rousseau's whale hunt? Current Anthropology 43:533–559

Anderson DG (1995) Paleoindian interaction networks in the Eastern Woodlands. In: Nassaney MS, Sassamn KE (eds) Native American interaction: Multiscalar analyses and interpretations in the Eastern Woodlands. University of Tennessee Press, Knoxville, TN, pp 1–26

Anderson DG, Faught MK (2000) Paleoindian artefact distributions: Evidence and implications. Antiquity 74:507–513

Anderson DG, Gillam JC (2000) Paleoindian colonization of the Americas: Implications from an examination of physiography, demography, and artifact distribution. Am Ant 65:43–66

Andrewartha HG, Birch LC (1961) The distribution and abundance of animals. University of Chicago Press, Chicago, IL

Bailey RC (1991) The behavioral ecology of Efe Pygmy men in the Ituri forest, Zaire. Museum of Anthropology University of Michigan, Ann Arbor, MI

Barnosky AD, Koch PL, Feranec RS, Wing SL, Shabel AB (2004) Assessing the causes of late Pleistocene extinctions on continents. Science 306:70–75

Barton CM, Schmich S, James SR (2004) The ecology of human colonization in pristine landscapes. In: Barton CM, Clark GA, Yesner DR, Pearson GA (eds) The settlement of the American continents: A multidisciplinary approach to human biogeography. University of Arizona Press, Tucson, AZ, pp 138–161

Beck C, Jones GT (1997) The terminal Pleistocene/early Holocene archaeology of the Great Basin. J World Prehist 11:161–236

Begon M, Harper JL, Townsend CR (1996) Ecology. Blackwell, Oxford

Bement LC, Carter BJ (2003) The Jake Bluff site: Clovis bison hunting adaptations at the brink of mammoth extinction on the southern high plains of North America. Paper presented at the 68th Annual Meeting of the Society for America Archaeology, Apr. 9–13, Milwaukee, WI

Binford LR (1978) Nunamiut ethnoarchaeology. Academic, New York

Binford LR (1991) A corporate caribou hunt. Expedition 33(1):33–43

Bird DW, Bird RB (2000) The ethnoarchaeology of juvenile foragers: Shellfishing strategies among the Meriam children. J Anthropol Archaeol 19, 461–476.

Bird RB, Bird DW, Smith EA, Kushnick GC (2002) Risk and reciprocity in Meriam food sharing. Evol Hum Behav 23:297–321

Boeri D (1983) People of the ice whale: Eskimos, white men, and the whale. Harcourt Brace Jovanovich, San Diego, CA

Brook BW, Bowman DMJS (2005) One equation fits overkill: Why allometry underpins both prehistoric and modern body size-biased extinctions. Popul Ecol 47:137–141

Brugal J, David F, Enloe JG, Jaubert J (1999) Introduction. In: Brugal J, David F, Enloe JG, Jaubert J (eds) Le bison: Gibier et moyen de subsistence des hommes du Paléolithique aux Paléoindians des Grandes Plaines. Éditions APDCA, Antibes, pp 7–11

Brunswig RH Jr, Fisher DC (1993) Research on the Dent mammoth site. Curr Res Pleistocene 10:63–65

Brush N, Smith F (1994) The Martins Creek mastodon: A Paleoindian butchery site in Holmes County, Ohio. Curr Res Pleistocene 11:14–15

Brush N, Newman M, Smith F (1994) Immunological analysis of flint flakes from the Martins Creek Mastodon site. Curr Res Pleistocene 11:16–18

Byers DA, Ugan A (2005) Should we expect large game specialization in the late Pleistocene? An optimal foraging perspective on Early Paleoindian prey choice. J Archaeol Sci 32:1624–1640

Byers DS (1955) Additional information on the Bull Brook site, Massachusetts. Am Ant 20:274–276

Cannon MD, Meltzer DJ (2004) Early Paleoindian foraging: Examining the faunal evidence for megafaunal specialization and regional variability in prey choice. Quat Sci Rev 23:1955–1987

Cashdan E (1985) Coping with risk: Reciprocity among the Basarwa of northern Botswana. Man 20:454–474

Charnov EL (1976) Optimal foraging: Attack strategy of a mantid. Am Nat 110:141–151

Clausen CJ, Cohen AD, Emiliani C, Holman JA, Stipp JJ (1979) Little Salt Spring, Florida: A unique underwater site. Science 203:609–614

Cleland CE (1965) Barren ground caribou (*Rangifer arcticus*) from an Early Man site in southeastern Michigan. Am Ant 30:350–351

Collins, MB (1999) Clovis blade technology. University of Texas Press, Austin, TX

Collins MB, Evans GL, Campbell TN, Winans MC, Mear CE (1989) Clovis occupation at Kincaid Shelter, Texas. Curr Res Pleistocene 6:3–4

Condon RG, Collings P, Wenzel G (1995) The best part of life: Subsistence hunting, ethnicity, and economic adaptation among young Inuit males. Arctic 48(1):31–46

Cox SJ, Sluckin TJ, Steele J (1999) Group size, memory, and interaction in the evolution of cooperation. Curr Anthropol 40(3):369–377

Crook, WW Jr, Harris, RK (1957) Hearths and artifacts of Early Man near Lewisville, Texas and associated faunal material. Bull Texas Arch Soc 28:7–97

Crook WW Jr, Harris RK (1958) A Pleistocene campsite near Lewisville, Texas. Am Ant 23:233–246

Dent RJ, Kauffman BE (1985) Aboriginal subsistence and site ecology interpreted from microfloral and faunal remains. In: McNett C (ed) Shawnee-Minisink: A stratified Paleoindian-Archaic site in the Upper Delaware Valley of Pennsylvania. Academic, New York, pp 55–79

Dillehay TD (ed) (2000) The settlement of the Americas: A new rehistory. Basic Books, New York

Dixon EJ (1999) Bones, boats, and bison: Archeology and the first colonization of Western North America. University of New Mexico Press, Albuquerque, NM

Driver JC (1999) Raven skeletons from Paleoindian contexts, Charlie Lake Cave, British Columbia. Am Ant 64:289–298

Duffy K (1984) Children of the forest. Dodd, Mead, New York

Dunbar JS (1991) Resource orientation of Clovis and Suwanee age sites Paleoindian sites in Florida. In: Bonnichsen R, Turnmire KL (eds) Clovis: Origins and adaptations. Center for the Study of the First Americans, Corvallis, OR, pp 185–213

Eisenberg L (1978) Paleo-Indian settlement pattern in the Hudson and Delaware River drainages. Department of Anthropology, Franklin Pierce College, Occasional Publications in Northeastern Anthropology, No. 4, Rindge, New Hampshire

Enloe JG (1999) Hunting specialization: Single-species focus and human adaptation. In: Brugal J, David F, Enloe J, Jaubert J (eds) Le bison: Gibier et moyen de subsistence des hommes du Paléolithique aux Paléoindians des Grandes Plaines. Éditions APDCA, Antibes, pp 501–510

Ferring CR (1995) The late Quaternary geology and archaeology of the Aubrey Clovis site, Texas: A preliminary report. In: Johnson E (ed), Ancient peoples and landscapes. Museum of Texas Tech University, Lubbock, pp 273–281

Ferring CR (2001) Archaeology of the Aubrey Clovis site. In: Ferring CR (ed) The archaeology and paleoecology of the Aubrey Clovis site (41DN479), Denton County, Texas. Center for Environmental Archaeology, University of North Texas, Denton, pp 121–202

Figgins JD (1933) A further contribution to the antiquity of man in America. Proceedings of the Colorado Museum of Natural History XII(12)

Fisher DC (1984) Taphonomic analysis of late Pleistocene mastodon occurrences: Evidences of butchery by North American Paleo-Indians. Paleobiology 10:338–357

Fisher JW (1993) Foragers and farmers: Material expressions of interaction at elephant processing sites in the Ituri Forest, Zaire. In: Hudson J (ed) From bones to behavior: Ethnoarchaeological and experimental contributions to the interpretation of faunal remains. Center for Archaeological Investigations, University of Southern Illinois Press, Carbondale, pp 247–262

Fitting JE, DeVisscher J, Wahla EJ (1966) The Paleo-Indian occupation of the Holcombe Beach. Anthropological Papers of the Museum of Anthropology, University of Michigan, No. 27, The University of Michigan, Ann Arbor

Fladmark K, Driver JC, Alexander D (1988) The Paleoindian component at Charlie Lake Cave (HbRF 39), British Columbia. Am Ant 53:371–384

Frison GC (1982) The Sheaman site: A Clovis component. In: Frison GC, Stanford DJ (eds) The Agate Basin Site: A record of the Paleoindian occupation of the northwestern High Plains. Academic, New York, pp 143–157

Frison GC (1989) Experimental use of Clovis weaponry and tools on African elephants. Am Ant 54(4):766–784

Frison GC, Todd LC (1986) The Colby mammoth site. University of New Mexico Press, Albuquerque, NM

Gould R (1980) Living archaeology. Cambridge University Press, Cambridge.

Graham RW, Kay M (1988) Taphonomic comparisons of cultural and noncultural faunal deposits at the Kimmswick and Barnhart sites, Jefferson County, Missouri. In: Laub RS, Miller NG, Steadman DW (eds) Late Pleistocene and early Holocene paleoecology and archaeology of the eastern Great Lakes region. Bulletin of the Buffalo Society of Natural Sciences, vol 33, Buffalo, NY, pp 227–240

Graham RW, Haynes CV, Johnson DL, Kay M (1981) Kimmswick: A Clovis-mastodon association in eastern Missouri. Science 213:1115–1117

Grayson DK (1984) Archaeological associations with extinct Pleistocene mammals in North America. J Arch Sci 11:213–221

Grayson DK (1988) Perspectives on the archaeology of the first Americans. In: Carlisle RC (ed) Americans before Columbus: Iceage origins. Ethnography Monographs 12, Department of Anthropology, University of Pittsburgh, Pittsburgh, PA, pp 107–122

Grayson DK (1993) The desert's past: A natural prehistory of the Great Basin. Smithsonian Institution Press, Washington, DC

Grayson DK (2001) The archaeological record of human impacts on animal populations. J World Prehist 15:1–67

Grayson DK, Delpech F (1998) Changing diet breadth in the early Upper Paleolithic of southwestern France. J Arch Sci 25:1119–1129

Grayson DK, Delpech F (2001) The Upper Paleolithic at Grotte XVI (Dordogne, France): Richness, evenness and cave bears. In: Hays MA, Thacker PT (eds) Questioning the answers: Resolving fundamental problems of the early Upper Paleolithic. BAR International Series1005, British Archaeological Reports, Oxford, pp 187–195

Grayson DK, Delpech, F (2002) Specialized early Upper Paleolithic hunters in southwestern France? J Arch Sci 29:1439–1449

Grayson DK, Meltzer DJ (2002) Clovis hunting and large mammal extinction: A critical review of the evidence. J World Prehist 16:313–359

Grayson DK, Meltzer DJ (2003) A requiem for North American overkill. J Arch Sci 30:585–593

Grayson DK, Delpech F, Regaud J.-P, Simek J (2001) Explaining the development of dietary dominance by a single ungulate taxon at Grotte XVI, Dordogne, France. J Arch Sci 28:115–125

Greaves RD (1997) Hunting and multifunctional use of bows and arrows. In: Knecht H (ed) Projectile technology. Plenum, New York, pp 287–320

Griffiths D (1980) Foraging costs and relative prey size. Am Nat 116(5):743–752

Gubser NJ (1965) The Nunamiut Eskimos: Hunters of caribou. Yale University Press, New Haven, CT

Gurven M, Hill K, Kaplan H, Hurtado A, Lyles R (2000) Food transfers among Hiwi foragers of Venezuela: Tests of reciprocity. Hum Ecol 28(2):171–218

Gustafson CE, Gilbow D, Daugherty RD (1979) The Manis mastodon: Early Man on the Olympic Peninsula. Can J Archaeol 3:157–164

Hames RB (1979) A comparison of the efficiencies of the shotgun and the bow in neotropical forest hunting. Hum Ecol 7(3):219–252

Hannus LA (1989) Flaked mammoth bone from the Lange/Ferguson site White River Badlands area, South Dakota. In: Bonnichsen R, Sorg MH (eds) Bone modification. Center for the Study of the First Americans, University of Maine, Orono, Maine, pp 395–412

Hannus LA (1990) The case for mammoth bone-butchering tools. In: Agenbroad LD, Mead J, Nelson L (eds) Megafauna and man: Discovery of America's heartland. Mammoth Site of Hot Springs, Hot Springs, SD, pp 86–99

Haury EW (1953) Artifacts with mammoth remains, Naco, Arizona. Am Ant 19(1):1–14

Haury EW, Sayles EB, Wasley WW (1959) The Lehner mammoth site, southeastern Arizona. Am Ant 25(1):2–42

Hawkes K (1991) Showing off: Tests of an hypothesis about men's foraging goals. Ethol Sociobiol 12:29–54

Hawkes K (1992) Sharing and collective action. In: Smith EA, Winterhalder B (eds) Evolutionary ecology and human behavior. Aldine de Gruyter, New York, pp 269–300

Hawkes K, O'Connell JF, Jones NGB (1991) Hunting income patterns among the Hadza: Big game, common goods, foraging goals and the evolution of the human diet. Phil Trans Royal Soc Lond 334:243–251

Hayden B (1994) Competition, labor, and complex hunter-gatherers. In: Burch E, Ellana L (eds) Key issues in hunter-gatherer research. Berg, Oxford, pp 223–239

Haynes CV Jr (1993) Clovis-Folsom geochronology and climatic change. In: Soffer O, Praslov ND (eds) From Kostenki to Clovis: Upper Paleolithic – Paleo-Indian adaptations. Plenum, New York, pp 219–236

Haynes CV Jr, Haury EW (1982) Archaeological investigations at the Lehner site, Arizona, 1974–1975. Natl Geogr Res Rep 14:325–334

Haynes G (2002a) The early settlement of North America: The Clovis era. Cambridge University Press, Cambridge

Haynes G (2002b) The catastrophic extinction of North American mammoths and mastodonts. World Arch 33:391–416

Haynes G, Stanford D (1984) On the possible utilization of *Camelops* by Early Man in North America. Quaternary Research 22:216–230

Heizer RF, Baumhoff MA (1970) Big game hunters in the Great Basin: A critical review of the evidence. Contributions of the University of California Archaeological Research Facility, vol 7, pp 1–12

Hemmings ET, Haynes CV Jr (1969) The Escapule mammoth and associated projectile points, San Pedro valley, Arizona. J Arizona Acad Sci 5:184–188

Hill K, Padwe J (2000) Sustainability of Ache hunting in the Mbaracayú Reserve, Paraguay. In: Robinson JG, Bennett EL (eds) Hunting for sustainability in tropical forests. Columbia University Press, New York, pp 79–105

Hill K, Kaplan H, Hawkes K, Hurtado AM (1985) Men's time allocation to subsistence work among the Ache of eastern Paraguay. Hum Ecol 13(1):29–47

Hodgeson AGO (1926) Some notes on the hunting customs of the Wandamba of the Ulanga valley, Tanganyika Territory, and other East African tribes. J Roy Anthropol Inst Great Britain Ireland 56:59–70

Hoffman CA (1983) A mammoth kill site in the Silver Springs Run. The Florida Anthropologist 36:83–87

Holliday VT, Vance Haynes CJ, Hofman JL, Meltzer DJ (1994) Geoarchaeology and geochronology of the Miami (Clovis) site, Southern High Plains of Texas. Quat Res 41:234–244

Jochim MA (1981) Strategies for survival: Cultural behavior in an ecological context. Academic Press, New York

Johnson D, Kawano P, Ekker E (1980) Clovis strategies of hunting mammoth. Can J Anthropol 1:107–114

Johnson E (1987) Vertebrate remains. In: Johnson E (ed) Lubbock Lake: Late Quaternary studies on the southern High Plains. Texas A&M Press, College Station, TX, pp 49–89

Johnson E (1991) Late Pleistocene cultural occupation of the Southern Plains. In: Bonnichsen R, Tunmire KL (eds) Clovis: Origins and adaptations. Center for the Study of the First Americans, Corvalis, OR, pp 215–236

Kaplan H, Hill K, Lancaster J, Hurtado AM (2000) A theory of human life history evolution: Diet, intelligence, and longevity. Evol Anthropol 9:156–185

Kelly RL, Todd LC (1988) Coming into the country: Early paleoindian hunting and mobility. Am Ant 53(2):231–244

Koch PL, Barnosky AD (2006) Late quaternary extinctions: State of the debate. Annu Rev Ecol Evol Syst 37:215–250

Kooyman B, Hills LV, McNeil P, Tolman S (2006) Late Pleistocene horse hunting at the Wally's beach site (DhPg-8), Canada. Am Ant 71:101–121

Kooyman B, Newman ME, Cluney C, Lobb M, Tolman S, McNeil P, Hills LV (2001) Identification of horse exploitation by Clovis hunters based on protein analysis. Am Ant 66:686–691

Krebs JR, McCleery RH (1984) Optimization in behavioural ecology. In: Krebs JR, Davies NB (eds) Behavioural ecology: An evolutionary approach. Blackwell, Oxford, pp 91–121

Krieger AD (1962) The earliest cultures in the western United States. Am Ant 28:138–143

Kurtén B, Anderson E (1980) Pleistocene mammals of North America. Columbia University Press, NY

Lance JF (1959) Faunal remains from the Lehner Mammoth Site. Am Ant 25(1):35–42

Laub RS, DeRemer MF, Duport CA (1988) The Hiscock site: A rich late Quaternary locality in western New York State. In: Laub RS, Miller NG, Steadman DW (eds) Late Pleistocene and early Holocene paleoecology and archaeology of the eastern Great Lakes region. Bulletin of the Buffalo Society of Natural Sciences, vol 33, Buffalo, NY, pp 67–81

Lee RB (1979) The !Kung San: Men, women, and work in a foraging society. Cambridge University Press, Cambridge

Leonhardy FC (1966) Domebo: A Paleo-Indian mammoth kill in the prairie-plains. Contributions of the Museum of the Great Plains, Lawton, OK

Leonhardy FC, Anderson AD (1966) The archaeology of the Domebo site. In: Leonhardy FC (ed) Domebo: A Paleo-Indian mammoth kill in the prairie-plains. Contributions of the Museum of the Great Plains, No. 1, Lawton, OK, pp 14–26

Lundelius EL Jr (1972) Vertebrate remains from the Gray Sand. In: Hester JJ (ed) Blackwater Locality No. 1: A stratified, Early Man site in eastern New Mexico. Fort Burgwin Research Center, Ranchos de Taos, NM, pp 148–163

Lyons SK, Smith FA, Brown JH (2004) Of mice, mastodons, and men: Human-mediated extinctions on four continents. Evol Ecol Res 6:339–358

Madsen DB, Schmitt DN (1998) Mass collecting and the diet breadth model: A Great Basin example. J Arch Sci 25:445–455

Marks SA (1976) Large mammals and a brave people: Subsistence hunters in Zambia. University of Washington Press, Seattle, WA

Marlowe F (2001) Male contribution to diet and female reproductive success among foragers. Curr Anthropol 42:755–760

Marlowe F (2003) A critical period for provisioning by Hadza men: Implications for pair bonding. Evol Hum Behav 24:217–229

Martin JE (1984) Fossil vertebrates and the paleoenvironment of the Lange/Ferguson Clovis kill site in the Badlands of South Dakota. Curr Res Pleistocene 1:69–71

Martin PS (1973) The discovery of America. Science 179:969–974

Martin PS (1984) Prehistoric Overkill: The global model. In: Martin PS, Klein RG (eds) Quaternary extinctions. The University of Arizona Press, Tucson, AZ, pp 354–403

Martin PS, Steadman DW (1999) Prehistoric extinctions on islands and continents. In: MacPhee R (ed) Extinctions in near time. Kluwer/Plenum, NY, pp 17–55

Meltzer DJ (1988) Late Pleistocene human adaptations in eastern North America. J World Prehist 2:1–52

Meltzer DJ (1989) Why don't we know when the first people came to North America? Am Ant 54:471–490

Meltzer DJ (1993) Is there a Clovis adaptation? In: Soffer O, Praslov ND (eds) From Kostenki to Clovis: Upper Paleolithic – Paleo-Indian adaptations. Plenum, NY, pp 293–310

Meltzer DJ (2004) Modeling the initial colonization of the Americas: Issues of scale, demography, and landscape learning. In: Barton CM, Clark GA, Yesner DR, Pearson GA (eds) The settlement of the American continents: A multidisciplinary approach to human biogeography. University of Arizona Press, Tucson, AZ, pp 123–137

Meltzer DJ, Smith BD (1986) Paleoindian and Early Archaic subsistence strategies in eastern North America. In: Neusius S (ed) Foraging, collecting, and harvesting: Archaic period subsistence and settlement in the eastern Woodlands. Center for Archaeological Investigations, Southern Illinois University, Carbondale, pp 3–31

Mithen S (1993) Simulating mammoth hunting and extinction: Implications for the late Pleistocene of the Central Russian Plain. In: Larsen G, Bricker HM, Mellars P (eds) Hunting and animal exploitation in the later Paleolithic and Mesolithic of Eurasia. Archaeological Papers of the American Anthropological Association, No. 4, pp 163–178

Ohtsuki R, Suzuki T (1990) Population ecology of human survival: Bioecological studies of the Gidra in Papua New Guinea. University of Tokyo Press, Tokyo

Overstreet DF (1996) Still more on cultural contexts of mammoth and mastodon in the southwestern Lake Michigan basin. Curr Res Pleistocene 13:36–38

Overstreet DF, Stafford TW Jr (1997) Additions to a revised chronology for cultural and non-cultural mammoth and mastodon fossils in the southwestern Lake Michigan Basin. Curr Res Pleistocene 14:70–71

Overstreet DF, Joyce DJ, Wasion D (1995) More on cultural contexts of mammoth and mastodon in the southwestern Lake Michigan basin. Curr Res Pleistocene 12:40–42

Palmer HA, Stoltman JB (1975) The Boaz mastodon: A possible association of man and mastodon in Wisconsin. Midcontinental J Archaeol 1(2):163–177

Peters RH (1986) The ecological implications of body size. Cambridge University Press, Cambridge

Ray CN (1930) Report on some recent researches in the Abilene section. Bull Texas Arch Paleontol Soc 2:45–58

Ray CN (1942) Ancient artifacts and mammoth's teeth of the McLean site. Bull Texas Arch Paleontol Soc 14:137–146

Ray CN, Bryan K (1938) Folsomoid point found in alluvium beside a mammoth's bones. Science 88:257–258

Rayl SL (1974) A Paleo-Indian mammoth kill site near Silver Springs, Florida. Unpublished M.A. thesis, Department of Anthropology, Northern Arizona University

Saunders JJ (1980) A model for man-mammoth relationships in late Pleistocene North America. Can J Anthropol 1(1):87–98

Saunders JJ (n.d.) Vertebrates of the San Pedro valley 11,000 BP. Manuscript

Sellards EH (1952) Early Man in America: A study in prehistory. University of Texas Press, Austin, TX

Sellards EH (1960) Some early stone artifact developments in North America. Southwestern J Anthropol 16:160–173

Slaughter BH (1966) The vertebrates of the Domebo local fauna, Pleistocene of Oklahoma. In: Leonhardy FC (ed) Domebo: A Paleo-Indian mammoth kill in the prairie-plains. Contributions of the Museum of the Great Plains, No. 1, Lawton, Oklahoma, pp 31–35

Smith EA (1991) Inujjuamiut foraging strategies: Evolutionary ecology of an arctic hunting economy. Aldine de Gruyter, New York

Smith FA, Lyons SK, Ernest SKM, Jones KE, Kaufman DM, Dayan T, Marquet PA, Brown JH, Haskell JP (2003) The body mass of late Quaternary mammals. Ecology 84:3403

Spiess AE, Curran ML, Grimes JR (1985) Caribou (*Rangifer tarandus* L.) bones from New England Paleoindian sites. North Am Archaeol 6:145–159

Stanford D (1987) The Ginsberg experiment. Nat Hist 9:10–14

Steadman DW (1988) Vertebrates from the late Quaternary Hiscock site, Genesee County, New York. In: Laub RS, Miller NG, Steadman DW (eds) Late Pleistocene and early Holocene paleoecology and archaeology of the eastern Great Lakes region. Bulletin of the Buffalo Society of Natural Sciences, vol 33, Buffalo, NY, pp 95–113

Steinhart EI (2000) Elephant hunting in 19th century Kenya: Kamba society and ecology in transformation. Int J Afr Hist Stud 33(2):335–345

Stephens DW, Krebs JR (1986) Foraging theory. Princeton University Press, Princeton, NJ

Stiner MC, Munro ND, Surovell TA, Tchernov E, Bar-Yosef O (1999) Paleolithic population pulses evidenced by small animal exploitation. Science 283:190–193

Stiner MC, Munro ND, Surovell TA (2000) The tortoise and the hare: Small game use, the broad spectrum revolution, and Paleolithic demography. Curr Anthropol 41(1):39–73

Storck PL, Spiess AE (1994) The significance of new faunal identifications attributed to an Early Paleoindian (Gainey Complex) occupation at the Udora site, Ontario, Canada. Am Ant 59:121–142

Stuart AJ (1999) Late Pleistocene megafaunal extinctions: A European perspective. In: MacPhee R (ed) Extinctions in near time. Kluwer/Plenum, NY, pp 257–269

Surovell TA (1999) Modeling occupation intensity and small game use in the Levant. In: Driver JC (ed) Zooarchaeology of the Pleistocene/Holocene boundary: Proceedings of a symposium held at the 8th Congress of the International Council for Archaeozoology (ICAZ), Victoria, British Columbia, Canada, August 1998. British Archaeological Reports International Series 800, Oxford, pp 31–36

Surovell TA (2000) Early Paleoindian women, children, mobility, and fertility. Am Ant 65(3):493–508

Surovell TA, Waguespack NM (2008) How many elephant kills are 14? Clovis mammoth and mastodon kills in context. Quat. Int. 191(1):82–97

Surovell TA, Waguespack NM (2007) Folsom hearth-centered use of space at Barger Gulch, Locality B. In: Brunswig RH, Pitblado BL (eds) Emerging frontiers in Colorado Paleoindian archaeology. University of Colorado Press, Boulder, CO, pp 219–259

Surovell TA, Waguespack NM, Brantingham PJ (2005) Global archaeological evidence for proboscidean overkill. Proc Natl Acad Sci USA 102:6231–6236

Tanaka J (1980) The San: Hunter-gatherers of the Kalahari. University of Tokyo Press, Tokyo

Tankersley KB, Schlecht KD, Laub RS (1998) Fluoride dating of mastodon bone from an Early Paleoindian spring site. J Arch Sci 25:805–812

Testart A (1982) The significance of food storage among hunter-gatherers: Residence patterns, population densities, and social inequalities. Curr Anthropol 23(5):523–537

Testart A (1986) Game sharing systems and kinship among hunter-gatherers. Man 22:287–304

Todd LC (1987) Analysis of kill-butchery bonebeds and interpretation of Paleoindian hunting. In: Nitecki MH, Nitecki DV (eds) The evolution of human hunting. Plenum, NY, pp 371–403

Townsend WR (2000) The sustainability of subsistence hunting by the Sirionο Indians of Bolivia. In: Robinson JG, Bennett EL (eds) Hunting for sustainability in tropical forests. Columbia University Press, NY, pp 267–281

Treves A, Naughton-Treves L (1999) Risk and opportunity for humans coexisting with large carnivores. J Hum Evol 36:275–282

Ugan A (2005) Does size matter? Body size, mass collecting, and their implications for understanding prehistoric foraging behavior. Am Ant 70:75–89

Waguespack NM (2002) Caribou sharing and storage: Refitting the Palangana Site. J Anthropol Archaeol 21:396–417

Waguespack NM (2003) Clovis hunting and the organization of subsistence labor. Unpublished Ph.D. dissertation, University of Arizona, Tucson, AZ

Waguespack NM (2005) The organization of male and female labor in foraging societies: Implications for early Paleoindian archaeology. Am Anthropol 107(4):666–676

Waguespack NM, Surovell TA (2003) Clovis hunting strategies, or how to make out on plentiful resources. Am Ant 68:333–352

Walker DN, Frison GC (1980) The late Pleistocene fauna from the Colby mammoth kill site, Wyoming. Contributions to Geology, University of Wyoming 19(1):69–79

Walker R, Hill K, Kaplan H, McMillan G (2002) Age-dependency in hunting ability among the Ache of eastern Paraguay. J Hum Evol 42:639–657

Webb SD, Milanich JT, Alexon R, Dunbar JS (1984) A *Bison antiquus* kill site, Wacissa River, Jefferson County, Florida. Am Ant 49:384–392

Wiessner P (2002) Hunting, healing, and *Hxaro* exchange: A long-term perspective on!Kung (Ju/'hoansi) large-game hunting. Evol Hum Behav 23:407–436

Winterhalder B (1981) Optimal foraging strategies and hunter-gatherer research in anthropology: Theory and models. In: Winterhalder B, Smith EA (eds) Hunter-gatherer foraging strategies: Ethnographic and archaeological analyses. University of Chicago Press, Chicago, IL

Winterhalder B (1983) Opportunity-cost foraging models for stationary and mobile predators. Am Nat 122(1):73–84

Winterhalder B, Baillargeon W, Cappelletto F, Randolph DI, Prescott C (1988) The population ecology of hunter-gatherers and their prey. J Anthropol Archaeol 7:289–328v

Winterhalder B, Lu F, Tucker B (1999) Risk-sensitive adaptive tactics: Models and evidence from subsistence studies in biology and anthropology. J Arch Res 7(4):301–348

Yates BC, Lundelius EL Jr (2001) Vertebrate faunal remains from the Aubrey Clovis site. In: Ferring CR (ed) The archaeology and paleoecology of the Aubrey Clovis site (41DN479), Denton County, Texas. Center for Environmental Archaeology, University of North Texas, Denton, pp 103–119

6. Ancient DNA and the Genetic Consequences of Late Pleistocene Extinctions

Alex D. Greenwood*
Department of Biological Sciences
Mills Godwin Building, Room 108E
Old Dominion University
Norfolk, Virginia 23529, USA
agreenwo@odu.edu

Keywords Ancient DNA • Pleistocene • megafauna • populations

Introduction

Most of what we can discern about the genetic history of living organisms can only be inferred through analysis of populations that currently exist or existed very recently. The question is whether the genetic endowments of modern populations of a species can be reliably decoded in such a way as to provide an accurate picture of the species' past population history. The field of phylogeography attempts to overlay population genetic associations of current populations and geographic areas to infer the histories of those populations (Emerson and Hewitt, 2005). However, without a temporal component to the analysis, any assumption that current distributions, diversity, or structuring are non-recent events may be erroneous. For example, if one looks at current distributions of a given species and the genetic associations within the existing gene pool, the question is whether this reflects a recent trend or is an ancient pattern established early in the formation of the species. If one were to go back in time hundreds or even thousands of years, would the pattern one sees today be the same or radically different? This issue is particularly acute for properly reconstructing the genetic history of megafaunal species (i.e., body sizes > 44 kg), because the majority of large species in many of the world's mammalian faunas suffered extinction as recently as 40–10 ka. Thus, for species that were extinguished and also for survivor taxa, the genetic consequences would have been dramatic – loss of different genetic lineages and/or reduction of genetic variation within and between populations.

These extinction events happened at different times in different places, and have left behind some tantalizing and at times contradictory clues as to what may have happened. For much of the time period in question there is an accessible climate record, as described in other chapters in this volume (e.g., Cione et al., Chapter 7). There is also an abundant fossil record for many extinct and extant species that span the relevant time periods. For example, mammoth bone collections are extremely large and accessible and new finds occur frequently. This is a great advantage for evaluating one particular aspect of these finds: preserved biomolecules, the study of which has only been feasible within the last 25 years (Willerslev and Cooper, 2005). Although the chemical boundaries of *post mortem* DNA preservation are not limitless, it is well within the time frame encompassing the Pleistocene/Holocene boundary. For some climates or well preserved samples, the boundary extends back even further in time (Lindahl, 1997; Loreille et al., 2001; Barnes et al., 2002). Thus, an area of research has opened up that allows one to study not just morphological variation over time but changes in DNA as well.

Observing the Pleistocene/Holocene transition at a species level, it appears as if there were unfortunate species that became extinct, thus ending their genetic histories and all future contributions to further generations (for reviews see Martin and Klein, 1984). However, the species that survived

* Address for correspondence: agreenwo@odu.edu

G. Haynes (ed.), *American Megafaunal Extinctions at the End of the Pleistocene*, 107–123.

were affected, as well. In some cases they retained the same distributions, shifted distributions, or even increased their distributions. Thus, it would not automatically be assumed that they had been negatively impacted by the end Pleistocene extinctions. However, recent studies of DNA from fossils from multiple species suggest a far more complicated picture than one of an abrupt extinction of particular species with an advantageous or neutral effect on the survivors. In addition, the discovery of species such as the woolly mammoth far into the Holocene on islands indicates the extinction event was far from simple (Vartanyan et al., 1993; Guthrie, 2004).

Studying DNA from fossils is a way to introduce a temporal component to population genetics that has until now been absent. A theme coalescing around this aspect of ancient DNA studies is that the late Pleistocene was a time of great transition for most species, not just those that went extinct (Leonard et al., 2000; Barnes et al., 2002; Shapiro et al., 2004; MacPhee et al., 2005). However, the sub-discipline of ancient DNA studies is plagued by problems and limitations that have made gaining insights into the past rare and sometimes very controversial. Many technical limitations still exist and much effort concentrates on individual specimens to address phylogenetic issues, or on projects whose central questions by their nature can be largely solved by looking at one or a few samples (Gill et al., 1994; Krings et al., 1997; Greenwood et al., 2001b; Orlando et al., 2003).

This chapter will provide a general, though not comprehensive, overview of ancient DNA research, including some relevant aspects of methodology. Emphasis will be placed on the difficulties of this area of research which should impress upon readers the challenges involved in attempting studies at the population level. This will set the context for a discussion of the several major population studies that cover both the region of interest of this volume and the time frame. The focus will be on what the genetics of American megafaunal survivor species tell us and some potential opportunities for further study.

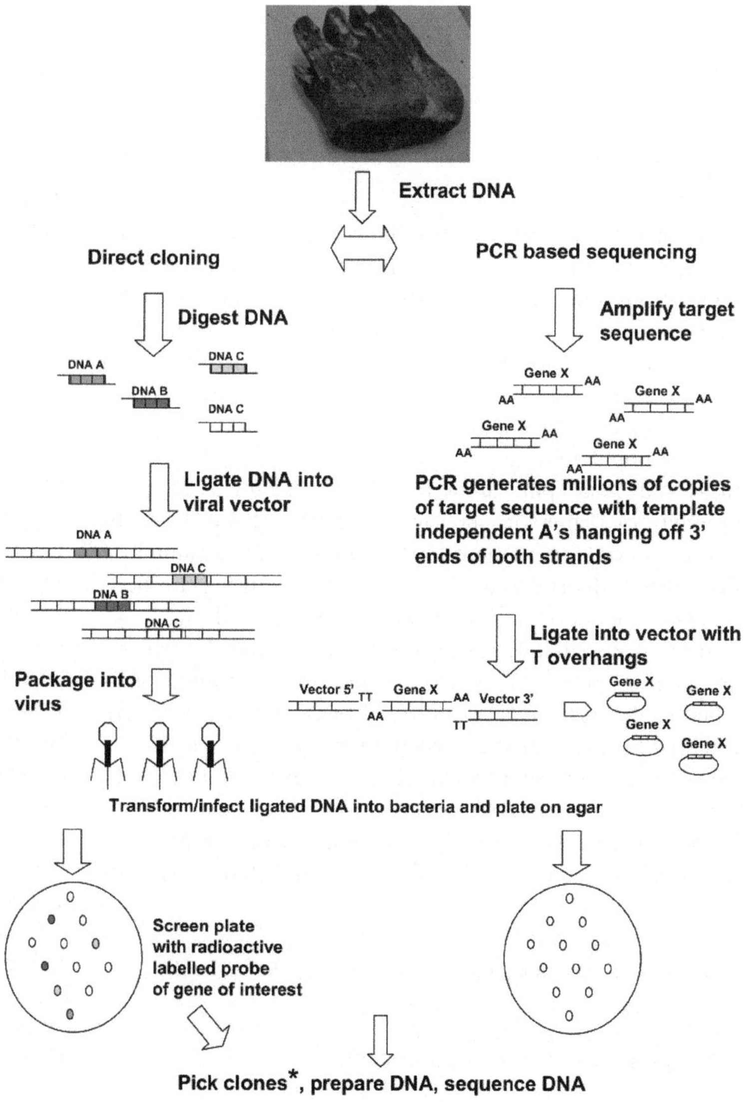

FIGURE 6.1. Direct cloning and screening of inserts with DNA probes versus cloning of PCR products. Both methods begin with the extraction of DNA from a given fossil; in this case a mammoth tooth is pictured. In the case of the non-amplification based approach, the DNA is digested with the same enzyme (s) used to cut an appropriate viral based vector such that the digested DNA will be able to "become a part" of the virus DNA by ligation. The sequences incorporated into the phage in this case are completely random. In the PCR based approach, the *Taq* polymerase enzyme normally leaves non-template derived overhangs of several A bases at the end of each strand which is used to ligate the DNA into the vector using T base overhangs on the vector. The direct method involves using a bacteria infecting virus to propagate the cloned DNA whereas the PCR approach uses bacteria. Both methods benefit from the fact that usually a single ligated molecule is taken up by a single transformed bacterium or infected by an individual phage. Thus, the DNA propagated as the phage or bacteria multiplies is a clone. At the end, in both cases, the cloned DNA is prepared for sequencing. An alternative in the case of PCR screening is to avoid cloning altogether and directly sequence the product. But this is not a favored technique due to several issues specific to ancient DNA that are discussed in the text and in Fig. 6.5.

As for dating and reporting ages of fossil specimens, in this chapter I use the notation "ka" to refer to thousands of years before now, and the notation "kya" to refer to duration in thousands of years. I have avoided using the commonplace "BP" notation (which refers to radiocarbon years Before Present, or before 1950 AD), because a similar notation ("bp") refers to something very different in genetic studies, namely base pairs.

Ancient DNA

The first arguably "ancient" DNA sequences reported in 1984 derived from an extinct zebra relative, the quagga (Higuchi et al., 1984). The approach at the time was to extract as much DNA as possible from a sample (ancient or modern), break it down into small fragments and randomly clone them into a viral vector that could be used to transmit them to bacteria. Individual viruses would receive a single "cloned" piece of DNA which would then be amplified as the viruses reproduced in the bacteria on an agar plate producing more viruses carrying the plasmid. This non-targeted approach meant one had no control over what specific sequence one incorporated and was very inefficient (Fig. 6.1).

Nonetheless, Higuchi et al. (1984) reported two clones of a mitochondrial gene sequence that allowed them to distinguish the quagga from other extant equids.

Exploring fossils for DNA was given a tremendous boost soon after the quagga report by the fortuitous development of the polymerase chain reaction (PCR) which allowed for the amplification of minute amounts of DNA. Prior to this development, methods requiring the presence of large amounts of starting DNA made analysis of older samples prohibitive or, at the very least, rare and to this day unverified (Pääbo, 1985). In addition to the incredible sensitivity of PCR, it allowed for the selective targeting of a sequence of interest instead of the hit and miss approach of random cloning of genomic DNA fragments. Since this development, the field of ancient DNA has progressed enormously, with a wide variety of studies on individual fossil species and genes now on record (Fig. 6.2). However, the tremendous sensitivity provided by PCR has brought its own share of problems to the field of ancient DNA. These problems will be discussed below.

Methodologies

Extracting DNA from a fossil, directly "cloning" the DNA by ligating it into a suitable vector, and screening the resultant clones for the sequences that were captured was the original pre-PCR technique employed in ancient DNA studies (Higuchi et al., 1984; Pääbo, 1985). This was a non specific approach to DNA analysis and was immediately dropped once PCR was developed. Ironically, the advent of high throughput sequencing of short DNA fragments by newer methods has brought a modified version of the original approach back to the forefront (Noonan et al., 2005, 2006; Poinar et al., 2006, Green et al., 2006).

The vast majority of ancient DNA studies use a similar protocol whereby a gene of interest, historically mitochondrial DNA (mtDNA), has been chosen as a target (Fig. 6.1). Mitochondria are organelles in the cytoplasm which convert the energy derived from food into adenosine triphosphate (ATP) which is used by the cell as an energy source for enzymatic activity. Each cell contains multiple mitochondria and each mitochondrion has multiple copies of its ca. 16,000 base pairs (bp) genome. In contrast, the nuclear genome is far larger, about 3 billion base pairs per haploid genome, and each sequence, for non-sex chromosomal sequences, exists in two copies per diploid genome (Fig. 6.3).

Although counterintuitive, this means that mtDNA is effectively more abundant than nuclear DNA since each sequence is represented multiple times as opposed to single

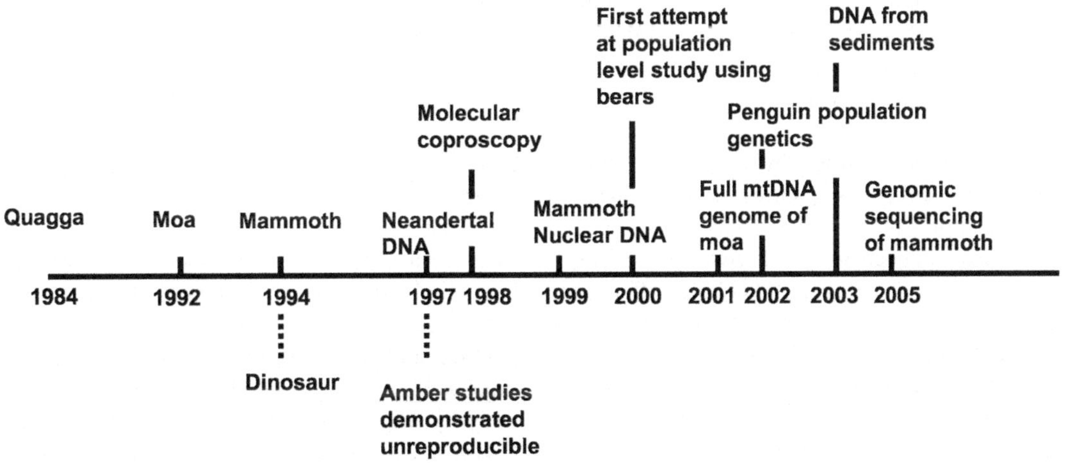

FIGURE 6.2. Selected milestones in ancient DNA research are shown in chronological order. Successes are shown above the timeline and failures below.

FIGURE 6.3. Each stretch of nuclear DNA exists in two copies per diploid (two full sets of chromosomes) genome. The exceptions are some sequences on the Y chromosome of males which exist in only one copy and only in males. In contrast, each cell has multiple mitochondria and each mitochondrion has multiple copies of its approximately 16 kb genome. So even though the nuclear genome is many orders of magnitude larger, there is effectively more of any mitochondrial sequence in a sample than specific nuclear DNA sequences.

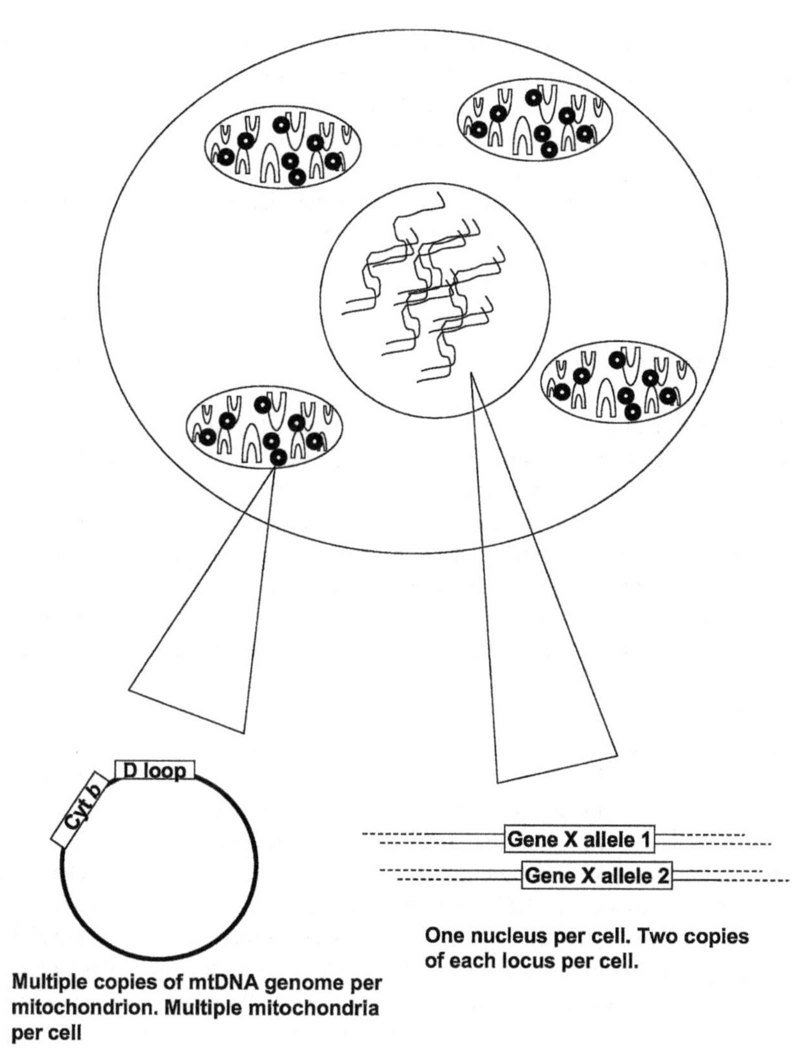

copy nuclear DNA sequences. Considering the *post mortem* degradation of nucleic acids that occurs, a given sample is more likely to "run out" of nuclear DNA that can be retrieved before it runs out of mtDNA. A comparison of mtDNA versus nuclear DNA amplifications from a woolly mammoth is shown in Fig. 6.4. Thus, the majority of ancient DNA work published has relied on mtDNA.

The target sequence is amplified by PCR, generating a product which can be directly sequenced. If one imagines starting with a few copies of a given template molecule that PCR amplifies, direct sequencing will give an "average" sequence for each base if the original templates differ. This has the drawback that if two or more molecules that differ exist in the original template, it may make it impossible to score the bases that vary since the software will detect a signal for more than one base at a given position (Fig. 6.5).

Alternatively, the PCR product generated can be cloned into a vector and a sub-sample of sequences determined from the millions of amplified fragments, given that each bacterial clone represents a single sequence from the millions copied. The advantage of cloning is that a single sequence is determined representing a unique sequence from the PCR. By sequencing multiple individual clones, one can determine if the original PCR product is homogeneous in composition or whether distinct molecules have been amplified, or in other words if there is a mixture of the target sequence and contaminant sequences. This is particularly critical in cases where PCR primers are not species specific and contaminants or artifacts might co-amplify with the sequence of interest. These can easily be sorted out by cloning and sequencing many individual clones, whereas direct sequencing would produce an unreadable sequence.

Once an mtDNA sequence has been determined, it can then be compared to known sequences from putative relatives by alignment much like morphological comparison. Performance of phylogenetic analysis can aid in determining the relationship of the sequence obtained to other distinct sequences.

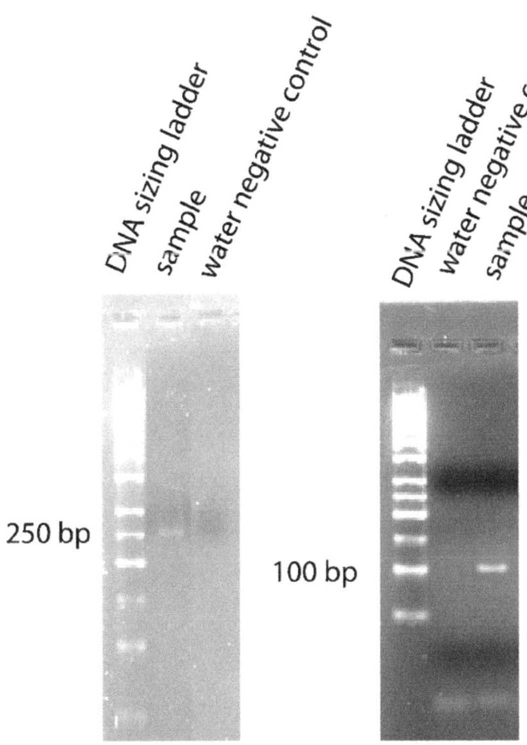

FIGURE 6.4. PCR products are generally resolved on agarose gels that are placed in a buffer and electric current applied. DNA has a net negative charge and migrates downwards towards the positive electrode. The agarose matrix allows smaller fragments to pass through quickly and larger fragments more slowly. By comparing the fragments to a known size standard (shown in the left most lane of each panel), one can determine the size of one's PCR amplified fragments. The agarose gels are stained with an agent (ethidium bromide) which allows for visualization of DNA under UV light. Amplified woolly mammoth mitochondrial DNA and nuclear DNA are shown. The left panel represents an approximately 250 bp mitochondrial DNA fragment amplified from a mammoth. The right panel is an approximately 100 bp nuclear DNA fragment amplified from a different mammoth. A size standard, the negative water control, and the mammoth samples themselves are indicated by lane. For the nuclear gene, the sequence of the product was confirmed by cloning and sequencing of the PCR product.

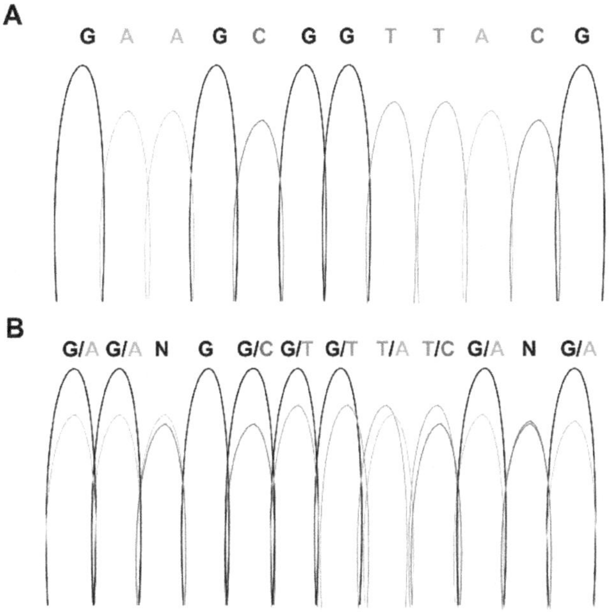

FIGURE 6.5. The figure illustrates a typical readout of a hypothetical sequencing reaction. Each base is marked with a different fluorescent dye which when scanned with a laser distinguishes the four bases. The bases are resolved in a gel matrix based on the same principles as those in Fig. 6.4. However, the gels allow for single base differences to be resolved. Panel A shows a clear legible sequence read. Panel B shows a direct sequence example where either due to damage or the presence of more than one sequence, the scoring of each individual base is hindered or rendered impossible. Panel B is shown to illustrate the perils of direct sequencing ancient DNA.

The Myriad Problems of Working with Ancient Biomolecules

The reality is that after decades of research this simple set of procedures is insufficient for producing authentic ancient sequences for a variety of reasons which are explored here in more detail.

Contamination

Most samples have been handled during excavation or curation in museums and thus are contaminated with human DNA (Sampietro et al., 2006). This is not a trivial issue, particularly for the study of human remains, as the sample itself is contaminated as opposed to the reagents used to process them. This means all experimental controls may be done with the utmost care but reproducible contamination will be detected due to the prior sample handling. Ideally one would excavate the samples in a way non-conducive to contamination introduction but it is impractical and would make using museum collections impossible.

In addition to sample contamination, most molecular biology labs are contaminated to some extent with previously amplified DNA products that can remain stable for long periods of time. Molecular laboratories that employ PCR will likely have aerosolized PCR products present which, depending on the sample being extracted, could contain more DNA copies of a given target than the sample itself. This is a particularly acute problem when one repeatedly characterizes the same region of DNA, such as mtDNA genes from multiple individuals, which also would lead to a build up of the same PCR product in the laboratory and increase the likelihood of cross contaminating samples. Although a variety of PCR controls should always be set up to detect such contamination, they may not always be successful. Thus, it is normal practice to separate rooms for DNA extractions and PCR setup for ancient samples from any modern molecular laboratory work. Other precautions are the use of isolation hoods with an ultraviolet light (UV) source designed to minimize cross contamination by aerosol-borne DNA. The UV source is on when sample processing is not in effect and should destroy DNA present in the hood or on the surfaces of anything placed within. A better alternative to separate laboratories is to separate sample storage and processing from PCR into different buildings. This, of course, adds the expense of duplicating laboratory facilities in another location, but is a useful step for minimizing spurious results.

Low DNA Concentration in the Samples

Most samples will have far less DNA than any contaminating source. *Post mortem* modification of DNA includes cytidine deamination, strand breakage, cross link formation, and, importantly, hydrolyzation of DNA into small fragments (Willerslev and Cooper, 2005). In general terms, the living cell affords DNA the protection necessary to survive by repairing it and preventing a hostile environment from gaining access. As this defence system is energy dependent, the protection is unavailable upon death of the organism. Subsequently, DNA is subject to the environment without any repair mechanisms to correct introduced damage. At the extreme end, there will be no retrievable DNA present, a not infrequent occurrence when studying fossils. More importantly, the low number of amplifiable molecules can lead to errors in sequence determination due to DNA damage.

DNA Damage

Once separated from the myriad mechanisms the cell uses to protect and repair DNA, the nucleic acids in a sub-fossil will be subject to an inexorable degradation which will ultimately result in the complete loss of all endogenous DNA. Thus, an ancient DNA result can be seen as a snap-shot of the process of degradation. If the process is too far along, the result will be impossible to interpret. Most of the DNA in a fossil is fragmented due to strand breaks caused by hydrolysis. One of the most common types of errors is caused by deamination of nucleotides (see Fig. 6.6) They can be broken into two classes, type 1 and type 2 transitions (T to C and G to A or C to T and A to G mutations respectively). The most common are type 2 errors.

The bias is irrespective of whether one looks at mtDNA or nuclear DNA (Binladen et al., 2006). A particularly alarming problem is that the sites in DNA that tend to be hotspots of damage-induced misincorporation tend to coincide with the sites of evolutionary change such that DNA damage can mimic expected changes over time. Thus, great care is essential in distinguishing DNA damage from a *bona fide* polymorphism. An example of historical interest is that the first published ancient DNA sequence (from quagga) exhibited two

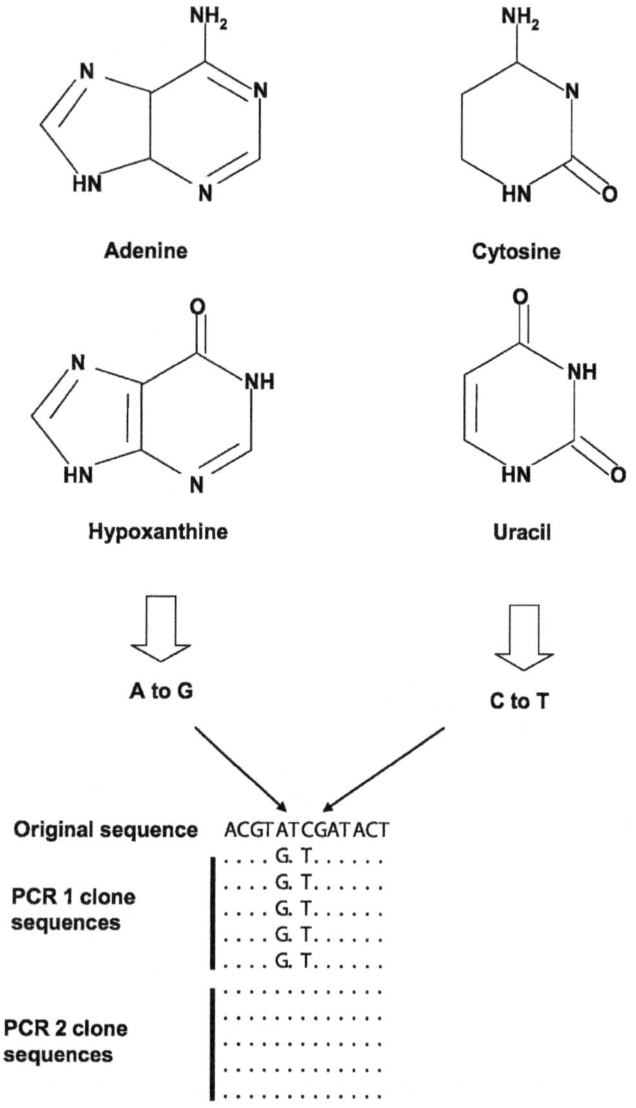

FIGURE 6.6. Two common types of DNA damage associated changes are shown. Oxidation can alter adenine to hypoxanthine which *Taq* DNA polymerase recognizes as a change from adenine to guanine. Similarly, cytosine can be altered to uracil, a base normally found in RNA, which is read as a T causing a C to T mutation. A typical sequencing error pattern is shown. Very often, ancient DNA extracts have such low concentration of target sequence, a mutated template amplified in the early cycles of PCR can dominate the reaction such that most or even all clones give an incorrect sequence because of DNA damage (i.e., PCR clone 1 sequences). A second PCR which will start from a different target molecule may give a different sequence, in this scenario the correct sequence. In such an instance, one must perform multiple rounds of PCR, cloning, and sequencing to determine the actual sequence from the background of damaged molecules on the assumption that multiple independent replications will ultimately reveal the true base composition of the original template since damage is random and should not be consistent from PCR to PCR. Thus, if two out of three PCR's yield a C at a given position, and one a T, the most parsimonious explanation is that the original DNA sequence at this position was a C.

sequencing errors representing both type 1 and 2 transitions (Hofreiter et al., 2001).

Nuclear Insertions of Mitochondrial Sequences (NuMts)

NuMts are pervasive elements that are copies of mtDNA sequences that have entered the nuclear DNA genome(Fig. 6.7).

They are a particularly problematic contamination source as they may be confused with *bona fide* ancient mitochondrial DNA sequence, but may represent modern nuclear DNA contamination. They may even be authentic ancient DNA but represent the incorrect target sequence. This has serious repercussions for phylogenetic analysis (Fig. 6.8).

One of the best examples of the consequences of this problem was the case of much publicized dinosaur DNA sequence which represented a small fragment of the mtDNA cytochrome *b* gene. It turned out upon phylogenetic analysis to be a human NuMt contamination and, thus, the main conclusion of the study (that dinosaur DNA had been retrieved) was falsified (Zischler et al., 1995). Similar problems have cropped up with other types of studies of very old DNA, such as sequences characterized from amber inclusions (Gutierrez and Marin, 1998). The various problems described and a lack of authentication of the sequences by those in the field have led to the general consensus that *bona fide* sequences older than the late Pleistocene have not been retrieved.

Some Proposed Solutions for Working with Difficult Samples

In an effort to combat these problems, standards have been proposed although not universally implemented by any research group, to avoid problems associated with low copy DNA amplification and to authenticate ancient DNA sequences.

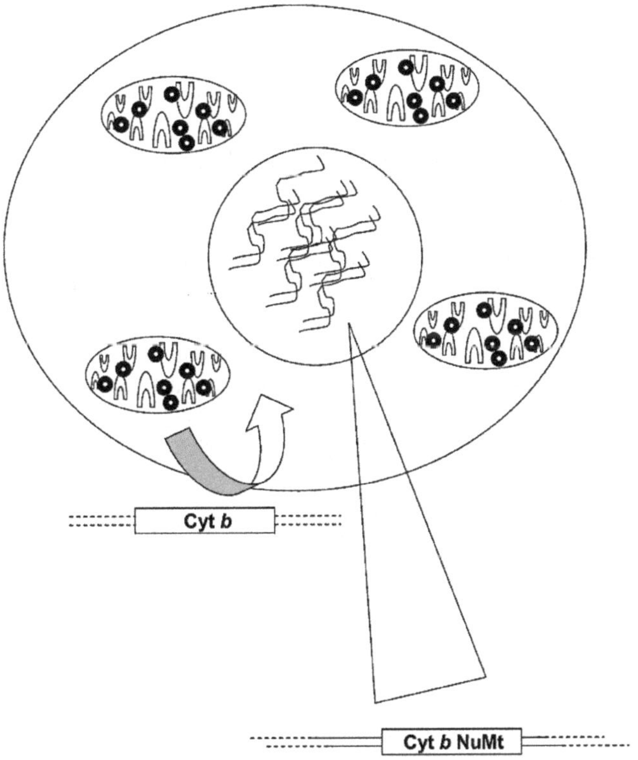

FIGURE 6.7. A common if unwelcome feature of mitochondrial DNA is that it can, from time to time, exit the mitochondria, integrate into the nuclear genome, and become transmitted vertically as a non-functional copy of the original mitochondrial sequence. This has many consequences for using such sequences for evolutionary analysis. First, because of differences in local cellular environment and differences in organellar versus nuclear DNA repair mechanisms, mitochondrial DNA evolves at a faster rate than nuclear DNA for most sequences. Nuclear integrations can be easily mistaken for organellar sequences. Younger integrations may be hard to distinguish from *bona fide* mitochondrial DNA. One consequence is shown in Fig. 6.8.

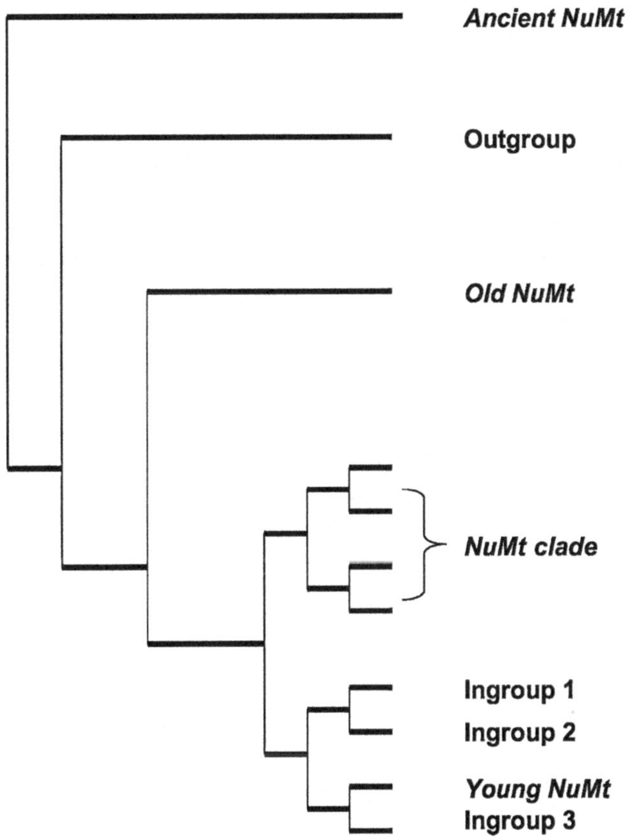

FIGURE 6.8. In the hypothetical tree, old NuMts can be near or even appear older than the outgroup sequence. However, younger NuMts may form distinct clades or appear randomly among non-NuMt clades making phylogenetic interpretation difficult. At worst, one may mistakenly identify a NuMt as mtDNA and obtain an incorrect phylogeny. This phenomeneon was responsible for the misidentification of dinosaur DNA and has been shown to be a serious problem for several other species including some primates and elephants.

Separation of Work on Modern DNA from Ancient DNA

This is a fairly obvious standard, but is not always employed. If one works on mammoths, one should not do mammoth DNA extractions in a laboratory working on the molecular biology of elephants. The modern DNA is an obvious source of contamination that could be hard to distinguish or separate from *bona fide* mammoth DNA. Most ancient biomolecular work requires a room or facility that is separate from modern DNA work and that employs ultraviolet light sources to destroy potential contamination. Much greater precaution is required for analysis of ancient human remains or human relatives such as Neandertals, where ancient DNA sequences may be more similar to human contamination than would be the case with nonhuman but related species. However, in general, the areas where ancient samples are stored, ancient DNA extracted, and PCR setup (but not amplification) performed are either in separate rooms or in a separate buildings away from laboratories employing molecular biology methods.

Molecular Preservation Analysis

Using preservation of ancient biomolecules other than DNA as a proxy for DNA preservation can be regarded as a pre-DNA analysis technique. Thus far, the techniques developed are the analysis of amino acid racemisation or flash pyrolysis followed by gas chromatography or mass spectrometry. Amino acids are chiral molecules and all living organisms use primarily L form amino acids as opposed to the D form used in building proteins. However, *post mortem*, as a function of time and temperature, the L form can be interconverted to the D form. Equilibrium is reached at a 1:1 ratio of D/L amino acids in a sample. This can be used as a proxy for preservation of DNA as the rate of racemisation and DNA depurination appear to be correlated (Poinar et al., 1996). However, the technique has not been applied widely and the range of racemisation ratios that truly indicate whether or not preservation of DNA is possible has not been determined. A similar problem exists for flash pyrolysis which provides a robust analysis of the molecular contents of a sample which can be compared between ancient and modern samples (Poinar and Stankiewicz, 1999). It indicates the relative amount of protein hydrolysis, which can be used as a proxy for overall biomolecular preservation in a given sample. This method has not been widely applied and is not generally available in molecular labs. Similarly, a correlation between DNA preservation and collagen preservation has been observed (Götherström et al., 2002). However, like amino acid racemisation, the boundaries need to be better defined before it can be used as a screening method for samples containing ancient DNA. Until then, it is still more economical to brute force screen multiple samples by extraction and PCR and pay the cost of working on a large number of samples that are devoid of DNA.

Sequencing of Multiple Clones

Because of DNA damage and NuMts, it is usually necessary to perform PCRs in short overlapping fragments (if mtDNA is chosen) and to clone and sequence multiple clones per PCR. In addition, clones from multiple PCRs are necessary (whether mtDNA or nuclear) to assure that mis-scoring of bases does not occur (Fig. 6.9).

Overlapping PCR fragments are useful as it is unlikely that two independent PCR primer pairs will detect the same NuMt. However, this is not always the case (Greenwood and Pääbo, 1999). Recent additional suggested tests include quantitative PCR analysis to determine the number of molecules

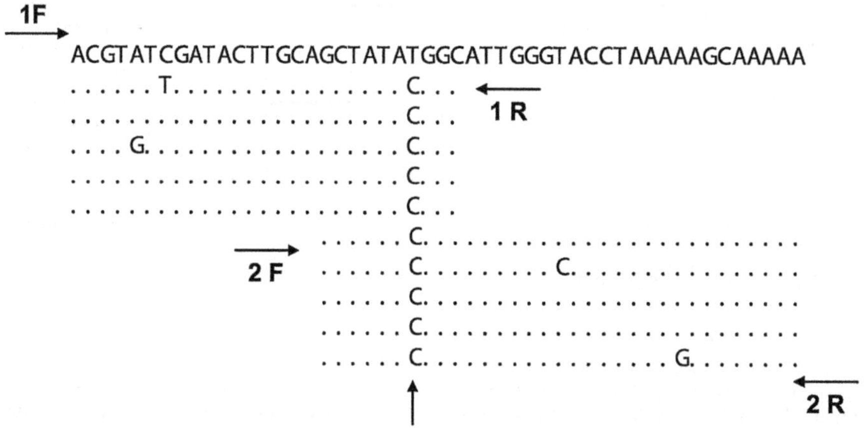

FIGURE 6.9. Overlapping PCR products generated by two independent sets of PCR primers can be used to attempt to exclude NuMts and to confirm sequences in the overlapping region. The region of overlap is determined from two completely independently derived PCR reactions and the chances that one has designed primers that pick up an NuMt, though not zero, are lower than with a single PCR primer pair. In the diagram, two overlapping PCR products show that a T to C change from the reference sequence appears in all clones from two independent PCR reactions with independent primers. This strongly suggests that C is the correct base at that position. Random damage induced among clone variation is also seen but does not interfere with determining the sequence as it appears randomly among clones as opposed to the consistent T to C change in the overlap.

of endogenous DNA per gram of tissue in a given sample. This can indicate whether or not a sample is likely to yield errors due to miscoding lesions (Poinar et al., 2003). Direct sequencing of PCR products should be avoided as miscoding lesions and contamination could mask the correct sequence at a given position and inflate the divergence between the sequence obtained and related sequences to which it is being compared. In the case of nuclear gene sequences, the presence of distinct alleles may not be detected by direct sequencing and valuable information may be lost (Greenwood et al., 1999).

Independent Reproduction of Results

This is a compelling way to ensure that the results obtained are authentic. In its best form, an independent sample (one never having been exposed to the first lab) is sent to a second lab for extraction, PCR amplification, and sequence determination. If the result is confirmed, there is a much higher confidence that the sequences determined are authentic. Two independent labs would not be expected to get the same results by chance or be exposed to identical contamination risks. This standard is not followed nearly as often as it should be. However, particularly in the case of human remains, contamination of the actual sample with DNA – i.e., by handling – will be reproducible and should be taken in context with the other authentication standards.

Given the preceding suggested standards, one should be very sceptical of reports of very old DNA (i.e., over 100 ka from climates non-conducive to DNA preservation) that have not been replicated, report odd phylogenetic placements, or report exceptionally long PCR product retrieval. In addition, reports of less stable molecules such as RNA from ancient samples or microbial sequences from samples where the microbe in question has free living relatives should also be viewed with extreme scepticism. Although less frequently than in the 1980s and 1990s, poorly executed ancient DNA studies (i.e., those ignoring most or all of the authentication standards) do still appear in the literature. Nonetheless, progress has been made and the first population genetic level studies and even genomic level studies are beginning to appear. In addition, a great deal of effort has been made in characterizing the types of damage one can expect from fossil DNA which facilitates the interpretation of the data. Given the low chance of success with any given sample, the rigorous methodology involved in authenticating sequences, and the expenses involved, most studies have focused on one or a few individual samples. With such low sample sizes, the vast majority of studies have focused on resolving phylogenetic issues associated with a sample (e.g., sloths, Neandertals [Greenwood et al., 2001b; Poinar et al., 2003; Krings et al., 1997]) or in characterizing the DNA damage observed in given samples (Mitchell et al., 2005; Binladen et al., 2006). While such studies have yielded interesting results, the scope of the types of questions asked has been limited. However, there have been a few exceptions.

Examples of Applications Beyond Phylogenetic Issues

Coprolites

Since DNA of organisms is preserved, it stands to reason that the contents of their diet will also be preserved if coprolites are found. This is exactly the case for a large sample of sloth coprolites from dry caves in both North America and Chile. Using a modified DNA extraction protocol that frees DNA from sugar-crosslinked proteins, Poinar et al. (1998) were able to simultaneously determine the plant content of a sloth coprolite and that the coprolite was defecated by a Shasta ground sloth (*Nothrotheriops shastensis*). Morphological studies of coprolites have been performed for decades; however, digestion and decay of the samples over time can obscure some of the morphological features of the ingested materials. In the Poinar et al. (1998) study, one family of plant was not identified by DNA analysis but observed via morphological analysis. However, four plant groups were identified by molecular methods that were not detected morphologically. Thus, the new approach allows one to identify the species that generated the coprolites and at the same time extend the range of plants that can be found in the sample. The technique was extended to look at changes in diet for sloths over 17.5 kya allowing for a temporal component to be added to the analysis (Hofreiter et al., 2000).

This type of work has been used to analyze the diet of humans from Hinds Cave in Texas from 2 ka (Poinar et al., 2001). A richer diet of plants and animals were detected from the paleofecal sample than morphological analysis yielded and the human DNA in the sample demonstrated a clear affinity to Native American mtDNA haplotypes. Thus, the origin and contents of human feces are also accessible to ancient DNA study.

Environmental Samplings

Ancient DNA retrieval from sediments has been performed (Willerslev et al., 2003). The motivating idea is that enough ancient biomolecules have been left behind in the soil to obtain information where no morphological information remains. Samples cored from permafrost in various localities in Siberia and from two temperate caves in New Zealand were analyzed for both plant and animal DNA sequences. Morphological data were almost entirely absent, and thus the study relied primarily on molecular data.

Multiple plant and animal species were detected, differing according to the environment and age (e.g., mammoth and bison sequences from Siberia and moa from New Zealand). Of note, the plant diversity changed dramatically in the late Pleistocene. Around the time of the Last Glacial Maximum (LGM) 25 ka, plant diversity was relatively low. It increased during the late Pleistocene subsequent to the LGM and then abruptly changed at the Pleistocene/Holocene boundary where grasses were replaced by sedges. Though the number of samples studied is relatively limited, the information content is interesting. Even with just seven cores, spanning 0 to 400 ka, a large amount of genetic data can be sampled in one experiment and changes over time for multiple plants and animals subsequently determined. Of interest, the record for plants extended as far back as 400 ka although for animals it was much more restricted. If other localities are amenable to this type of study it could greatly enhance the understanding of the many changes that were occurring in plant and animal life at this time.

Disease

Many studies of ancient pathogens have been unconvincing. The studies of tuberculosis from ancient faunal and human remains, for example, suffer from the fact that there are free living bacteria that are genetically similar. In addition, the authentication criteria were rarely or unevenly followed. In the case of a study claiming to have detected *Yersinia pestis* (the plague bacteria), attempts at independent confirmation failed (Gilbert et al., 2004). Finally, in most studies, many samples were tested for the presence of pathogens but not tested for host DNA to confirm that the results were in keeping with the overall biomolecular preservation of the samples. For example, it would be highly unusual to detect a pathogen that exists at less than single copy DNA concentrations in a given tissue but fail to detect nuclear or mtDNA from the host in the same sample. In some reports, however, this was the case. In other cases, testing for the presence of endogenous human DNA was not done at all. In addition, the overwhelming use of direct sequencing in these studies may mask DNA damage induced polymorphisms which could obscure the true source of the sequences by magnifying or minimizing differences between the sequences obtained and the reference sequences used. Finally, the lack of knowledge of the microbial content of surrounding soil environments will always prove challenging to those looking for microbes with potential free living relatives. For example, retrieving a previously uncharacterized sequence from an ancient sample could merely represent a previously uncharacterized modern contaminant. Until these issues are better resolved, a large segment of the ancient DNA community will remain sceptical of claims that reported sequences from microbes in sub-fossils are authentic. Encouragingly, those studying tuberculosis from ancient remains are beginning to take steps towards authenticating their results. Testing of multiple markers to narrow down the specific bacterial species being analyzed and replicating of results in independent laboratories are starting to appear in the literature (Donoghue et al., 2004). If it becomes a general practice, then the analysis of ancient pathogens could be a major benefit to the study of disease-causing agents. However, somewhat discouragingly, a recent study that attempted to identify tuberculosis and syphilis DNA from 59 samples using established criteria failed to identify any pathogen DNA (Barnes and Thomas, 2006) – this in spite of the fact that 20% of the individuals from the collections tested were known to have died from syphilis and tuberculosis respectively. Thus, this field of study remains both difficult and controversial.

Ironically, the disease studies that best meet the established criteria for authenticity as "ancient" have surprisingly involved an RNA virus, influenza A. RNA is much less stable than double stranded DNA and is not expected to exist in even the youngest of ancient remains. The samples studied were arguably not truly ancient as they dated from the 1918 flu pandemic and involved paraffin fixed tissue samples for the most part. However, influenza sequences were also obtained from corpses interred in graves in permafrost in Alaska. The entire 1918 flu genome has been painstakingly reconstructed and has yielded marvellous insights into the biology of this critical pathogen (Taubenberger et al., 2005).

Population Studies: Genetics and American Megafaunal Extinctions

Ancient DNA has been applied to population level analysis for several species including penguins and pocket gophers (Hadly et al., 1998; Lambert et al., 2002; Shepherd et al., 2005). However, very few have dealt with the time period around the Pleistocene-to-Holocene transition. It should be evident at this point that, given the intrinsic difficulties of working with ancient DNA from one or a few specimens, population studies are extremely difficult and time consuming, particularly so in light of the rigor with which each individual sample must be examined to exclude contamination and NuMts while ensuring that data are reproducible. It is therefore unsurprising that the number of studies addressing population level questions is small. However, there are several examples in the literature pertinent to the question of American megafaunal extinctions.

Genetic Consequences of Late Pleistocene Extinctions: Brown Bears

Bears, both brown bears (*Ursus arctos*) and European cave bears (*Urus spelaeus*), have been a focal point of a great deal of ancient DNA research. *Ursus arctos* is the focus of this section as the time points of study for cave bears is prior to the Pleistocene/Holocene transition and they were exclusively European in their distribution. However, several population level studies have been done and readers should refer to them if interested (Lorielle et al., 2001; Orlando et al., 2002; Hofreiter et al., 2002).

Brown bear phylogeography has been extensively studied with over 300 bears genotyped for mitochondrial control and/or hypervariable region sequences (Waits et al., 1998). Overall, modern brown bear populations appear to group into three clades with one containing two subclades (clades or groups 2, 3a, 3b and 4). From the modern distribution of the clades in Alaska and Canada, one would surmise that the clades represent distinct conservation units that are ancient in origin. However, ancient DNA studies of brown bears suggest this is not the case at all (Fig. 6.10).

The first indication that something is amiss with the hypothesized evolutionary scenarios regarding bears came from a limited study of seven sub-fossils from Beringia

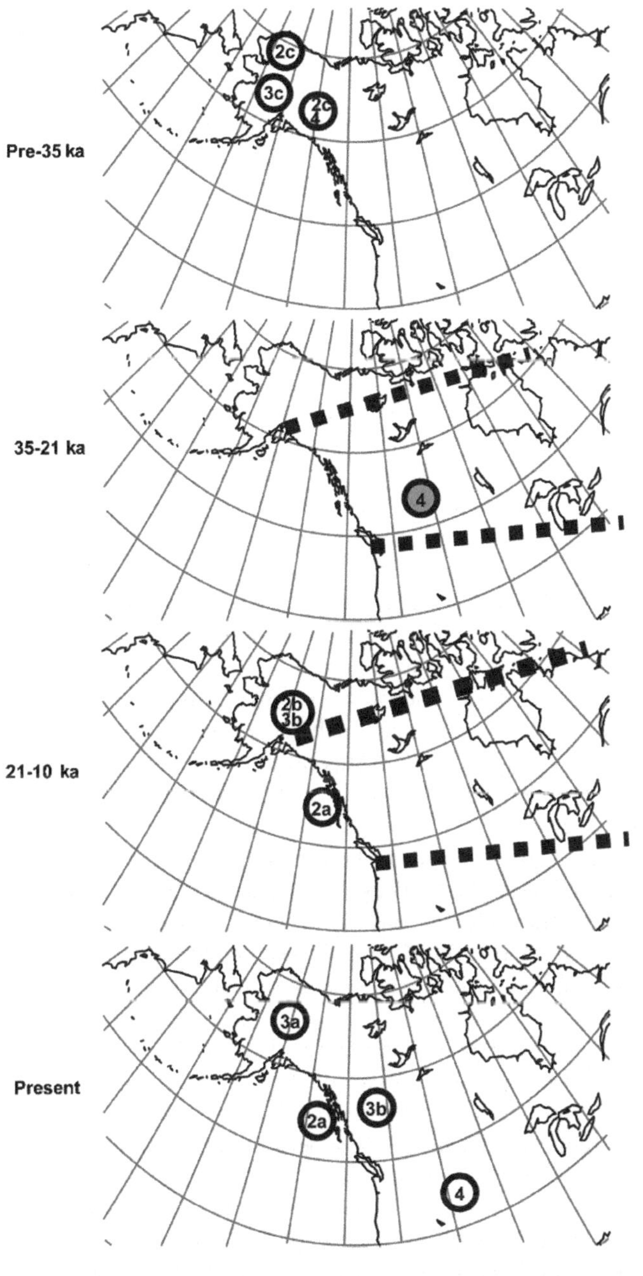

FIGURE 6.10. Brown bear mitochondrial DNA clade distributions before 35 ka, between 35–21 ka, 21–10 ka, and the present are shown in individual panels. The focus area for ancient bears is within Beringia and North America. Circles are used to highlight clades on the map and do not indicate clade boundaries. The clade 4 sample circled in grey indicates a recently discovered ancient clade 4 individual from Canada near Edmonton dated at 26 ka. Dashed lines indicate very roughly the positions of the ice coverage from 21 ka to 18 ka. The base map was created using the Paleobiology Database plotting software (Alroy, 2006).

spanning a time of 14 to 42 ka. It was suggested that brown bears entered Beringia 50–70 ka and then entered the central part of North America around 13 ka. The seven samples fell into three groups but with temporal differences. Thus, clade 4 was only found at 43 ka. Clade 3 was found in the youngest samples at around 14–17 ka, and clade 2 was found in both 35–45 ka specimens and in the current distribution but restricted to the ABC Islands. Clades such as clade 2 and 4 which are currently absent in Alaska preceded the LGM whereas clade 3b which still exists in the area had established itself by at least 15 ka. The partitioning of the different clades was a recent event that preceded the end Pleistocene extinctions by about 5 kya.

However, seven samples are too few from which to draw many conclusions, especially in light of the limited sampling area and the comparison of only seven samples to over 300 modern representatives. This was partially rectified by another study that included 71 bear fossils, 36 of which yielded ancient DNA and spanned a 60 kya record. Prior to 35 ka, several extinct clades were present. There was then a gap in the record from 35 to 21 ka when bears were absent possibly due to competitive exclusion with the short faced bear, *Arctodus*. From 21 to 10 ka a new subclade of clade 3 (3b) dominated, which then changed in the Holocene to the two current subclades 3a and 3b.

What the results suggest is that throughout the time examined, bear populations were highly genetically structured (Fig. 6.10). The current clade structure is a recent phenomenon. One addendum to the scenario is that recently a clade that only appeared in the pre-35 ka Beringia populations and then was absent until the Holocene where it is found in Canada, clade 4, has been found in a 26 ka specimen from Canada. This would explain the origin of modern clade 4 animals which was difficult to explain from the larger study. It also illustrates that although the population study represents a great effort, there are still many issues of insufficient sampling that hinder the development of a complete story of Pleistocene bear genetic diversity and structure. Moreover, since the work has focused on Beringia, the origin of the different clades and the magnitude of the changes in other regions where the Holarctic species brown bear existed is unclear. However, the conclusions demonstrate that current distributions and trends may not reflect ancient events.

Genetic Consequences of Late Pleistocene Extinctions: Steppe Bison

Extant bears are carnivores and/or omnivores and do not reach population densities as high as many herbivores. However, the end Pleistocene extinctions affected a great many megafaunal herbivores such as woolly mammoths. Given the differences in population sizes and life history traits, it would be of interest to know what happened to survivor species with an exclusively herbivorous life style. Steppe bison (*B. priscus* and *B. bison* were considered the same species in the studies covered here due to genetic similarity) were abundant in Beringia with a fossil record extending back at least 300 kya (McDonald, 1981). Different populations of bison were occasionally separated from one another by regional ice barriers. Of particular interest, due to overhunting of bison in the United States 150 years ago, bison suffered a severe population bottleneck. Thus, a historical event with known genetic repercussions is an excellent control for events that happened much deeper in the past for which no firsthand documentation exists.

A total of 442 fossil bison representing Alaska, Canada, the lower 48 states of the United States, Siberia, and China were sampled. Of the 442 samples, 220 samples were radiocarbon dated. A total of 685 base pairs (bp) of the mitochondrial DNA control region was amplified and sequenced from the samples. Three hundred fifty two of 442 samples yielded DNA and of those only 328 yielded the entire 685 bp fragment which was derived from amplification and sequencing of smaller overlapping PCR products. Sequence data were independently reproduced in a separate laboratory for 16 samples (Shapiro et al., 2004).

The results suggested the most recent common ancestor of modern bison lived 136 ka. Gene flow among populations in Beringia was apparent from ~60 ka until 25 ka. Subsequently, the Laurentide and Cordilleran ice sheets interrupted north-south gene flow (Fig. 6.11).

When the ice free corridor began to appear near the end of the Pleistocene, gene flow resumed. However, all modern bison are related to those of a single clade, deriving from populations to the south of the ice sheets in Canada and with a most recent common ancestor estimated to have lived between 22–15 ka (Shapiro et al., 2004). Thus, Holocene bison genetic diversity is restricted to the last 12 ka which overlaps with the wave of extinctions of megafauna that was in effect at the Pleistocene/Holocene transition.

Using statistical modelling, the demographic history of bison was calculated over the span of radiocarbon dates obtained. The model suggests bison populations were expanding until 37 ka when they began to decline. As a control, bison from the last 1 ka were analyzed separately, demonstrating expansion followed by an abrupt decline associated with overhunting in the late 1800s. The age of 37 ka is consistent with environmental changes at the time including reduction of the steppe environment and expansion of forests during a warm cycle (Anderson and Lozhkin, 2001). A re-analysis of the data suggests that although there was a decline that preceded the end Pleistocene extinctions, in fact there was an abrupt decline in genetic diversity of bison around 10 ka that was not observed in the original analysis (Drummond et al., 2004). The timing in this case does coincide with human arrival and the extinction of end Pleistocene non-survivor species. The Pleistocene/Holocene transition bottleneck in bison may reflect effects of the cause of the end Pleistocene extinctions on this survivor species.

Several aspects of the study are of particular interest. Almost none of the Pleistocene genetic diversity has survived in modern bison. According to the analysis of Drummond et al. (2004), there was an abrupt decline at the Pleistocene/Holocene transition accelerating a decline that apparently was in progress since at least 37 ka. This study also shows that the end Pleistocene extinctions were likely complex as the data for bison do not completely correlate with that of brown bears. The bear genetic structure changed approximately 5 kya earlier than the bottleneck in bison. However, like the bear work, the bison study exemplifies the fact that measuring current genetic diversity and genetic structure of species may not be representative of the diversity that existed as recently as 10 ka, since for many species, particularly megafauna, the diversity was largely erased or, in the case of mammoths for example, wiped out entirely.

Genetic Consequences of Late Pleistocene Extinctions: Muskox

Another survivor species is the muskox (*Ovibos moschatus*). Having originated in Eurasia, the muskox achieved a Holarctic distribution by the early Pleistocene. Its current natural distribution is restricted to the Arctic archipelago and Greenland (Fig. 6.12). The restricted range is also reflected in the genetic diversity of modern *Ovibos* which is very limited according to the few analyses on record.

Modern muskox genetic diversity is notable by its absence. Of 35 individuals tested for mitochondrial hypervariable region sequences, only eight distinct haplotypes were observed and these differed by a maximum of 1.3% (Groves, 1997). The percentage of variable sites was only 1.4% whereas in many other species it was much higher, e.g., *Bos taurus* at 7% (see table 4 in Groves, 1997). This result was obtained despite the fact that two putative sub-species of muskox (*O. moschatus moschatus* and *O. m. wardii*) were examined. Nuclear DNA yields a similar result as both highly variable microsatellite loci and sequence analysis of a major histocompatibility gene demonstrate highly constrained genetic diversity (Holm et al, 1999; Mikko et al., 1999). The conclusion from studies of modern muskoxen is that they lack significant genetic diversity.

A number of scenarios could explain this finding. Muskoxen could have always maintained a low effective population size and therefore suffered from restricted genetic diversity as a consequence. Generally low population densities would be expected in the extreme Holarctic climate for which they are specialized. There could have been a long and slow decline of genetic diversity since the Early Pleistocene. Alternatively, there could have been one or several abrupt population crashes that have restricted muskox genetic diversity to its present levels.

Several lines of evidence suggest that muskox diversity may have coincided with the decline and extinction of other megafaunal species. An analysis of the distribution of 188 radiocarbon dates, based on fossils from the Taimyr peninsula and covering woolly mammoth (*Mammuthus primigenius*),

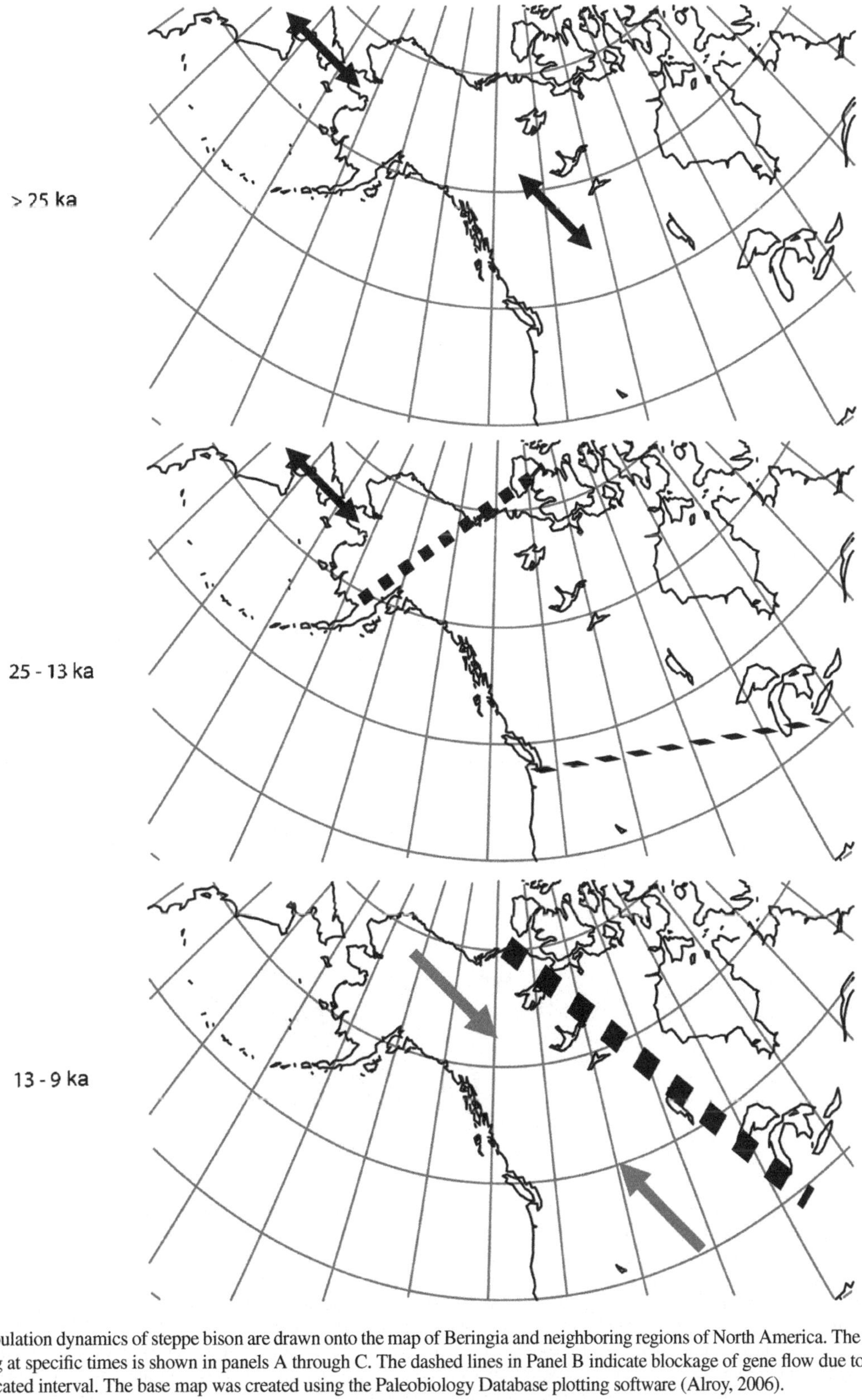

FIGURE 6.11. Population dynamics of steppe bison are drawn onto the map of Beringia and neighboring regions of North America. The regions where gene flow was occurring at specific times is shown in panels A through C. The dashed lines in Panel B indicate blockage of gene flow due to ice sheets during a portion of the indicated interval. The base map was created using the Paleobiology Database plotting software (Alroy, 2006).

steppe bison (*Bison "priscus"*), moose (*Alces alces*), reindeer (*Rangifer tarandus*), horse (*Equus caballus*), wolf (*Canis lupus*) and muskox, suggest that a major event occurred at the Pleistocene/Holocene transition which affected all large species, not just those that failed to survive in Taimyr beyond the end Pleistocene (MacPhee et al., 2002). The record for all species is fairly continuous until 9–10 ka, at which point there is a gap for all taxa represented by significant numbers of dated

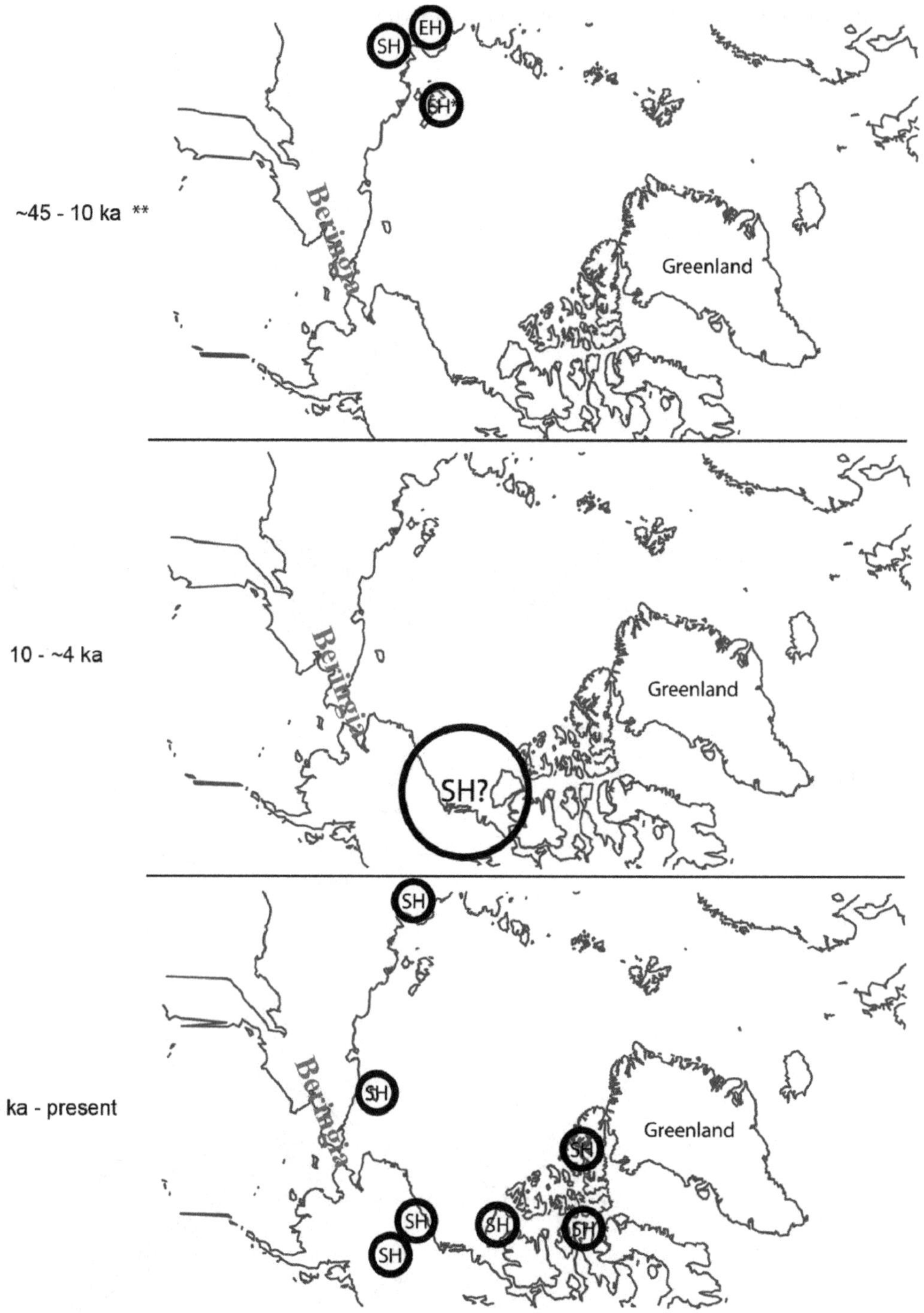

FIGURE 6.12. *Ovibos* haplotype dynamics in Eurasia and North America are shown. From approximately 45 to 10 ka, both a mitochondrial haplotype that is now extinct (extinct haplotype, EH) and mitochondrial haplotypes that currently exist (surviving haplotype, SH) in the muskox standing herd occurred in Eurasia. Because of ice sheet coverage, the North American muskox fossil record is extremely sparse probably representing exclusion of large populations from these areas prior to the Holocene. The asterisk "*" indicates that the one sample from the Bol'shoi Islands that yielded DNA was only partially characterized for mitochondrial sequence and for one region of the genome, not reproducible. Thus, it is questionable as to which haplotype existed in this locality. The double asterisk "**" indicates that the earliest dated muskox finds in Taimyr are circa 10 ka. However, DNA was not retrieved from the youngest Pleistocene samples and the last observed occurrence of the EH was circa 15 ka. The haplotypes from the few 10,000 year old samples is unknown. From 10 ka to approximately 4 ka, there is a gap in the fossil record for muskoxen in Taimyr. It also exists for other species and thus likely represents their extirpation from this region (see MacPhee et al., 2002). No DNA was retrieved from the limited number of fossils dating to this period in North America. However, considering no sample has yielded EH from North America, it is possible that during this time the animals in this location were SH. From 3,790 to the present muskoxen populations were on the rise in North America as evident from their increased representation in the fossil record for this time period. Extant populations are largely restricted to Alaska, Canada, and Greenland and all bear SH's. Muskoxen also briefly reappeared after a fossil record gap of about 7kya years in the Taimyr Peninsula and on Wrangel Island at about 3,790 before disappearing for good from the region. These animals also carried SH. The base map was created using the Paleobiology Database plotting software (Alroy, 2006).

fossils (mammoth, horse, bison, and muskox). The gap in the record extends over a period of roughly 4 kya; thereafter, horse and muskox reappear, whereas mammoth and bison do not and are presumed to have become at least regionally extinct. Subsequently, muskox and horse permanently disappear from the record *ca.* 2 ka, in line with their absence from these areas in historical times. Thus, of the three alternatives suggested previously, the best supported is that muskoxen were eliminated from the Taimyr Peninsula at the same time as other megafaunal species. The later dates may represent recolonization events from an as yet unknown refugial population.

To further probe this scenario, MacPhee et al. (2005) undertook additional genetic studies of several of the muskox fossils that span both sides of the extinction gap, as well as ancient musk-ox samples from localities in Russia beyond Taimyr and several localities in the New World (Alaska and western Arctic Archipelago of Canada). Cyt *b* and hypervariable region sequences were determined for 16 samples including a nearly complete cyt b sequence from one of the Taimyr samples (Fig. 6.12). The results demonstrated that all Taimyr samples predating 10 ka contained haplotypes distinct from all other muskoxen tested to date. All other samples yielded sequences identical or nearly identical to those of modern muskoxen. Although sample sizes were low, the diversity among Taimyr musk-oxen appears to have been greater than that represented in modern muskox populations as recently as 18 ka. The late dated Taimyr muskoxen were identical to modern. However, the modern haplotype was found in non-Taimyr samples older than 10 ka (i.e., Wrangel Island at 22 ka). Thus, it appears that muskoxen suffered a loss of genetic diversity associated with the end Pleistocene extinction event. However, it should be noted that, given the limited number of genetic data points, one cannot distinguish between an abrupt decline 10 ka or a more gradual decline starting 5 kya earlier. Still, the radiocarbon data for Taimyr suggest that the most abrupt decline occurred during the Pleistocene/Holocene transition (MacPhee et al., 2002). Regardless of the timing, several haplotypes, including the most common modern haplotype, existed before the end Pleistocene, but only the modern haplotype survived into the Holocene.

Thus, muskoxen represent a surviving group that was affected by the transition from the Pleistocene to the Holocene much like bison and bears. The current genetic diversity, or lack thereof in muskoxen, does not represent the total diversity that existed relatively recently and again suggests that the cause of the end Pleistocene extinctions had a broader impact beyond the species that became extinct.

Genetic Consequences of Late Pleistocene Extinctions: The Animals that Went Extinct

The late Quaternary population histories of American megafaunal species that became extinct remain largely unknown. Although multiple mitochondrial sequences (mainly for cyt *b*) from woolly mammoths have been determined, little differentiation has been observed (Rogaev et al., 2006). However, cyt *b* is not ideal for such a study as it does not evolve fast enough to be useful for the time frame of interest and because of the problems associated with NuMts in elephants (Greenwood and Pääbo, 1999). Nuclear DNA sequences might resolve the issue but have not yet been attempted at a population scale. However, several reports on mammoth nuclear DNA suggest that such a study is feasible (Greenwood et al., 1999, 2001a; Poinar et al., 2006; Capelli et al., 2006; Rompler et al., 2006). Additional reports of large scale nuclear DNA sequencing from a Neandertal sample also suggest that genomics research may widen the number of available markers for population genetic studies of extinct species (Green et al., 2006; Noonan et al., 2006).

While several reports on the ancient DNA (including nuclear DNA) of species of giant ground sloth have been published, no population genetic studies have been attempted on this extinct group. This is also true for other American megafaunal species, although some Eurasian species have been briefly surveyed (e.g., *Megaloceros*, Lister et al., 2005). More effort has gone into determining the genetic histories of fauna in the Holocene including several studies of penguin and vole population genetics. Thus, an area of future study will be to determine if taxa that became extinct at the end of the Pleistocene shared any genetic or populational features that might have predisposed them toward collapse.

Concluding Remarks

While it is clearly a difficult task to retrieve ancient DNA from Holocene or late Pleistocene samples, given the right conditions and using proper precautions, insights into the population histories of species are both possible and important. It is clear that all models of the end Pleistocene extinctions must factor in the somewhat discordant genetic data regarding survivor species, all of which indicate that current genetic structure is of recent origin and that as recently as 10 ka the genetic structure measurably differed. Unanswered questions are whether species that did not survive the end Pleistocene extinctions demonstrated a similar pattern of genetic changes or whether something unique happened to them. But for the moment, it appears that many if not all megafaunal species were affected by different factors for an extended period of time in the late Pleistocene both before and after humans presumably arrived. For reasons still unknown, some species survived in an altered state whereas others were prevented from passing their genes on to subsequent generations.

Acknowledgments. I wish to thank S. Kate Lyons, Martha Stokes and Kelly Wyatt for helpful comments on early versions of the manuscript. Martha Stokes generated the agarose gel data in Fig. 6.4. I also wish to thank Ross MacPhee for our continuing collaborations and his support.

References

Alroy J (2006) Plotting software. Paleobiology Database. Available at: http://paleodb.org/cgi-bin/bridge.pl?user=Guest&action=display MapForm

Anderson PM, Lozhkin AV (2001) The Stage 3 interstadial complex (Karginskii/middle Wisconsinan interval) of Beringia: Variations in paleoenvironments and implications for paleoclimatic interpretations. Quat Sci Rev 20(1):93–125

Barnes I, Matheus P, Shapiro B, Jensen D, Cooper A (2002) Dynamics of Pleistocene population extinctions in Beringian brown bears. Science 295(5563):2267–2270

Barnes I, Thomas MG (2006) Evaluating bacterial pathogen DNA preservation in museum osteological collections. Proc Roy Soc B, 273(1587):645–653

Binladen J, Wiuf C, Gilbert MT, Bunce M, Barnett R, Larson G, Greenwood AD, Haile J, Ho SY, Hansen AJ, Willerslev E (2006) Assessing the fidelity of ancient DNA sequences amplified from nuclear genes. Genetics 172(2):733–741

Capelli C, MacPhee RD, Roca AL, Brisighelli F, Georgiadis N, O'Brien SJ, Greenwood AD (2006) A nuclear DNA phylogeny of the woolly mammoth (*Mammuthus primigenius*). Mol Phylogenet Evol 40(2):620–627

Donoghue HD, Spigelman M, Greenblatt CL, Lev-Maor G, Bar-Gal GK, Matheson C, Vernon K, Nerlich AG, Zink AR (2004) Tuberculosis: From prehistory to Robert Koch, as revealed by ancient DNA. Lancet Infect Dis. 4(9):584–592

Drummond AJ, Rambaut A, Shapiro B, Pybus OG (2005) Bayesian coalescent inference of past population dynamics from molecular sequences. Mol Biol Evol. 22(5):1185–1192

Emerson BC, Hewitt GM (2005) Phylogeography. Curr Biol 15(10):R367–371

Gilbert MT, Cuccui J, White W, Lynnerup N, Titball RW, Cooper A, Prentice MB (2004) Absence of *Yersinia pestis*-specific DNA in human teeth from five European excavations of putative plague victims. Microbiology 150(pt 2):341–354

Gill P, Ivanov PL, Kimpton C, Piercy R, Benson N, Tully G, Evett I, Hagelberg E, Sullivan K (1994) Identification of the remains of the Romanov family by DNA analysis. Nature Genet 6(2):130–135

Götherström A, Collins MJ, Angerbjörn A, Lidén L (2002) Bone preservation and DNA amplification. Archaeometry 44(3):395–404

Green RE, Krause J, Ptak SE, Briggs AW, Ronan MT, Simons JF, Du L, Egholm M, Rothberg JM, Paunovic M, Pääbo S (2006) Analysis of one million base pairs of Neanderthal DNA. Nature 444(7117): 330–336

Greenwood AD, Lee F, Capelli C, DeSalle R, Tikhonov A, Marx PA, MacPhee RD (2001a) Evolution of endogenous retrovirus-like elements of the woolly mammoth (*Mammuthus primigenius*) and its relatives. Mol Biol Evol 18(5):840–847

Greenwood AD, Castresana J, Feldmaier-Fuchs G, Pääbo S (2001b) A molecular phylogeny of two extinct sloths. Mol Phylogenet Evol 18(1):94–103

Greenwood AD, Capelli C, Possnert G, Pääbo S (1999) Nuclear DNA sequences from late Pleistocene megafauna. Mol Biol Evol 16(11):1466–1473

Greenwood AD, Pääbo S (1999) Nuclear insertion sequences of mitochondrial DNA predominate in hair but not in blood of elephants. Mol Ecol 8(1):133–137

Groves P (1997) Intraspecific variation of mitochondrial DNA of muskoxen based on control region sequences. Can J Zool 75: 568–575

Guthrie RD (2004) Radiocarbon evidence of mid-Holocene mammoths stranded on an Alaskan Bering Sea island. Nature 429(6993): 746–749

Gutierrez G, Marin A (1998) The most ancient DNA recovered from an amber-preserved specimen may not be as ancient as it seems. Mol Biol Evol 15(7):926–929

Hadly EA, Kohn MH, Leonard JA, Wayne RK (1998) A genetic record of population isolation in pocket gophers during Holocene climatic change. Proc Natl Acad Sci USA 95(12):6893–6896

Higuchi R, Bowman B, Freiberger M, Ryder OA, Wilson AC (1984) DNA sequences from the quagga, an extinct member of the horse family. Nature 312(5991):282–284

Hofreiter M, Jaenicke V, Serre D, Haeseler Av A, Pääbo S (2001) DNA sequences from multiple amplifications reveal artifacts induced by cytosine deamination in ancient DNA. Nucl Acids Res 29(23): 4793–4799

Hofreiter M, Poinar HN, Spaulding WG, Bauer K, Martin PS, Possnert G, Pääbo S (2000) A molecular analysis of ground sloth diet through the last glaciation. Mol Ecol 9(12):1975–1984

Hofreiter M, Capelli C, Krings M, Waits L, Conard N, Munzel S, Rabeder G, Nagel D, Paunovic M, Jambresic G, Meyer S, Weiss G, Pääbo S (2002) Ancient DNA analyses reveal high mitochondrial DNA sequence diversity and parallel morphological evolution of late Pleistocene cave bears. Mol Biol Evol 19(8):1244–1250

Holm, LE, Forchhammer, MC, Boomsma, JJ (1999) Low genetic variation in muskoxen (*Ovibos moschatus*) from western Greenland using microsatellites. Molecular Ecology 8: 675–679.

Krings M, Stone A, Schmitz RW, Krainitzki H, Stoneking M, Pääbo S (1997) Neandertal DNA sequences and the origin of modern humans. Cell 90(1):19–30

Lambert DM, Ritchie PA, Millar CD, Holland B, Drummond AJ, Baroni C (2002) Rates of evolution in ancient DNA from Adelie penguins. Science 295(5563):2270–2273

Leonard JA, Wayne RK, Cooper A (2000) Population genetics of Ice Age brown bears. Proc Natl Acad Sci USA 97(4):1651–1654

Lindahl T (1997) Facts and artifacts of ancient DNA. Cell 90(1):1–3.

Lister AM, Edwards CJ, Nock DA, Bunce M, van Pijlen IA, Bradley DG, Thomas MG, Barnes I (2005) The phylogenetic position of the 'giant deer' *Megaloceros giganteus*. Nature 438(7069): 850–853

Loreille O, Orlando L, Patou-Mathis M, Philippe M, Taberlet P, Hanni C (2001) Ancient DNA analysis reveals divergence of the cave bear, *Ursus spelaeus*, and brown bear, *Ursus arctos*, lineages. Curr Biol 11(3):200–203

MacPhee RDE, Tikhonov AN, Mol D, De Marliave C, Van der Plicht H, Greenwood AD, Flemming C, Agenbroad L (2002) Radiocarbon chronologies and extinction dynamics of the late Quaternary megafauna of the Taimyr Peninsula, Russian Federation. J Archaeol Sci 29:1017–1042

MacPhee RD, Tikhonov AN, Mol D, Greenwood AD (2005) Late Quaternary loss of genetic diversity in muskox (*Ovibos*). BMC Evol Biol 5:49

Martin PS, Klein RG (1984) Quaternary extinctions: A prehistoric revolution. University of Arizona Press, Tucson

Matheus P, Burns J, Weinstock J, Hofreiter M (2004) Pleistocene brown bears in the mid-continent of North America. Science 306(5699):1150

McDonald JN (1981) North American bison: Their classification and evolution. University of California Press, Berkeley, CA

Mikko S, Røed K, Schmutz S, Andersson L (1999) Monomorphism and polymorphism at Mhc DRB loci in domestic and wild ruminants. Immunol Rev. 167:169–178

Mitchell D, Willerslev E, Hansen A (2005) Damage and repair of ancient DNA. Mut Res 571(1–2):265–276

Noonan JP, Hofreiter M, Smith D, Priest JR, Rohland N, Rabeder G, Krause J, Detter JC, Pääbo S, Rubin EM (2005) Genomic sequencing of Pleistocene cave bears. Science 309(5734):597–599

Noonan JP, Coop G, Kudaravalli S, Smith D, Krause J, Alessi J, Chen F, Platt D, Pääbo S, Pritchard JK, Rubin EM (2006) Sequencing and analysis of Neanderthal genomic DNA. Science 314(5802): 1113–1118

Orlando L, Bonjean D, Bocherens H, Thenot A, Argant A, Otte M, Hanni C (2002) Ancient DNA and the population genetics of cave bears (*Ursus spelaeus*) through space and time. Mol Biol Evol 19(11):1920–1933

Orlando L, Leonard JA, Thenot A, Laudet V, Guerin C, Hanni C (2003) Ancient DNA analysis reveals woolly rhino evolutionary relationships. Mol Phylogenet Evol 28(3):485–499

Pääbo S (1985) Molecular cloning of ancient Egyptian mummy DNA. Nature 314(6012):644–645

Poinar HN, Hoss M, Bada JL, Pääbo S (1996) Amino acid racemization and the preservation of ancient DNA. Science 272(5263):864–866

Poinar HN, Hofreiter M, Spaulding WG, Martin PS, Stankiewicz BA, Bland H, Evershed RP, Possnert G, Pääbo S (1998) Molecular coproscopy: Dung and diet of the extinct ground sloth *Nothrotheriops shastensis*. Science 281(5375):402–406

Poinar H, Kuch M, McDonald G, Martin P, Pääbo S (2003) Nuclear gene sequences from a late Pleistocene sloth coprolite. Curr Biol 13(13):1150–1152

Poinar HN, Stankiewicz BA (1999) Protein preservation and DNA retrieval from ancient tissues. Proc Natl Acad Sci USA 96(15):8426–8431

Poinar HN, Kuch M, Sobolik KD, Barnes I, Stankiewicz AB, Kuder T, Spaulding WG, Bryant VM, Cooper A, Pääbo S (2001) A molecular analysis of dietary diversity for three Archaic Native Americans. Proc Natl Acad Sci USA 98(8):4317–4322

Poinar HN, Schwarz C, Qi J, Shapiro B, Macphee RD, Buigues B, Tikhonov A, Huson DH, Tomsho LP, Auch A, Rampp M, Miller W, Schuster SC (2006) Metagenomics to paleogenomics: Large-scale sequencing of mammoth DNA. Science 311(5759):392–394

Rogaev EI, Moliaka YK, Malyarchuk BA, Kondrashov FA, Derenko MV, Chumakov I, Grigorenko AP (2006) Complete mitochondrial genome and phylogeny of Pleistocene mammoth *Mammuthus primigenius*. Pub Lib Sci Biol 4(3):e73

Römpler H, Rohland N, Lalueza-Fox C, Willerslev E, Kuznetsova T, Rabeder G, Bertranpetit J, Schöneberg T, Hofreiter M (2006) Nuclear gene indicates coat-color polymorphism in mammoths. Science 313(5783):62

Sampietro ML, Gilbert MT, Lao O, Caramelli D, Lari M, Bertranpetit J, Lalueza-Fox C (2006) Tracking down human contamination in ancient human teeth. Mol Biol Evol 23(9):1801–1807

Shapiro B, Drummond AJ, Rambaut A, Wilson MC, Matheus PE, Sher AV, Pybus OG, Gilbert MT, Barnes I, Binladen J, Willerslev E, Hansen AJ, Baryshnikov GF, Burns JA, Davydov S, Driver JC, Froese DG, Harington CR, Keddie G, Kosintsev P, Kunz ML, Martin LD, Stephenson RO, Storer J, Tedford R, Zimov S, Cooper A (2004) Rise and fall of the Beringian steppe bison. Science 306(5701):1561–1565

Shepherd LD, Millar CD, Ballard G, Ainley DG, Wilson PR, Haynes GD, Baroni C, Lambert DM (2005) Microevolution and mega-icebergs in the Antarctic. Proc Natl Acad Sci USA 102(46): 16717–16722

Taubenberger JK, Reid AH, Lourens RM, Wang R, Jin G, Fanning TG (2005) Characterization of the 1918 influenza virus polymerase genes. Nature 437(7060):889–893

Vartanyan SL, Garrut VE, Sher AV (1993) Holocene dwarf mammoths from Wrangel Island in the Siberian Arctic. Nature 382:337–340

Waits LP, Talbot SC, Ward RH, Shields GF (1998) Mitochondrial DNA phylogeography of the North American brown bear and implications for conservation. Conserv Biol 12:408–417

Willerslev E, Hansen AJ, Binladen J, Brand TB, Gilbert MT, Shapiro B, Bunce M, Wiuf C, Gilichinsky DA, Cooper A (2003) Diverse plant and animal genetic records from Holocene and Pleistocene sediments. Science 300(5620):791–795

Willerslev E, Cooper A (2005) Ancient DNA. Proc Roy Soc B 272(1558):3–16

Zischler H, Höss M, Handt O, von Haeseler A, van der Kuyl AC, Goudsmit J (1995) Detecting dinosaur DNA. Science 268(5214): 1192–1193

7 Did Humans Cause the Late Pleistocene-Early Holocene Mammalian Extinctions in South America in a Context of Shrinking Open Areas?

Alberto L. Cione,* Eduardo P. Tonni, and Leopoldo Soibelzon
División Paleontología de Vertebrados
Facultad de Ciencias Naturales y Museo
Universidad Nacional de La Plata, Paseo del Bosque
1900 La Plata, Argentina
acione@museo.fcnym.unlp.edu.ar,
eptonni@museo.fcnym.unlp.edu.ar,
lsoibelzon@museo.fcnym.unlp.edu.ar

Keywords South America • mammals • extinction • pseudoextinction • human impacts

Introduction

The last important – and possibly the most spectacular – turnover in South American mammal history occurred around the Pleistocene-Holocene boundary, when 100% of megamammal species and about 80% of large mammal species became extinct. In this paper, we consider as "megamammals" those with body mass over 1,000 kg, and "large mammals" those over 44 kg (Tables 7.1 and 7.2). With the exception of a few smaller mammals, no other animals or plants disappeared. Consequently, this extinction event was distinct from mass extinctions (see comments in Cione et al., 2003).

South American communities had included at least some megamammals since the latest Paleogene, but gigantism in mammals markedly emerged during the Ensenadan age (early Pleistocene; Fig. 7.1) when many large mammal genera first appeared (Ameghino, 1889; Pascual et al., 1965; Cione and Tonni, 2005). Most genera, with different species, survived during the late Pleistocene when the remarkable figure of 37 megamammal species can be documented (Table 7.1). Many appear to have persisted in the early Holocene. However, none is living today. Moreover, presently there are no megamammals in Europe and Australia. In Asia today there are only two megamammals and in Africa four megamammal species and the giraffe, close to a ton in body mass. The largest terrestrial mammal in the Neotropics is the tapir *Tapirus bairdii*, some individuals of which slightly surpass 300 kg (Nowak and Paradiso, 1983).

The large mammal extinction of the latest Cenozoic occurred on different continents and islands at different times (Steadman et al., 2005). For explaining these extinctions several hypotheses have been proposed, most related to cold climate, disease, or human activities. We agree with Barnosky et al. (2004) in that the accumulated evidence suggests it is time to move beyond casting the Pleistocene extinction debate as a dichotomy of climate versus humans. In this context, we have proposed what we call the Broken Zig-Zag hypothesis (Cione et al., 2003). During most of the middle and late Pleistocene, dry and cold climate caused open areas to predominate in South America. Nearly all megamammals and large mammals that became extinct were adapted to this kind of environment. The periodic though relatively short interglacial increases in temperature and humidity led to dramatic shrinking of open areas and extreme reduction of the mammalian biomass (albeit not of species richness) adapted to open habitats. However, during the longer glacial periods, open-area mammal populations recovered. This alternation of low and high biomass of mammals from open and closed areas is what we refer to as the Zig-Zag. Remarkably, the extinction rate of large mammals and megamammals during more than a half million years was not high, until rising abruptly during the latest Pleistocene.

In this paper, (1) we analyze the biostratigraphic and chronologic pattern of the large continental mammals of South America

* Address for correspondence: acione@museo.fcnym.unlp.edu.ar

G. Haynes (ed.), *American Megafaunal Extinctions at the End of the Pleistocene*, 125–144.

TABLE 7.1. Mammal taxa present in putative Lujanian (_sensu_ Cione and Tonni, 1999) beds in South America and that became extinct (modified from Cione et al., 2003). Asterisks indicate taxa that occur in archeological sites.

Megamammals	Large mammals
_Cuvieronius humboldti_*	_Antifer niemeyeri_
Cuvieronius hyodon	_Arctotherium bonariense_
Doedicurus clavicaudatus *	_Arctotherium brasiliense_
Eremotherium carolinense	_Arctotherium tarijense_
Eremotherium laurillardi	_Brasiliochoerus stenocephalus_
Eremotherium mirabile	_Equus (A. merhippus) andium_
Eremotherium rusconii	_Equus (A.) insulatus_
Glossotherium (Oreomylodon)	_Equus (A.) lasallei wagneri_
Glossotherium lettsomi	_Equus (A.) neogeus_*
Glossotherium (Pseudolestodon)	_Equus (A.) santa-elenae myloides_
_Glossotherium robustum_*	_Eulamaops paralellus_
Glossotherium tropicorum	_Eutatus seguini_*
Glyptodon clavipes	_Eutatus punctatus_
Glyptodon perforatus	_Glyptotherium_ sp.
Glyptodon reticulatus	_Hippidion principale_*
_Glyptotherium cf. cylindricum_*	_Holmesina occidentalis_
Hemiauchenia paradoxa *	_Holmesina paulacoutoi_
Lestodon armatus	_Hoplophorus euphractus_
Lestodon trigonidens	_Lama gracilis_
_Macrauchenia patachonica_*	_Morenelaphus lujanensis_
Megalonyx sp.	_Mylodopsis ibseni_
_Megatherium americanum_*	_Neochoerus aesopy_
Megatherium medinae	_Neochoerus sirasakae_
Mixotoxodon larensis	_Neosclerocalyptus paskoensis_*
Mylodon darwinii	_Neuryurus_ n. sp.
_Mylodon listai_*	_Nothropus priscus_
Neothoracophorus depressus	_Nothrotherium roverei_
Panochthus frenzelianus	_Ocnopus gracilis_
Panochthus morenoi	_Ocnotherium giganteum_
Panochthus tuberculatus	_Onohippidion saldiasi_*
Plaxhaplous canaliculatus	_Palaeolama niedae_
Stegomastodon platensis	_Palaeolama weddelli_
Stegomastodon guayasensis	_Pampatherium humboldti_
Stegomastodon waringi	_Pampatherium typum_
Toxodon burmeisteri	_Paraceros fragilis_
_Toxodon platensis_*	_Parapanochthus jaguaribensis_
Xenorhinotherium bahiense	_Propraopus grandis_
	Propraopus humboldti
	Propraopus magnus
	Scelidodon cuvieri
	Scelidodon chiliensis
	Scelidodon reyesi
	Scelidotherium leptocephalum
	Smilodon populator
	Tapirus cristatellus
	Trigonodops lopesi
Medium sized mammals	**Small mammals**
Canis dirus	_Eligmodontia_ n. sp.
_Dusicyon avus_*	_Microcavia robusta_
Protocyon orcesi	_Desmodus draculae_
Protocyon troglodytes	
Protopithecus brasiliensis	
Valgipes deformis	

TABLE 7.2. Large mammals and megamammals in the Guerrero Member of the Luján Formation and correlative units.

Antifer neumeieri	_Megatherium americanum_
Arctotherium bonariense	_Megatherium lundi_
Arctotherium tarijense	_Megatherium tarijense_
Ctenomys lujanensis	_Morenelaphus lujanensis_
Cuvieronius humboltii	_Mylodon darwinii_
Dusicyon avus	_Mylodon listai_
Doedicurus clavicaudatus	_Neochoerus aesopy_
Equus (A.) andium	_Neosclerocalyptus paskoensis_
Equus (A.) neogeus	_Neothoracophorus depressus_
Eremotherium sp.	_Neuryurus_ n. sp.
Eulamaops paralellus	_Onohippidium salidiasi_
Eutatus seguini	_Palaeolama_ sp.
Eutatus punctatus	_Pampatherium typum_
Glossotherium lettsomi	_Panochthus morenoi_
Glossotherium myloides	_Panochthus tuberculatus_
Glossotherium robustum	_Paraceros fragilis_
Glyptodon clavipes	_Plaxhaplous canaliculatus_
Glyptodon perforatus	_Propraopus grandis_
Glyptodon reticulatus	_Scelidodon_ sp.
Glyptotherium cylindricum	_Scelidotherium leptocephalum_
Hemiauchenia paradoxa	_Smilodon populator_
Hippidion principale	_Stegomastodon platensis_
Holmesina sp.	_Stegomastodon waringi_
Lama gracilis	_Toxodon burmeisteri_
Lestodon armatus	_Toxodon platensis_
Lestodon trigonidens	
Macrauchenia patachonica	

during the late Pleistocene-earliest Holocene with new evidence, (2) discuss why megamammals and large mammals were more liable to become extinct than those that survived, (3) discuss the timing of human entry into South America, and (4) examine the possible role of humans in this extinction.

The Broken Zig-Zag Hypothesis

We have proposed that megamammal and large mammal extinction in South America during the late Pleistocene-earliest Holocene was caused by human foragers (Cione et al., 2003). However, we believe that this event would have been favoured by a particular circumstance: total biomass (not diversity) and distribution of open-area-adapted mammals began to be extremely reduced in response to the periodic shrinking of this kind of environments (the Zig-Zag) which was stimulated by the last (present) interglacial's periodic rising temperature and humidity. Humans certainly did not exterminate all the extinct taxa (e.g., the large carnivores), but killed off many and provoked changes that occasioned the disappearance of the remaining ones (see also Kay, 2002).

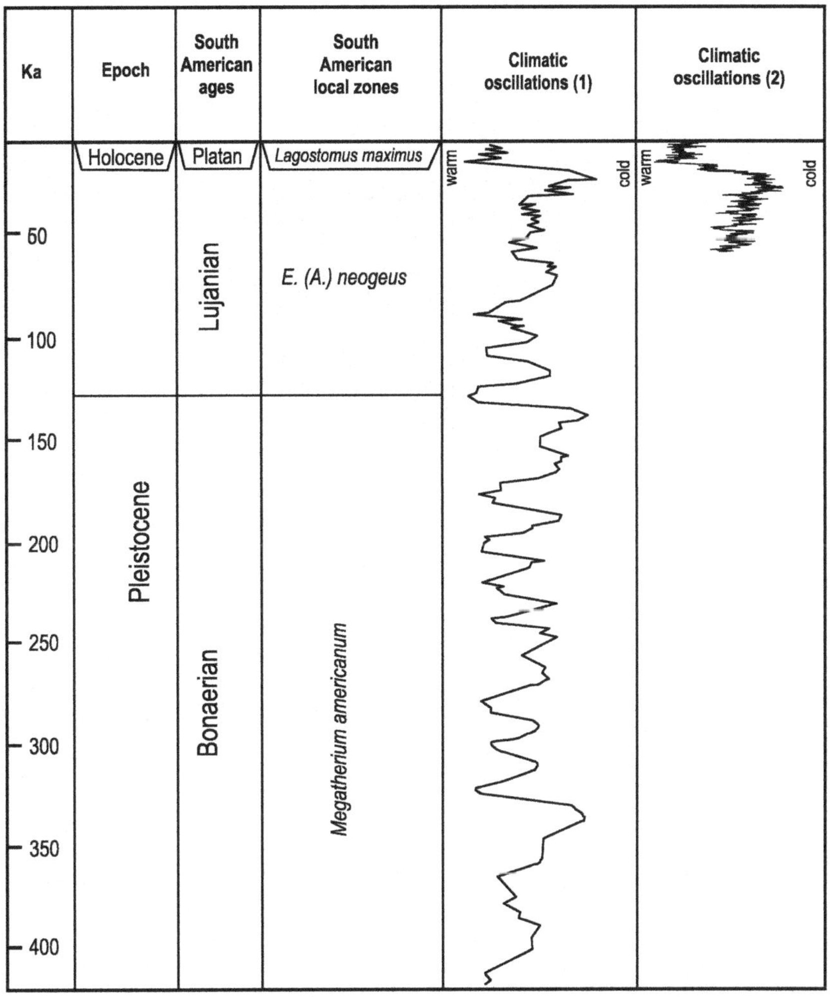

FIGURE 7.1. Chronological chart of the middle Pleistocene-Recent in southern South America depicting mammal zones and South American ages (modified from Cione and Tonni, 2005) and the climatic oscilations represented by (1) ^{18}O of Vostok, Antarctica (Petit et al., 1999) and (2) data from Dronning Maud Land, Antartica (Steig, 2006).

The Zig-Zag

Studies based on geochemical proxies in glacial ice cores from Greenland, Antarctica, and South America show that temperatures strongly fluctuated during at least the last 400 ka and that the present interglacial is not substantially different from the earlier ones, of which there were over 20 during the middle-late Pleistocene (Fig. 7.1; McCulloch et al., 2000; Blunier and Brook, 2001; Steig, 2006). The periodic changes in Pleistocene climate led to dramatic modifications in the distribution and biomass of the biota in South America and other continents. We use the expression Zig-Zag to stress the periodicity of biotic trends.

To determine the impact of changes, we evaluated the possible modification of vegetation physiognomy. In modern South America, 18% of the land is characterized as open areas, while medium vegetated areas make up 15%, and closed areas make up 67% (Cione et al., 2003). In contrast, by using the reconstruction of South America during the LGM, we calculated that open areas would have encompassed 31% of the territory, medium areas 54%, and closed areas 15% (Cione et al., 2003). Simberloff (1986 *fide* Raup, 1992: 136) had calculated that areas of wet forests were reduced by 84% during this time. We understand that the most difficult definition is that of the "medium vegetated areas." In this term, *cerrado*s, *chaco*, *monte*, and other relatively closed areas are included along with some more open areas (see Clapperton, 1993a, his Fig. 23.10). For this, we consider that the geological, floristic, and faunistic evidence from the present densely vegetated areas suggests that open areas were even larger than depicted by Clapperton (1993a, his Fig. 23.10). Similar results were obtained by Vivo and Carmignotto (2004) based on the distribution of plant formations (Fig. 7.2; see also Johnson, 2002).

We commented above that during the late Pleistocene the mammal diversity was higher especially because many different species of large and megamammals were thriving in

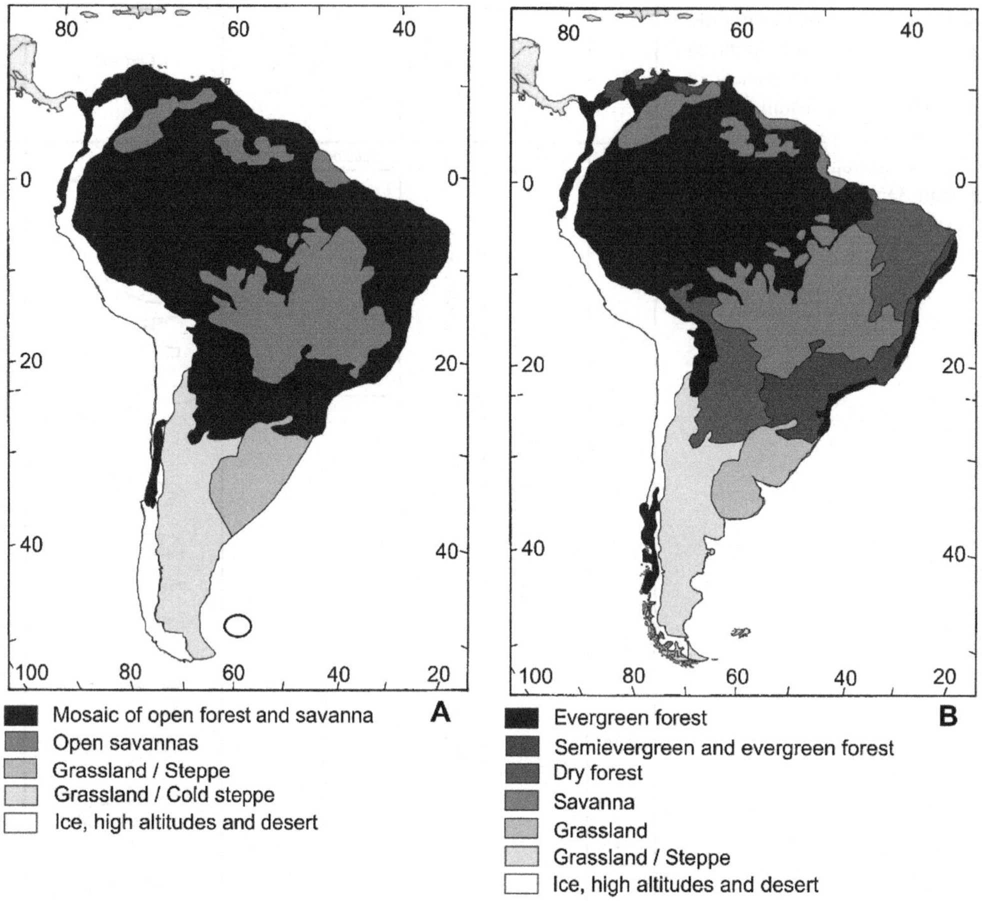

FIGURE 7.2. Map depicting plant distribution in South America (modified from Vivo and Carmignotto, 2004). A. Last Glacial Maximum (the emerged shelf is depicted). B. Present day.

the Pampean area and the rest of South America. However, although the taxonomic diversity was elevated, the number of individuals per species and the total biomass in each location probably was not high because most of the South American glacial ecosystems should not have been very productive. During the glacial times, while climate was colder and drier in extraglacial areas, open areas expanded and animals and plants that were adapted to these environments augmented their distribution and biomass (Tonni and Cione, 1997; Tonni et al., 1999a; Cione et al., 2003). During the shorter interglacial periods, when temperatures were higher than present (Thompson, 2000; Blunier and Brook, 2001; Vivo and Carmignotto, 2004), an expansion of the forested areas may be expected, and with this an increase in the biomass of forest dwelling animals, including those that survived the Pleistocene-Holocene extinction, such as tapirs (Pardiñas et al., 1996; see above).

Remarkably, the fossil record does not suggest that these cyclic environmental changes either produced depletion of niches or caused massive extinction of mammals (Cione et al., 1999; Tonni et al., 1999b; Nabel et al., 2000). In the short interglacial periods, mammals adapted to open areas had not necessarily been starving, but populations were surely remarkably reduced and under ecological stress. Some populations possibly were close to the minimal viable number and many surely became isolated so that genetic flux greatly diminished among them, reducing variability. The persistence of South American mammal species through a prolonged cycle of important environmental changes could be explained according to the Plus Ça Change Model, in which morphological stasis over geological timescales tends to arise not from the stability of physical environments, but from their instability (Sheldon, 1998).

Implications of the Hypothesis to Be Tested

This hypothesis is supported by the climatic evolution of the continent, the vegetational history, the positive biostratigraphical evidence, the chronology of extinctions, the paleobiogeography of mammals, the adaptation to open environments of those mammals that became extinct which also were probably k-adapted, the selective disappearance of all megamammals and most large mammals, the almost nil extinction of middle sized and small mammals and other vertebrates, the protected environments where the few surviving large mammals live today, and the fact that American mammals did not coevolve with humans. Morever, the only certain new biological or geological event for the middle-late Pleistocene to Holocene that occurred in the time of extinction was the entrance of humans into the continent.

Methods

Dates given here are mostly uncalibrated (BP: radiocarbon years before present; see below) and some calibrated (cal bp).

Simulations of South America during the LGM with a regional climate model coupled with a potential vegetation model produce a quantitative picture of LGM climate and vegetation distributions that is consistent with the geological proxy data. The model indicates a smaller Amazonian rainforest through the glacial time, associated with drier conditions (Cook and Vizy, 2006). For comparison of the general vegetation of the Recent and the Last Glacial Maximum (LGM) we had designed two maps where three arbitrary categories of vegetational structure were used: open, medium, and closely vegetated areas (see Cione et al., 2003; their Figs. 2 and 3). For this chapter we have modified the biome maps of South America (Fig. 7.2) designed by Vivo and Carmignotto (2004) following the data cited in Cione et al. (2003) and some new references mentioned below.

In Colombia, pollen indicates that at 18,000 BP a generally cool and dry environment is reflected in biome assignments of cold mixed forests, cool evergreen forests, and cool grassland/shrub, the latter extending to lower altitudes than presently recorded (Marchant et al., 2004). From 11,150 to 9,100 cal bp, grass savanna dominated the landscape while gallery forest along the drainage system was poorly developed. Water availability was lower than today and the length of the dry season longer. From 9,100 to 7,330 cal bp gallery forest expanded pointing to wetter conditions (Wille et al., 2003).

In São Francisco de Assis in the western Rio Grande do Sul State in southern Brazil, the region was naturally covered by *Campos* (grassland) throughout the recorded glacial and Holocene period under conditions that were cold and relatively dry, and warm and dry, respectively. Initial expansion of gallery forest after 5,170 cal bp indicates a change to wetter climatic conditions. Maximum extent of gallery forest after 1550 cal bp reflects the wettest recorded period (Behling, 2002; Behling et al., 2002, 2005).

In the Pampean area of Argentina, several paleosols were dated at about 10,000 BP, at the Holocene Climatic Optimum (HCO), and at about 2,000 BP (Tonni et al., 2001, 2003), indicating wetter conditions.

During the LGM, the prominence of hyper-humid vegetation in southern Chile (north Patagonian and Subantarctic forests and parkland) implies sustained, periodic northward migration of the southern westerlies. Significant cooling events (inferred from the expansion of grass-dominated Subantarctic Parkland) occurred between 17,000 and 35,000 BP. At the end of the last glacial, glacial vegetation was abruptly replaced by more temperate Valdivian and Lowland Deciduous Forests at 17,000 BP. A brief climate reversal, centered on 14,000–12,000 BP, interrupted the unidirectional glacial-interglaciation transition. "The structure and variability of southern Chilean vegetation and climate closely resemble changes in Antarctic ice core data and in marine surface offshore" (Heusser et al., 2006:484).

Results

The Pleistocene-Holocene Chronology

The Pliocene to Recent continental sequence of southern South America is the most complete for studying the land mammal evolution in the continent for the period. More specifically, the Pampean stratigraphic sequence gives the basic framework for the South American late Cenozoic chronological scale (see articles in Tonni and Cione, 1999.; Tonni et al., 2003; Cione and Tonni, 2005).

It is well known since the 19th century that South American mammal faunas underwent marked turnovers in taxonomic composition during the Neogene (Ameghino, 1889; Pascual et al., 1965; Marshall et al., 1984; Tonni et al., 1992; Cione and Tonni, 1995, 2001, 2005; see papers in Tonni and Cione, 1999). The periodic alternation of glacial and interglacial epochs during the middle-late Pleistocene dramatically modified the distribution, composition, and biomass of plant and animal communities in South America as in other regions of the world (Tonni and Fidalgo, 1978; Tonni and Politis, 1980; Graham, 1986; Tonni and Cione, 1997; Pardiñas 1999; Pardiñas et al., 2004; Tonni et al., 1999a; Whitlock and Bartlein, 1997; Haynes, 2002). However, these remarkable changes did not necessarily provoke large extinctions although many new species appeared (Table 7.3).

The global inception of the glacial periods that characterize the middle and late Pleistocene is chronologically related to an important turnover in the South American mammal fauna. It is represented by the boundary between the *Mesotherium cristatum* and *Megatherium americanum* biozones in the stratigraphic sections of the Pampean area and in Bolivia (Cione and Tonni, 1999; Cione et al., 1999; MacFadden, 2000; Soibelzon et al., 2005; Cione and Tonni, 2005; Fig 7.1). Both units provide the biostratigraphic basis for the Ensenadan and Bonaerian ages. According to magnetostratigraphic analyses in the Pampean area and Bolivia, the base of the Bonaerian is younger than 0.78 Ma, probably ca. 0.5 Ma (Nabel et al., 2000; Cione and Tonni, 2005; Fig. 7.1).

A much smaller turnover has allowed the definition of a biostratigraphic unit in the Pampean area: the *Equus (Amerhippus) neogeus* biozone (Fig. 7.1). This unit is the basis of the Lujanian age and is dated from about 130,000 to about 7,000 BP. It is characterized by the first record of

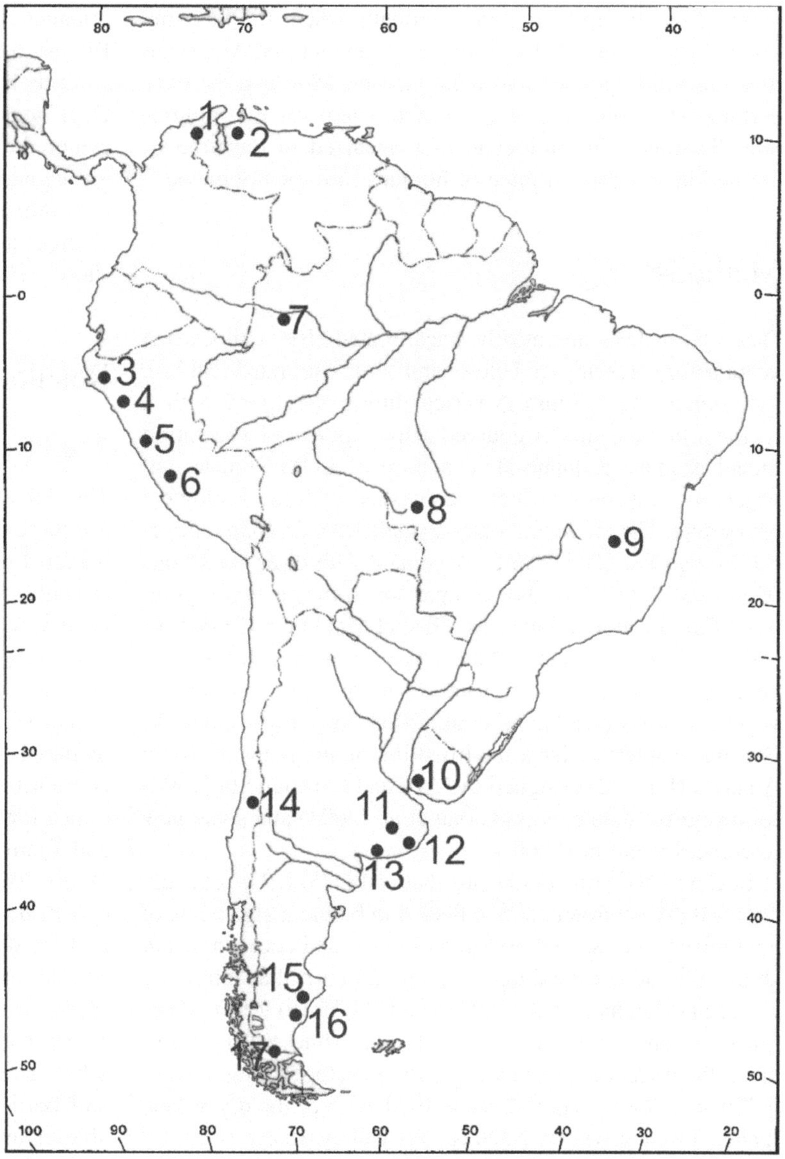

FIGURE 7.3. Map of South America depicting most of the sites mentioned in the text. 1. Taima Taima; 2. Muaco; 3. Talara; 4. Cupisnique; 5. Huargo; 6. Ayacucho complex; 7. Río Juruá; 8. Cuiabá; 9. Minas Gerais; 10. Sauce; 11. Tapalqué and Campo Laborde; 12. Arroyo Seco; 13. Cerro La China; 14. Tagua Tagua; 15. Los Toldos and Piedra Museo; 16. Las Buitreras; 17. Cueva del *Mylodon*, Cueva Lago Sofía, Cueva del Medio, Cueva Fell, Pali Aike.

TABLE 7.3. Large mammal and megamammal pattern of extinction during different ages of the middle Pleistocene to Holocene of South America.

	Ensenadan	Bonaerian	Lujanian
Total number of genera	99	75	68
Number of extinct genera	66	25	43
Percentage of taxa per total number of large and mega	67%	33%	63%
Duration	1.2 Ma	0.5 Ma	0.01 Ma
Percentage of taxa per duration	55%	50%	358%

some mammal taxa and the extinction of other (Pardiñas et al., 1996; Cione and Tonni, 2005). Overlying the *Equus (Amerhippus) neogeus* biozone is the *Lagostomus maximus* biozone which constitutes the basis for the Platan age, dated from about 7,000 BP to the 16th century AD (Fig. 7.1). In this zone, only autochthonous Recent mammals occur, except for the peculiar occurrence of the canid *Dusicyon avus* during this period (see below). Consequently, the deposition of sediments representing this stage appears to have occurred after the mammal extinctions in South America. In the *Lagostomus maximus* biozone there are no introduced mammals from the Old World, such as cows, sheep, and horses. They occur in the *Bos taurus-Ovis aries* biozone (Deschamps, 2005), which represents the most recent sediments.

Mammal Diversity during the Middle-Late Pleistocene and Shifts in Distributions

An outstanding feature of the South American Neogene terrestrial fauna is the presence of a very diverse endemic fauna of gigantic mammals (Ameghino, 1889; Simpson, 1980).

The large size of individuals is striking in the Ensenadan, but also continues to be marked in the Bonaerian and Lujanian (Tonni et al., 1992). During the Neogene, until the establishment of the Isthmus of Panama at about 3 Ma, these very large mammals were mainly endemic notoungulates, litopterns, and xenarthrans. However, mostly during the Pliocene Chapadmalalan and Marplatan ages and the Pleistocene Ensenadan age, many other taxa of Holarctic origin appeared in South America, most of them represented in the first record by endemic genera and species (Cione and Tonni, 1995, 1996). Many of the Holarctic mammals that invaded South America were large (e.g., several felids, ursids, tayassuids, equids, camelids, and cervids), and some others fall in the category of megamammals (gomphotheriids, some ursids).

The Quaternary was a time of extensive evolution among mammals and many of them show adaptations to peculiarly Quaternary environments (Lister, 2004). The South American middle Pleistocene-earliest Holocene (Bonaerian-Lujanian) mammal diversity was significatively higher than that of the middle Holocene to Recent, because it can be presumed that most of the extant mammals (243 genera; 73 genera if we exclude Chiroptera, Rodentia, Lagomorpha, Marsupialia, and Insectivora; Nowak and Paradiso, 1983) were present together with those mammals that became extinct in the continent (49 genera, see Table 7.1). In the particular case of the Pampean area, the faunal composition during most of the middle and late Pleistocene was also different from the Recent one because many of the extant mammals occurring there during this interval (except for interglacial times) corresponded to those presently living in arid and/or colder areas to the south and west. At least 24 species of mammals of the Platan or the Recent were documented as inhabiting the Pampean region during at least part of the late Pleistocene (see Cione et al., 1999, 2003; Cione and Tonni, 2005). Except for the absence of the extinct taxa, few "non-analogue" associations (see Graham et al., 1996) were documented in the Bonaerian, Lujanian, and Platan from the Pampean area (Cione and López Arbarello, 1994; Pardiñas, 1999; Pardiñas et al., 2004). During interglacial times mammals from warmer climates occupied the region, and during the glacial times Patagonian and western taxa occurred there (Pardiñas et al., 1996; Tonni et al., 1999a; Verzi et al., 2002). Actually, there is a very poor record of interglacial times probably because they did not last as long as the glacial intervals, and also because they permitted the development of the kind of soils that foster bone destruction (see Retallack, 1998; Tonni et al., 1999b).

Eighty three extinct species and 46 extinct genera of large mammals and megamammals occur in putative Lujanian beds of South America (from 130,000 to about 7,000 BP; Table 7.1). There is some stratigraphic uncertainity in several parts of South America outside Argentina and some records could correspond to beds of Bonaerian age. This means that some of these species might not have been involved in the massive extinction at the end of Lujanian. However, the Lujanian of the Pampean region appears to be characterized primarily by new occurrences, with few extinctions in the top of the Bonaerian (Cione et al., 1999; Cione and Tonni, 2005; Table 7.3), perhaps because some species developed flexible adaptations enabling them to inhabit broad niches and to survive major environmental changes (Lister, 2004), although some evolved fixed adaptations to specialist habits. The large mammals that became extinct at the end of the Pleistocene and beginning of the Holocene were adapted to open environments. Certainly, they could find such environments somewhere in the continent both during the glacial and interglacial periods.

The total mammal generic diversity during Lujanian times could have been as high as 286 genera (49 extinct genera plus 243 Recent genera; Tables 7.1, 7.2). The total large mammal (and megamammal) diversity present during Lujanian times could have been as high as 83 species distributed into 48 genera. Most of the Lujanian taxa that became extinct were large mammals and megamammals.

The well studied Guerrero Member of the Luján Formation in the Pampean area, dated between 21,000 to 10,000 BP (Tonni et al., 2003), and other units in different parts of South America with radiometric dates include at least 52 species distributed into 38 genera (49 large and megamammal species plus 3 smaller mammal species and 35 large and megamammal genera plus 3 smaller mammal genera) (see below).

Was the South American Fauna Declining during the Pleistocene?

Present evidence indicates that the large-mammal and megamammal fauna was very well diversified at the end of the Pleistocene and did not suffer from any kind of declining trend in previous time intervals. Some South American clades were declining in diversity since the Great American Biotic Interchange. Litopterns and notoungulates as well as marsupials and large ground birds probably suffered from competition with Holarctic taxa. However, xenarthrans continued being much diverse although many new genera of Holarctic origin integrated in the ecosystems, with eight families during the Chapadmalalan to Ensenadan (Cione and Tonni, 1995). Consequently, the fauna was changing but was highly diverse and abundant. There is no evidence that the extinction rate rose during the the glacial periods of the middle and late Pleistocene (see Tonni et al., 1992; Cione and Tonni, 2005). Moreover, the number of megamammals in South America was the largest in the world at this time.

Youngest Records of Extinct Mammals in South America

The youngest beds where now-extinct mammals (with the exception of *Dusicyon avus*) have been found in South America are Lujanian in age. Several of these Lujanian localities appear to be early Holocene in age. In North America, there is agreement that the extinctions took place before the beginning of the Holocene, before 10,000 BP (Roy, 2001; Haynes, 2002).

In the river valleys of the South American Pampean region, no extinct mammal has been found in the typical Río Salado Member of the Luján Formation or correlative units, except for the peculiar case of *Dusicyon avus* (see below). The *Lagostomus maximus* biozone (and the Platan) begins in the base of this member (Cione and Tonni, 1995). The Río Salado Member is correlated with at least the middle Holocene Las Escobas Formation marine beds which were dated between ca. 6,500 to ca. 2,000 BP. Their deposition coincides with the high sea-level of the Holocene Climatic Optimum (HCO) (Tonni and Fidalgo, 1978; Fidalgo, 1992; Aguirre and Whatley, 1995). Eleven ^{14}C ages based on molluscan shells and total organic matter were obtained from transitional beds between the Guerrero and Río Salado members of the Luján Formation in sections at the Arroyo Tapalqué (central Buenos Aires province; Fig. 7.3) ranging between 9,710 ± 110 BP and 8,810 ± 140 BP (Figini et al. 1995; Zárate et al. 1995; Cione et al., 2001). In the divides of the Pampean region, no extinct mammal occurs in the uppermost beds of La Postrera Formation (of middle to late Holocene age).

Certainly, the best-known mammal faunas in the South American upper Pleistocene are those found in the Guerrero Member of the Luján Formation, the La Postrera Formation in the Pampean area, and the correlative Dolores Formation of Uruguay (Ameghino, 1889; Prado et al., 1987; Alberdi et al., 1988; Tonni et al., 1992; Martínez, 1997; Cione et al., 1999; Ubilla and Perea, 1999). According to stratigraphic relationships and isotopic dating, the Guerrero Member was deposited between 21,000 BP and earliest Holocene (Tonni et al., 2003 and papers cited therein). The localities most accurately dated are Tapalqué and Paso Otero (Tonni et al., 2003; Fig. 7.3).

In addition to those records, in the uppermost Pleistocene beds (dated between 20,000 and 10,000 BP) of several sites in Argentina, Chile, Peru, Uruguay, and Venezuela (Fig. 7.3), the extinct species shown in Table 7.4 were documented. Many of these records were based on dates directly from bones, and dating carried out on the basis of other materials is sufficient to support that timing to the last 20,000 BP.

There are several other records that are doubtful, at best. The identification of *Lama* cf. *owenii* is here dismissed (Nami and Nakamura, 1995). The Bonaerian genus *Antifer* was incorrectly cited for the latest Pleistocene of Buenos Aires and Chile by Aramayo (1997) and Casamiquela (1999). The *Mylodon* species that is recorded in Patagonia appears to be *M. listai* and not *M. darwinii* (Gustavo Scillato-Yané, personal communication). *Panthera onca mesembrina* is here considered as indistinguishable from the living jaguar (*Panthera onca*).

Table 7.5 shows 16 different extinct taxa documented in dated South American units of putative early Holocene age. Other extinct taxa were reported in the Holocene of West Indian islands (ground sloths; Steadman et al., 2005) and Alaskan Bering Sea islands (mammoths; Guthrie, 2003), among others.

The youngest dated remains of extinct fauna in South America (excepting *Dusicyon avus*) occur in the La Moderna site, Campo Laborde, and Arroyo Seco 2 sites (Fig. 7.3), all in the Pampean area. The fossils recorded at La Moderna include the glyptodonts *Neosclerocalyptus* (=*Sclerocalyptus*) and *Glyptodon*, and an indeterminate mylodontid. They occur in a bed (UL a' of Zetti et al., 1972; see discussion in Politis and Gutiérrez, 1998) transitional between two members of the typical Luján Formation: the Guerrero Member and the Holocene Río Salado Member. According to Politis et al. (1995) and Politis and Gutiérrez (1998), bones of the glyptodont *Doedicurus clavicaudatus* yielded dates between 7,500 to 7,000 BP. Other authors suggested that some young contaminants could have rejuvenated the samples (Borrero et al., 1998). However, Politis and Gutiérrez (1998) even mention that a sample that gave an older date (LM-2-4; 12,350 ± 370 BP) was redated with a better treatment and gave younger dates (7,010 ± 100 and 7,510 ± 370 BP; Beukens, 1992, unpublished report *fide* Politis and Gutiérrez, 1998). Recently, Politis et al. (2003) reported new radiocarbon dates from the soil humates fraction of La Moderna sediments that support the current accepted chronology of ca. 7,000–7,500 BP.

In the Arroyo Seco 2 site, a diverse extinct fauna was dated from 12,240 ± 110 BP to 7,320 ± 50 BP (Politis et al., 1995, 2004). It includes *Megatherium americanum*, cf. *Mylodon*, *Glossotherium robustum*, *Equus (Amerhippus) neogeus*, *Hippidion* sp., *Toxodon platensis*, *Macrauchenia patachonica*, and *Hemiauchenia* sp. (Fidalgo et al., 1986; Tonni, 1990; Politis et al., 1995; Politis and Gutiérrez, 1998). Only *Megatherium americanum*, *Hippidion* sp., and *Equus (Amerhippus) neogeus* were argued to have been human food resources at this locality (Politis and Gutiérrez, 1998). In the same site, a human burial accompanied by a grave offering dated at 7,800 to 6,300 BP included a *Glyptodon* osteoderm close to the human skull with which it seemed to be contemporaneous (Politis and Gutiérrez, 1998).

At the Campo Laborde archeological site, three AMS ^{14}C dates were obtained from *Megatherium americanum* bones. These dates correspond to the early Holocene, between 7,700 and 8,800 BP (Messineo and Politis, 2006).

A Holocene date (8,639 ± 450 BP) based on bones of *Mylodon*, *Lama*, and horse found at Pali Aike, Santa Cruz, Argentina was considered as a minimal age for that site (see Borrero, 1997).

Gomphotheriidae and Scelidotherinae remains were found in a bed overlying mylodontine remains dated 8,660 ± 150 BP in Tafí del Valle (Tucumán; Collantes et al., 1993).

In northern Uruguay, Suárez (2003) reports *Glyptodon* remains at a cultural component dated between 9,120 ± 40 BP and 8,570 ± 150 BP.

Bones of *Dusicyon avus*, without evidence of human association, were recovered from levels dated at 4,865 ± 65

TABLE 7.4. Additional records in the uppermost Pleistocene beds (dated between 20,000 and 10,000 BP) of several sites in Argentina, Chile, Peru, Uruguay and Venezuela.

Taxon	Locality	RCYBP	Reference
Mylodon listal	Cueva del Milodón, southern Chile		Borrero (1997)
Megatherium tarijense	Ayacucho complex, Peru	18,000–15,000	Hoffstetter (1986)
Megatherlidae indet.	Ayacucho complex, Peru	12,200	Mac Neish et al. (1970)
Glyptotherium cf. cylindricum	Taima Taima, Venezuela	12,580 ± 150	Carlini and Zurita (2006)
Glyptotherium cf. cylindricum	Muaco, Venezuela	13,390 ± 130	Carlini and Zurita (2006)
Eutatus seguini	Cerro La China, Cueva Tixi, Buenos Aires, Argentina		Flegenheimer and Zárate (1997), Mazzanti (1997)
Scelidodon	Talara, Peru	13,616 ± 600 and 14,418 ± 500	Hoffstetter (1970), Marshall et al. (1984)
Scelidotherium	Huargo, Peru	13,400 ± 700	Hoffstetter (1986)
Glossotherium aff. G. lettsomi	Santa Elina Rockshelter, Cuiaba, Brazil	23,320 ± 1000 and 10120 ± 60	Vialou (2003)
Holmesina	Talara, Peru		Hoffstetter (1970), Marshall et al. (1984)
Eremotherium	Cupisnique, Peru		Marshall et al. (1984)
Lestodon	Sauce, Uruguay		Arribas et al. (2001)
Macrauchenia	Cueva del Mylodon, southern Chile.		Borrero (1997)
Macrauchenia	Taima Taima, Venezuela		Gruhn and Bryan (1984)
Onohippidion saldiasi	Cueva del Mylodon, Piedra Museo, Los Toldos, Las Bultreras, Cueva Lago Sofia, Cueva del Medio, Fell, Pali Aike, Cerro Sota; southern Chile and Argentina		Alberdi and Prado (1992), Miotti (1993), Borrero, (1997), Alberdi et al. (2001)
Onohippidion saldiasi	Huargo, Peru	13,400 ± 700	Hoffstetter (1986)
Lama gracills	Los Toldos, Piedra Museo, southern Patagonia		Miotti and Cattáneo (1997), Borrero (1997), Borrero et al. (1998)
Equus (Amerhippus) neogeus	Tagua Tagua, Chile		Casamiquela (1999)
Equus andium	Ayacucho complex, Peru		Hoffstetter (1986)
Equus	Taima Taima, Venezuela		Gruhn and Bryan (1984)
Hippidion	Taima Taima, Venezuela		Gruhn and Bryan (1984)
Cuvieronius humboldti	Monte Verde, Chile	11,900 ± 200	Borrero (1997)
Stegomastodon	Talara, Perú		Hoffstetter (1970)
Stegomastodon	Taima Taima, Venezuela		R. Casamiquela in Gruhn and Bryan (1984)
Palaeolama	Cupinisque, Peru		F. Pujos, personal communication
Palaeolama	Monte Verde, Chile		Casamiquela and Dillehay (1989 fide Borrero, 1997: 91)
Antifer niemeyeri	Tagua Tagua, Chile		Casamiquela (1999)
Dusicyon avus	Cueva las Buitreras, Cueva Tixi, Argentina		Mazzanti (1997), Borrero (1997), Borrero et al. (1998)
Smilodon	Cueva Lago Sofia, southern Argentina	11,210 ± 50	Borrero (1997)
Arctotherium tarijense	Pali Aike Nacional Park, Chile		Soibelzon (2002), Prevosti et al. (2003)
Arctotherium tarijense	Rio Negro, Uruguay	11,600 + 130	Ubilla and Perea, (1999), Soibelzon et al. (2005)

TABLE 7.5. Extinct mammal taxa documented in dated South American units of putative early Holocene age.

Taxa	Locality	Reference
Doedicurus clavicaudatus, Sclerocalyptus, Glyptodon, Scelidotherinae indet., *Megatherium americanum, probably Glossotherium robustum, Equus (Amerhippus) neogeus, Hippidion, Toxodon platensis, Macrauchenia patachonica, Hemiauchenia,* Gomphotherildae, *Dusicyon avus.*	Pampean area, Tucumán, and Patagonia of Argentina and southern Chile	Fidalgo et al., 1986; Collantes et al., 1993; Politis et al., 1995; Borrero, 1997; Borrero et al., 1998; Politis and Gutiérrez 1998.
Eremotherium mirabile, Stegomastodon waringi, Arctotherium wingei	Cedeño District, Venezuela	Linares (1993), Linares pers. comm

BP in the Cueva Tixi and also in youger beds at Zanjón Seco 2, both in the Pampean area (Politis et al., 1995; Mazzanti and Quintana, 1997). The late disappearance of the species *Dusicyon avus* was not considered as a real extinction but a hybridation with other canids (Berman and Tonni, 1987). A date of 4,300 BP for a carapace of a glyptodont (Rosello et al., 1999) was considered unsupported by Cione et al. (2001).

In summary, we counted at least 52 species and 38 genera of confirmed extinct mammals in beds deposited between 20,000 to 10,000 (probably 7,000) BP in South America. These taxa certainly disappeared in the last massive extinction. There are many other mammals documented in Lujanian beds of South America for which we do not have certainity if they reached this period (Table 7.1). Future work will confirm or reject this hypothesis.

Mammal Biogeography during the Last Glacial and Postglacial

South America is a continent marked by the Andean influence and the remarkable latitudinal extension (see Cabrera and Willink, 1973). Most of the knowledge about latest Pleistocene mammals is based on fossils from southern South America. However, increasing records and new systematic studies allow recognition of patterns of distribution of many species (e.g., Cione et al., 2005; Scillato-Yané et al., 2005; Soibelzon et al., 2005). Some species were wide-ranging in the continent (e.g. *Smilodon*, *Panthera*, *Arctotherium*, *Equus*, *Hippidion*, *Hemiauchenia*). We devised several geographic patterns for several large mammal and megamammal taxa (Fig. 7.4). Notwitstanding the different distribution patterns, all became extinct.

Oldest Human Presence in South America

In South America, archeological evidence indicates that the continent was occupied by several distinct groups at the end of the Pleistocene (around 13,000–11,000 BP; Borrero et al., 1998). Some of the dated sites with humans for this period are: Taima Taima (ca. 13,000 BP) and El Jobo complex (both sites in the Estado de Falcón,Venezuela; see Gruhn and Bryan, 1984; Politis and Gutiérrez, 1998); Tibitó, Bogotá basin, Colombia, dated 11,740 ± 110 BP (see Gruhn, 1997); four different cultural complexes in Peru dated between 10,000 and 12,000 BP (Cardich, 1977); Quereo, near Los Vilos, Chile, dated 11,600 ± 190 BP (see Gruhn, 1997); Tagua Tagua, Chile, dated 10,200 BP and 9,900 BP (Núñez et al., 1994); Monte Verde, a site with mastodont and paleocamelid remains located in southern Chile dated ca. 12,000 BP (Dillehay, 1989; however, also see Fiedel, 1999); Pedra Pintada, Amazonia, Brazil, dated ca. 11,100 BP (Roosevelt et al., 1996); Lapa do Boquete, Minas Gerais, Brazil, dated 12,070 BP to 11,000 BP (see Gruhn, 1997); Abrigo de Santana do Riacho, Minas Gerais, Brazil, dated 11,960 ± 250 BP (see Gruhn, 1997). For supposed dates of ca. 13,000 BP in southern Brazil, see Schmidt Dias and Jacobus (2002); Los Toldos (11 level of Los Toldos, Santa Cruz, Argentina, dated 12,600 ± 600 BP; Cardich, 1977); Cueva del Lago Sofía, Santa Cruz, Argentina, dated 12,990 ± 490 BP and 11,570 ± 60 BP (Politis et al., 1995); Tres Arroyos, Tierra del Fuego, Argentina, dated 11,800 ± 250 BP (Politis et al., 1995); Cerro La China and Cerro El Sombrero, Argentina, dated ca. 11,000-10,000 BP (Flegenheimer and Zárate, 1997); and the Piedra Museo site, Santa Cruz, Argentina, which corresponds to a time between 12,890 and 10,300 BP (Miotti, 1993; Miotti and Cattáneo, 1997; Borrero et al., 1998; Ramírez Rozzi et al., 2001).

In the Holocene, there are many localities with cultural remains or human bones, some of which are of putative early Holocene age and include extinct fauna (see above).

The oldest widely accepted human presence in North America south of Alaska is dated at about 11,500 BP (Pielou, 1992: 112; Ward, 1997; Haynes, 2002; but see Fiedel, 1999).

Discussion

Climate When Humans Entered South America

After the Last Glacial Maximum (LGM; 20,000-18,000 BP), temperatures began to rise (Clapperton, 1993a,b; Petit et al., 1999; Thompson, 2000; Blunier and Brook, 2001; Steig, 2006). However, climate appears to have continued to be relatively dry in South America for several thousand years and humans seem to have entered southern South America when climate was still relatively dry (Fig. 7.1; Iriondo, 1997, 1999; Carignano, 1999; Tonni et al., 1999a). Ice cores from South America, Greenland, and Antarctica indicate that there was a peak of temperature followed by a decrease at about 12,700 and 10,300 BP (the Antarctic Cold Reversal and the Younger Dryas chronozone; see McCulloch et al., 2000). After this, there was a steep increase in temperature near the Pleistocene-Holocene boundary (Fig. 7.1; Thompson, 2000; McCulloch et al., 2000). In the Pampean area, at about 10,000 BP, development of dated paleosols indicate a change to wetter conditions (Fidalgo et al., 1975; Tonni and Fidalgo, 1978; Zárate et al., 1995; Prieto, 1996; Tonni et al., 1999a; Cione et al., 2001: Fig. 5). This series of climatic events is in agreement with ice cores and other temperature evidence (Thompson, 2000; McCulloch et al., 2000). Consequently, a substantial change to warmer and wetter climate occurred in South America at about 10,000 BP. During the Holocene Climatic Optimum, humid conditions were also documented in Argentina and Brazil (see Tonni et al., 2001; Vivo and Carmignotto, 2004).

According to the ice record in Greenland and Antarctica and the local evidence, the present interglacial is not essentially different enough from the preceding ones to have caused extinctions.

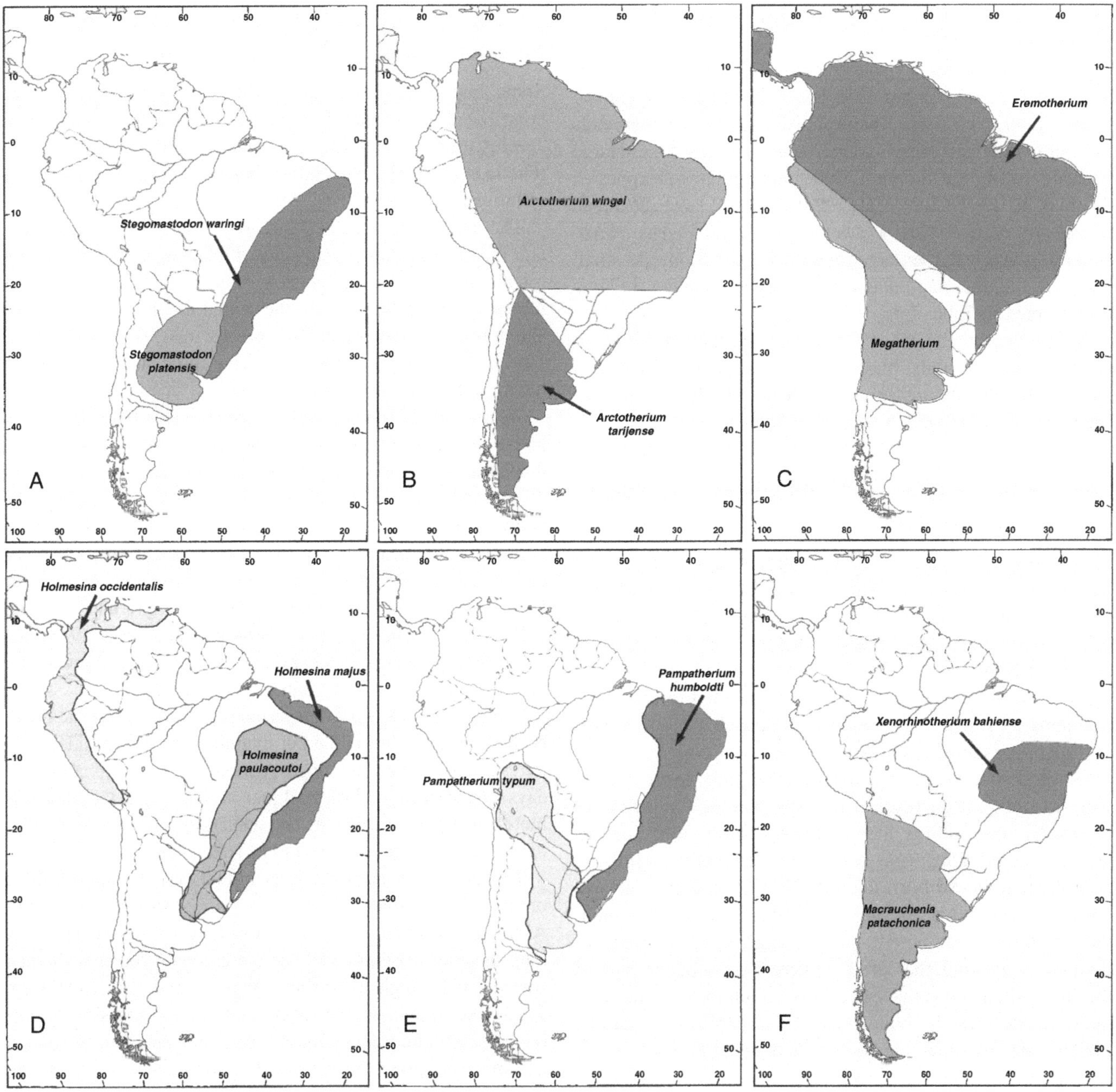

FIGURE 7.4. Distribution of some extinct large and megamammals in South America during the late Pleistocene. A. *Stegomastodon waringi* and *S. platensis* (modified from Cione et al., 2005); B. *Arctotherium tarijense* and *A. wingei* (Soibelzon et al., 2005); C. *Eremotherium* and *Megatherium* (Cartelle, 1994; Pujos and Salas, 2004); D. *Holmesina majus*, *H. occidentalis* and *H. paulacoutoi* (modified from Scillato-Yané et al., 2005); E. *Pampatherium typum* and *P. humboldti* (modified from Scillato-Yané et al., 2005); F. *Xenorhinotherium bahiense* and *Macrauchenia patachonica* (Melo et al., 2005; Carlini and Tonni, 2000).

Which Animals Became Extinct?

The extinctions were notably size-selective. In North America, some very large mammals (below the size of megamammals, but weighing almost a ton) survived the great extinction (e.g., bison, polar bear). In South America there are no very large mammals, because no megamammal survived.

Regarding the extinct taxa in South America, more than 70% of megamammal species and 55% of the large mammal species were xenarthrans (this figure results from the most recent revisions; see Scillato-Yané et al., 1995; Carlin and Scillato-Yané, 1999). No megaxenarthran and only one large xenarthran survived. On the contrary, no small xenarthran became extinct, notwithstanding that there were many medium-sized and small xenarthrans present. As well, the last notoungulates, litopterns, and gomphotheriids disappeared.

Of the 14 species of large mammals that presently occur in South America, 12 are of Holarctic origin and two are of South American origin. The proportion of extinct mammals of South American origin is large because almost all the large xenarthrans and all litopterns and notoungulates disappeared (Tonni et al., 1992; Cione et al., 1999). Only one large xenarthran (*Priodontes maximus*) and a caviomorph (*Hydrochoerus hydrochaeris*) survived the large mammal extinction. A small number of extinct mammals with a body mass under 44 kg is documented in Table 7.6.

Consequently, the latest Quaternary extinction in South America was strongly biased towards large mammals and megamammals (about 90% of the species; Table 7.1), including those of both South American and Holarctic origin.

South American Extinct Mammals were Adapted to Open Environments

Although the 83 species of large and mega-mammals that became extinct in South America include species with different feeding habits, nearly all seem to have been adapted to open environments (Webb and Marshall, 1982; Marshall et al., 1984; Bond et al., 1995; Tonni and Scillato-Yané, 1997; Table 7.1). Moreover, many were even adapted to arid environments. A possible exception is the capybara *Neochoerus aesopy* of presumed amphibious habit.

Glyptodonts, tardigrades, toxodonts, equids, and camelids were grazers; litopterns and perhaps gomphotheriids were browsers; dasypodiids are omnivores; smilodons, ursids, and the canid *Dusicyon avus* were carnivores. Although gomphotheres have been traditionally considered as browsers, grass phytolith assemblages in feces from three individual animals contained similarly high concentrations of pooids, suggesting that grasses were a significant part of the diet. Abundant pooid phytoliths, in addition to diatoms, indicate that these individuals grazed in cool and moist late Pleistocene environments (Gobetz and Bozarth, 2001).

TABLE 7.6. Extinct mammals with a body mass under 44 kg.

Taxon	Locality	South American age/stage	Reference
Canis dirus	Peru	Lujanian	Hoffstetter (1970)
Dusicyon avus	Argentina	Lujanian-Platan	Berman and Tonni (1987)
Protocyon orcesi	Ecuador	Bonaerian-Lujanian	Hoffstetter (1970)
Protocyon troglodytes	Brazil	Lujanian	Cartelle (1994)
Valgipes deformis	Brazil	Lujanian	Cartelle (1994)
Eligmodontia n. sp.	Argentina	Lujanian	Pardiñas, 1999
Microcavia robusta	Argentina	Lujanian	Quintana (1996)
Ctenomys lujanensis	Argentina	Lujanian	Morgan (1999)
Desmodus draculae	Brasil	Lujanian	Morgan (1999)
Protopithecus brasiliensis	Brasil	Lujanian	Cartelle (1994)

Constraints Related to Life-History Traits in Megamammals and Large Mammals

Some important megamammal life-history traits may have contributed to the extinction process because they made some species more susceptible to extinction than other mammals (Cione et al, 2003), especially when considered in the context of human hunting.

The age at which sexual maturity is attained is an important factor to be considered since it informs about how long a female must survive in the ecosystem to reach the age of first reproduction. In this regard, the sexual maturity of almost all extant female megammamals is greater than 10 years (Nowak, 1999); hence, a female must survive more than 10 years in nature to have its first offspring.

Johnson (2002) demonstrated correlations of both body mass and reproductive rate with extinction in mammals, and showed that extinction eliminated species at nearly the same threshold of reproductive rate in all groups, regardless of differences between groups in distributions of body size. He found that over all groups, the mean reproductive rate at which the probability of extinction reached 0.5 was 0.98 offspring per female per year. The correlation does not necessarily imply that a female will produce young every year (Nowak and Paradiso, 1983), but does seem to suggest that an average of less than one young per year gives a taxon < 50% chance of surviving.

The length of the gestation period is positively correlated with body mass in extant eutherian mammals. If we consider only terrestrial mammals, the longest period is that of elephants (668 days). At least in the last third part of that period the female is more vulnerable to predators and hunters due to reduced abilities to take evasive actions related to its added mass.

The length of lactation in mammals ranges almost three orders of magnitude, from 4 or 5 days to over 900 days. Although extremely short lactation lengths (<10 d) are rare, long lactation lengths (>500 d) are common for large-bodied species with single offspring (Hayssen and van Tienhoven, 1993). For mammals, the length of lactation is positively correlated with adult female mass. The maintenance of the parent and offspring bond during lactation and after weaning is crucial to offspring growth, development, and protection against predators and other threats, as well as necessary for learning from their parents the location of food, nest sites, etc. (Hayssen and van Tienhoven 1993). Hence, a young offspring needs its parents (at least its mother) to survive the first years.

Gregarious behavior of potential prey was explained by Hamilton (1971) on the basis of risk-sharing: the probability of being picked up by a predator is small when one is part of a large aggregate of prey. These defenses were evolved for "natural" predators (i.e., members of the Carnivora) but the Pleistocene humans did not behave the same as natural predators. For example, this is the case of armored xenarthrans, which humans would have hunted differently from carnivores which depended on claws and teeth, weak weapons against the xenarthran defenses. In our view, technology (i.e., the use of fire and rocks and others primitive weapons) and superior intellect constitute the difference between humans and other predators, hence some behavioral defenses of megamammals against natural predators would not be as effective as against humans.

There is correlation between population density and body size, resulting from the simple fact that for animals with similar metabolic processes (e.g., homeotherms), existing at the same trophic level (e.g., herbivores), equal amounts of biomass can be consumed by many individuals (Waguespack and Surovell, 2003). In this sense, it is important to note the amazing differences in population density between taxa with different body masses (see Table 1 in Wagespack and Surovell, 2003). In addition, several lines of theory link abundance with species richness. Clades whose species tend to be locally abundant will produce new species at a faster rate than clades of rare species (Isaac et al., 2005). If speciation rate is directly proportional to global population size (Hubbell, 2001) and abundance is a measure of global population size (Gaston, 1994), then abundance provides a natural buffer against extinction and leads to the accumulation of species (Isaac et al., 2005). Also, Kay (2002) suggests that large mammals and megamammals were few because they were limited in number by predators and, especially during the glacial periods, by the low basic plant productivity (see also Pielou, 1992).

American Mammals were Not Adapted to Humans

Many of the South American late Pleistocene mammals were of Holarctic origin. However, it is certain that neither these animals nor those of South American origin evolved in contact with humans. Quite certainly, these animals had not developed preventive reflexes against human hunting practices. There is no doubt that Paleoindians both in North and South America were at least occasionally big-game hunters, notwithstanding that they also used other resources for subsistence (Martin and Klein, 1984; Neves and Cornero, 1997; Borrero et al., 1998; Politis and Gutiérrez, 1998; Haynes, 2002).

Which of the Mammals Survived in South America?

No megamammal survived in South America and the only relatively large mammals that avoided extinction are inhabitants of areas of relatively difficult access for humans, such as (1) forests and cerrados (tapirs, *Tapirus terrestris*; large armadillos, *Priodontes giganteus*; peccaries, *Catagonus wagneri*; and some deers, *Blastoceros dichotomus; Hippocamelus bisulcus*); (2) mountainous areas (spectacled bear, *Tremarctos ornatus*; other tapirs, *Tapirus pinchaque* and *Tapirus bairdii*; vicuña, *Lama vicugna*; and deer, *Hippocamelus antisensis*); and (3) wetlands (carpinchos, *Hydrochoerus hydrochaeris*; cervids, *Ozotoceros bezoarticus*) (Johnson, 2002; Cione et al., 2003; Barnosky et al., 2004). *Priodontes giganteus* also has nocturnal habits. Other large tetrapods that survived are the more fully aquatic vertebrates such as river dolphins, manatees, seals, sea lions, and caimans (see Cione et al., 2000). The other survivors are some eurytopic species such as large carnivores (*Panthera onca*, *Puma concolor*). The guanaco is a special case because it is one of the most frequently cited taxa in archeological sites, but once again we could find an explanation in its life-history traits: the age of maturity in females is one year, the gestation period is 335 days, the reproductive rate is one offspring a year, the lactation period is six to 12 weeks and the parental care lasts one year, and the population density is 2.13 individual per square km (Sarno and Franklin, 1999; Nowak and Paradiso, 1983; Montes et al., 2000). Consequently, guanaco does not appear to be a species easy to extinguish. Besides, guanacos are fast, numerous, and have montane populations. In addition, there is evidence in the Andean region of early domestication of the guanacos (Pires Ferreira et al., 1976; F. Pujos, personal communication). The jaguar and puma include a large variety of prey in their diets and certainly were not especially adapted to feed on only those big herbivores that became extinct.

Man as a Hunter

Waguespack and Surovell (2003; Chapter 5) analyzed the zooarcheological record of North America and arrived at the conclusion that North American Paleoindians were big game specialists. The ratio between hunt effort and food obtained is obviously higher for a megammamal than for others mammals. Recently some authors (Grayson and Meltzer, 2003; Cannon and Meltzer, 2004) have expressed doubts about the role of man as a megafaunal specialist hunter. However, this is not necessary for explaining the South American extinction.

Lyons et al. (2004) analyzed the presently threatened and endangered mammals and found that those mammals menaced by human hunting were significantly larger than those subject to different extinction pressures. Thus, the prediction that early human hunters preferentially targeted large mammals is substantiated by modern hunting practices. In the scenario presented here a blitzkrieg is not necessary to explain the extinction. Only a moderate and occasional hunting of females (solitaries or with offspring), juveniles, or offspring is necessary to produce extinction in a few thousand years. Certainly, most xenarthrans would not be difficult to kill by trained hunters. Moreover, we consider that there was no time after the entrance of humans to allow the megamammals to evolve biological compensations for avoiding extinction.

The archeological evidence indicates that the overhunting was focused on guanacos and deer, which, paradoxically, are those mammals that survived the extinction. The scarcity or infrequent occurrence of megammamals in archeological sites more likely implies that these mammals were less abundant in the area, not that they were ignored or inaccessible to humans (see also Kay, 2002). Note that horses are infrequent in the archeological sites but were not more difficult to hunt than guanaco or deer; hence, a possible explanation is a reduced population. Another explanation is provided by Gary Haynes's observations (in Fiedel and Haynes, 2004) on modern killsites (both cultural and non-cultural) in Africa. Haynes observed that modern death sites are rarely preserved (less than 0.01% or less of total number being killed or dying naturally). Indirect evidence of hunting of some large mammals not recorded in archeological contexts were obtained from blood residue on lithic points (Kooyman et al., 2001). In South America, more than 15 extinct large mammals and megamammals were reported in association with Paleoindian sites (see above). However, many marks on bones should be taken with precaution as a direct evidence of hunting (F. Martin, 2003; pers. comm.)

How Many Humans were Required for the Extinction?

Simulations of human foraging on multiple prey yield the idea that overkill of slowbreeding prey is more likely when hunter populations became large (see Barnosky et al., 2004). However, according to the model of Alroy (2001), the number could be small (see also Kay, 2002), especially if we consider the putative reduced number of megamammals and large mammals in the continent (see above; see also Haynes, Chapter 3).

Large Extinct Carnivores

Large carnivores such as the short-faced bears of genus *Arctotherium* and the large saber-tooth cat *Smilodon populator* also became extinct (Soibelzon, 2002). Using phylogenetic comparative methods, Cardillo et al. (2004) show that extinction risk in the mammal order Carnivora is predicted more strongly by biology than exposure to high-density human populations. What was the human impact on carnivores competing for prey and carcasses? As Van Valkenburgh (2001) explains for Africa, if early *Homo* was consuming large prey regularly, it is probable that prey were often acquired by confrontational scavenging, as well as perhaps hunting, and this would be possible only because *Homo* had sufficient intelligence to overcome the superior strength, speed, and weaponry of the many other predators. Moreover, if we consider that humans who first inhabited America had advanced weaponry technology, it is easy to imagine that the resource competition with the medium to large carnivore/omnivore fauna was probably unequal. In this regard, we wonder what was the responsibility of man in the extinction of the medium to large carnivores/omnivores.

On the other hand, the absence or reduced number of top predators in the ecosystem could have large effects on the structure of the entire community. Both empirical and theoretical studies have shown that the loss of a top predator can trigger a cascade of secondary extinctions (Borrvall and Ebenman, 2006, and papers cited therein).

Wilmers and Getz (2005) and Wilmers and Post (2006) demonstrated that the presence of top predators provide insurance against the effects of climatic change. Wilmers et al. (2006) studied the consequences of the absence of wolves in two ecosystems and concluded that the top predators have buffering effects on changes in populations numbers (of other taxa) caused by climatic variations. Regardless of the importance of specific predators in particular systems, Wilmers et al.'s (2006) findings strongly support the hypothesis that intact food webs including top predators appear to be more resistant to stress. Bottom-up factors thus appear to be more important when food chains are shortened and top-down control is reduced (Sala, 2006).

We do not reject the possibility that some herbivores could also have been exterminated by cascade effects (see Owen-Smith, 1987).

Timing of Extinction in the Americas

From a biogeographic point of view, if humans first entered South America from North America, and if the extinction was caused by them, it would have to be accomplished earlier in North America than in South America. Not surprisingly, the youngest records of extinct mammals in North America are dated at 10,370 BP (i.e., latest Pleistocene; Beck, 1996) while in South America, the youngest records appear to correspond to the early Holocene. However, the dispersal of humans in South America could have been very fast. Youngblood (2002) calculated that humans could have traveled from Beringia to central Chile in 250 years. Available evidence suggests that the extermination process lasted longer in South America than in North America. This long lasting process could have been related to the peculiar climatic history and biocenotic evolution of each continent.

In Summary

The distribution and biomass of the flora and fauna had strongly fluctuated many times during the middle-late Pleistocene in a Zig-Zag but we have not identified important extinctions until the end of the Pleistocene and the beginning of the Holocene. However, almost all living mammal species make their first appearance in the Pleistocene (Lister, 2004).

The studies based on geochemical and dust proxies in Greenland, Antarctica, and South American glaciers attest that the present interglacial is not substantially different from the preceding ones.

The arrival of humans in South America was the sole new biological or geological event that occurred in the present interglacial.

The late Pleistocene and earliest Holocene South American mammal fauna was very different from the Recent, not only by the presence of many extinct very large mammals but also because of the different distribution of many animal and plant species.

All megamammals (37 species) and most large mammals (46 species) present during the late Lujanian (latest Pleistocene-earliest Holocene) became extinct in South America. Several did not overlap their geographical distributions.

Many of the extinct mammals were large and scarce. Females probably reached sexual maturity late, had a very long gestation period (one year or more), and prolonged parental care, implying one offspring in two or three years, and had a total low number of offspring during life.

Pleistocene megamammals and extinct large mammals were mostly adapted to open areas. The shrinking of open areas in the interglacials (including the present) provoked population fragmentation. Thus, the populations that originally inhabited the large open areas became isolated in large "islands" (see Haynes, Chapter 3, and the discussion of sources and sinks). The consequences of population isolation is gene flow interruption. The main disadvantage associated with gene flow interruption is the loss of genetic variability which offers some population "resistance" to new diseases, environmental severities, etc., and the emergence of genetic problems associated with inbreeding. On the other hand, natural barriers prevented dispersal of species to counteract local extinctions.

The few large mammals that survived (14 species) are adapted to forests, mountainous areas, or wetlands, or at least some populations inhabit these kinds of environments. Some have nocturnal habits. Other large vertebrates usually not considered in the extinction discussion, such as pinnipeds, manatees, and caimans, inhabit protected aquatic environments.

Shortly after humans entered South America, climate changed and open areas began to shrink, especially after the beginning of the Holocene.

It has been documented that humans hunted many of the mammals that became extinct. The ratio between hunt effort and food obtained is obviously higher for a megammamal than other mammals.

The extinction appears to have been more concentrated in taxa of South American origin. However, this is mainly apparent because many of the large mammals were xenarthrans (and litopterns and notoungulates).

Contrasting with North America, recent dating suggests that several megamammals became extinct in South America during the early Holocene (perhaps as late as 7,000 BP). As humans entered South America ca. 13,000-11,000 BP, the process of extinction endured several thousand years (perhaps up to 6,000 years). So, chronological evidence does not support the blitzkrieg model for South America.

The extinction event in South America was the largest in the world at the time and perhaps the youngest one at a large scale.

We believe that in the particular situation of a dramatic shrinking of geographic range and the number of large, mostly slow and naive mammals, a relatively small number of active foragers with specialized weapons would have been able to exterminate many of them after no more than several thousand years of cohabitation. Perhaps only a moderate and occasional hunting of females (solitaries or with offspring), juveniles, or offspring would be necessary to produce their extinction in few thousand years. Quite certainly, when populations diminished to a threshold, they became no longer viable. With the demise of many mammals and the taking of other actions such the use of fire, humans altered the ecologic equilibrium of the continent, provoking other extinctions in cascade. The large carnivores such as *Smilodon populator* probably became extinct when their preferred prey victims disappeared.

Finally, it is plausible that without the appearance of humans in the continent, the extraordinarily varied fauna of large mammals and megamammals that characterized South America during the Pleistocene should have recovered and survived, as it had occurred after each of the previous interglacial periods.

Acknowledgments. The authors would like to thank the following institutions and workers: For the invitation to participate in this book: Gary Haynes. For the valuable reviewing that permitted great improvement of the manuscript: Gary Haynes and two anonymous referees. For valuable information: Luis Borrero, María Susana Bargo, Mariano Bond, Marcel Cardillo, Alfredo Carlini, Bo Ebenman, Stuart Fiedel, Jean-Michel Gaillard, Donald Grayson, Nick Isaac, Charles Kay, Jennifer Leonard, Omar Linares, Adrian Lister, Kate Lyons, Fabiana Martin, Gustavo Politis, Francisco Prevosti, François Pujos, Gustavo Scillato-Yané, Tony Stuart, Sergio Vizcaíno, Nicole M. Waguespack, Lars Werdelin, and Alfredo Zurita. For financial support: Consejo Nacional de Investigaciones Científicas y Técnicas, Comisión de Investigaciones Científicas de la Provincia de Buenos Aires, Agencia Nacional de Promoción Científica y Tecnológica, and Universidad Nacional de La Plata.

References

Aguirre ML, Whatley RC (1995) Late Quaternary marginal marine deposits and palaeoenvironments from northeastern Buenos Aires province, Argentina: A review. Quat Sci Rev 14:223–254

Alberdi MT, Menegaz A, Prado JL, Tonni EP (1988) La fauna local Quequén Salado-Indio Rico (Pleistoceno tardío) de Buenos Aires, Argentina. Aspectos paleoambientales y bioestratigráficos. Ameghiniana 25:225–236

Alberdi MT, Prado JL (1992) El registro de *Hippidion* Owen 1869 y *Equus* (*Amerhippus*) Hoffstetter 1950 (Mammalia, Perissodactyla) en América del sur. Ameghiniana 29:265–284

Alberdi MT, Miotti L, Prado JL (2001) *Hippidion saldiasi* Roth 1899 (Equidae, Perissodactyla) at the Piedra Museo site (Santa Cruz, Argentina): Its implication for the regional economy and environmental reconstruction. J Archaeol Sci 28:411–419

Alroy J (2001) A multispecies overkill simulation of the end-Pleistocene megafaunal mass extinction. Science 292:1893–1896

Ameghino F (1889) Contribución al conocimiento de los mamíferos fósiles de la República Argentina. Actas de la Academia Nacional de Ciencias en Córdoba 6:1–1027

Aramayo SA (1997) Cronología radiocarbónica de localidades fosilíferas pleistocenas y holocenas de la costa sud-sudeste de la provincia de Buenos Aires, Argentina. Atas do VI Congresso da Associação Brasileira do Estudos do Quaternário, Curitiba, Brazil, pp 305–308

Arribas A, Palmquist P, Pérez-Claros JA, Castilla R, Vizcaíno SF, Fariña R (2001) New evidence on the interaction between humans and megafauna in South America. XVII Jornadas Sociedad Española de Paleontología, Valencia, Spain, pp 228–238

Bard E, Hamelin B, Fairbanks RG, Zindler A (1990) Calibration of the ^{14}C timescale over the past 30,000 years using mass spectometric U-Th ages from Barbados corals. Nature 345:405–410

Barnosky AD, Koch PL, Feranec R, Wing SSL, Shabel AB (2004) Assessing the causes of late Pleistocene extinctions on the continents. Science 306:70–75

Beck M (1996) On discerning the cause of late Pleistocene megafaunal extinctions. Paleobiology 22:91–103

Behling H (2002) South and southeast Brazilian grasslands during Late Quaternary times: A synthesis. Palaeogeogr Palaeocl Palaeoecol 177:19–27

Behling H, Arz H, Pätzold J, Wefer G (2002) Late Quaternary vegetational and climate dynamics in southeastern Brazil, inferences from marine cores GeoB 3229–2 and GeoB 3202–1. Palaeogeogr Palaeocl Palaeoecol 179:227–243

Behling H, DePatta Pillarb V, Girardi Bauermann S (2005) Late Quaternary grassland (Campos), gallery forest, fire and climate dynamics, studied by pollen, charcoal and multivariate analysis of the São Francisco de Assis core in western Rio Grande do Sul (southern Brazil). Rev Palaeobot Palynol 133:235–248

Berman D, Tonni EP (1987) *Canis (Dusicyon) avus* Burmeister,1864 (Carnivora, Canidae) en el Pleistoceno tardío y Holoceno de la provincia de Buenos Aires. Aspectos sistemáticos y bioestratigráficos relacionados. Ameghiniana 24:245–250

Blunier T, Brook EJ (2001) Timing of millenial-scale climate changes in Antarctica and Greenland during the last glacial period. Science 291:109–112

Bond M, Cerdeño E, López GM (1995) Los ungulados nativos de América del Sur. In: Alberdi MT, Leone G, Tonni EP (eds) Evolución biológica y climática de la región pampeana durante los últimos cinco millones de años. CSIC Monografías 12:257–277

Borrero LA (1997) La extinción de la megafauna en la Patagonia. Anales del Instituto de la Patagonia 23:89–102

Borrero LA, Zárate M, Miotti L, Massone M (1998) The Pleistocene-Holocene transition and human occupations in the southern cone of South America. Quat Intl 49/50:191–199

Borrvall C, Ebenman B (2006) Early onset of secondary extinctions in ecological communities following the loss of top predators. Ecol Lett 9:435–442

Broecker W (2001) Was the medieval warm period global? Science 291:1497–1499

Cabrera AL, Willink A (1973) Biogeografía de América Latina. Programa de Desarrollo Científico y Tecnológico, Serie Biología 13:1–120

Cannon MD, Meltzer DJ (2004) Early Paleoindian foraging: Examining the faunal evidence for large mammal specialization and regional variability in prey choice. Quat Sci Rev 23:1955–1987

Cardich A (1977) Las culturas Pleistocénicas y Postpleistocénicas de Los Toldos y un bosquejo de la prehistoria de Sudamérica. Obra del Centenario del Museo La Plata 2:149–172

Cardillo M, Purvis A, Sechrest W, Gittleman JL, Bielby J, Mace GM (2004) Human population density and extinction risk in the world's carnivores. PLoS Biology 2:909–914

Carignano C (1999) Late Pleistocene to recent climate change in Córdoba Province, Argentina: Geomorphological evidence. Quat Intl 57/58:117–134

Carlini AA, Scillato-Yané G (1999) In: Tonni E. P., Cione A. L. (eds), Quaternary vertebrate palaeontology in South America. Quat South Am Antarctic Peninsula 12:149–175

Carlini AA, Zurita AE (2006) *Glyptotherium* Osborn (Mammalia, Xenarthra, Cingulata) en el Pleistoceno Tardío de Venezuela. Abstracts 9° Congreso Argentino de Paleontología y Bioestratigrafía, p 98

Carlini AA, Tonni EP (2000) Mamíferos fósiles del Paraguay. SEAM. BGR, La Plata.

Cartelle C (1994) Tempo passado. Mamiferos do Pleistoceno em Minas Gerais. Palco Ed., Brasilia

Casamiquela R (1999) The Pleistocene vertebrate record of Chile. In: Tonni EP, Cione AL (eds) Quaternary vertebrate paleontology in South America. Quat South Am Antarctic Peninsula 12:91–107

Cione AL, Azpelicueta MM, Bond M, Carlini AA, Casciotta JR, Cozzuol MA, de la Fuente M, Gasparini Z, Goin FJ, Noriega J, Scillato-Yané GJ, Soibelzon LH, Tonni EP, Verzi D, Vucetich MG (2000) Miocene vertebrates from Paraná, Eastern Argentina. In: Aceñolaza FG, Herbst R (eds) El Neógeno de Argentina INSUGEO, serie Correlación Geológica 14:191–237, Tucumán

Cione AL, López Arbarello A (1994) La ictiofauna del Bajo San José (provincia del Buenos Aires, Pleistoceno temprano). Su significación paleoambiental y paleobiogeográfica. VI Congreso Argentino de Paleontología y Biostratigrafía, Trelew, Argentina, pp 22–23

Cione AL,Tonni EP (1995) Chronostratigraphy and "Land-mammal ages" in the Cenozoic of southern South America: Principles, practices, and the "Uquian" problem. J Paleontol 69:135–159

Cione AL,Tonni EP (1996) Reassesment of the Pliocene-Pleistocene continental time scale of southern South America. Correlation of the Chapadmalalan with Bolivian sections. J South Am Earth Sci 9:221–236

Cione AL,Tonni EP (1999) Biostratigraphy and chronological scale of uppermost Cenozoic in the pampean area, Argentina. In: Tonni EP, Cione AL (eds) Quaternary vertebrate palaeontology in South America. Quat South Am Antart Peninsula 12:23–51

Cione AL, Tonni EP (2001) Correlation of Pliocene to Holocene southern South American and European mammal-bearing units. Boletin della Società Paleontologica Italiana 40:167–174

Cione AL, Tonni EP (2005) Bioestratigrafía basada en mamíferos del Cenozoico superior de la región pampeana. In: Barrio R, Etcheverry RO, Caballé MF, Llambías E (eds) Geología y Recursos Minerales de la Provincia de Buenos Aires. Relatorio del XVI Congreso Geológico Argentino. La Plata 11:183–200

Cione AL, Tonni EP, Bond M, Carlini AA, Pardiñas UF, Scillato-Yané GJ, Verzi D, Vucetich MG (1999) Occurrence charts of Pleistocene mammals in the pampean area, eastern Argentina. In: Tonni EP, Cione AL (eds) Quaternary vertebrate palaeontology in South America. Quat South Am Antart Peninsula 12:53–59

Cione AL, Tonni EP, Figini AJ (2001) Did the megafauna range to 4,300 yr BP in South America? Radiocarbon 43:69–75

Cione AL, Tonni EP, Soibelzon LH (2003) The broken zig-zag: Late Cenozoic large mammal and turtle extinction in South America. Revista del Museo Argentino de Ciencias Naturales "Bernardino Rivadavia" 5:1–19

Cione AL, Tonni EP, Dondas A (2005) A mastodont (Mammalia, Gomphotheriidae) from the Argentinian continental shelf. Neues Jahrbuch für Geologie und Paläontologie, Mitteilungen 2005:614–630

Clapperton C (1993a) Nature and environmental changes in South America at the Last Glacial Maximum. Palaeogeogr Palaeocl Palaeoecol 101:189–208

Clapperton C (1993b) Quaternary geology and geomorphology of South America. Elsevier, Amsterdam

Coltorti M, Ficcarelli G, Jahren H, Moreno Espinosa M, Rook L, Torre D (1998) The last occurrence of Pleistocene megafauna in the Ecuadorian Andes. J South Am Earth Sci 11:581–586

Collantes M, Powell J, Sayago JM (1993) Formación Tafí del Valle (Cuaternario superior), provincia de Tucumán (Argentina): Litología, paleontología y paleoambientes. Actas XII Congreso Argentino de Geologia y II Congreso Exploración de Hidrocarburos 1:200–206

Cook KH, Vizy EK (2006) Delayed onset of the South American summer monsoon during the Last Glacial Maximum. PAGES News, pp 17–18

Deschamps CM (2005) Late Cenozoic mammal bio-chronostratigraphy in southwestern Buenos Aires province, Argentina. Ameghiniana 42:733–750

Dillehay T (1989) The cultural relationships of Monte Verde: A late Pleistocene settlement site in the Subantarctic forest of south central Chile. In: Bryan AL (ed) New evidence for the Pleistocene peopling of the Americas, Center for the Study of Early Man, University of Maine, Orono, pp 319–337

Ferigolo J (1999) Late Pleistocene South American land-mammal extinctions: The infection hypothesis In: Tonni EP, Cione AL (eds) Quaternary vertebrate palaeontology in South America. Quat South Am Antart Peninsula 12:279–310

Ficcarelli G, Azzaroli A, Bertini A, Coltorti M, Mazza P, Mezzabotta C, Moreno Espinosa M, Rook L, Torre D (1997) Hypothesis on the cause of extinction of the South American mastodonts. J South Am Earth Sci 10:29–38

Fidalgo F (1992) Provincia de Buenos Aires. Continental. In: Iriondo M (ed) El Holoceno de Argentina. Cadinqua, Buenos Aires

Fidalgo F, De Francesco F, Pascual R (1975) Geología superficial de la llanura bonaerense. Relatorio VI Congreso Geologico Argentino, pp 103–138

Fidalgo F, Meo Guzmán LM, Politis GG, Salemme M, Tonni EP (1986) Investigaciones arqueológicas en el sitio 2 de Arroyo Seco (Pdo. de Tres Arroyos-Provincia de Buenos Aires, República Argentina). In: Bryan AL (ed) New evidence for the Pleistocene peopling of the Americas. Center for the Study of Early Man, University of Maine, Orono, pp 221–269

Fiedel SJ (1999) Older than we thought: Implications of corrected dates for Paleoindians. Am Ant 64:95–115

Fiedel S, Haynes G (2004) A premature burial: Comments on Grayson and Meltzer's "Requiem for overkill". J Archaeol Sci 31:121–131

Figini A, Fidalgo F, Huarte R, Carbonari J, Gentile R (1995) Cronología radiocarbónica de los sedimentos de la Formación Luján en Arroyo Tapalqué, provincia de Buenos Aires. IV Jornadas Geologicas y Geofisicas Bonaerenses, Junín, Argentina, pp 119–126

Flegenheimer N, Zárate M (1997) Considerations on radiocarbon and calibrated dates from Cerro La China and Cerro El Sombrero, Argentina. Curr Res Pleistocene 14:27–28

Gaston KJ (1994) Rarity. Chapman & Hall, London

Gobetz K, Bozarth S (2001) Implications for Late Pleistocene mastodon diet from opal phytoliths in tooth calculus. Quat Res 55:115–122

Graham RW (1986) Plant-animal interactions and Pleistocene extinctions. In: Elliott DK (ed) Dynamics of extinction. Wiley, New York, pp 131–154

Graham RW, Lundelius EL Jr (1984) Coevolutionary disequilibrium and Pleistocene extinctions. In: Martin PS, Klein RG (eds) Quaternary extinctions: A prehistoric revolution. University of Arizona Press, Tucson, pp 223–249

Graham RW, Lundelius EL Jr, Graham M, Schroeder EK, Toomey RS III, Anderson E, Barnosky AD, Burns JA, Churcher C, Grayson D, Guthrie RD, Harington CR, Jefferson GT, Martin LD, McDonald HG, Morlan RE, Semken HA Jr, Webb SD, Werdelin L, Wilson MC (1996) Spatial response of mammals to late Quaternary environmental fluctuations. Science 272:1601–1606

Grayson DK, Meltzer DJ (2003) Clovis hunting and large mammal extinction: A critical review of the evidence. J World Prehistory 16:313–359

Gruhn R (1997) The South American context of the Pedra Pintada site in Brazil. Curr Res the Pleistocene 14:29–31

Gruhn R, Bryan A (1984) The record of Pleistocene megafaunal extinction at Taima-Taima, northern Venezuela. In: Martin P, Klein R (eds) Pleistocene extinctions: A prehistoric revolution. University of Arizona Press, Tucson, pp 128–137

Guthrie RD (2003) Rapid body size decline in Alaskan Pleistocene horses before extinction. Nature 426:169–171

Hamilton WD (1971) Geometry for the selfish herd. J Theoret Biol 31:295–311

Haynes G (2002) The catastrophic extinction of North American mammoths and mastodonts. World Archaeol 33:391–416

Hayssen V, van Tienhoven A (1993) Asdell's patterns of mammalian reproduction. A compendium of species specific data. Comstock, Ithaca, NY

Heusser L, Heusser C, Pisias N (2006) Vegetation and climate dynamics of southern Chile during the past 50,000 years: Results of ODP site 1233 pollen analysis. Quat Sci Rev 25:474–485

Hoffstetter R (1970) Vertebrados Cenozoicos y mamíferos Cretácicos de Perú. Actas IV Congreso Latinoamericano de Zoología 2:971–983

Hoffstetter R (1986) High Andean mammalian faunas during the Plio-Pleistocene. In: Vuilleumier F (ed) High altitude tropical biogeography. Oxford University, New York

Hubbell SP (2001) The unified neutral theory of biodiversity and biogeography: Monographs in population biology. Princeton University Press, Princeton, NJ

Iriondo M (1997) Models of deposition of loess and loessoids in the upper Quaternary of South America. J South Am Earth Sci 10:71–79

Iriondo M (1999) Climatic changes in the South American plains: Records of a continent-scale oscillation. Quat Intl 57/58:93–112

Isaac NJB, Jones KE, Gittleman JL, Purvis A (2005) Correlates of species richness in mammals: Body size, life history, and ecology. The American Naturalist 165:600–607

Johnson CN (2002) Determinants of loss of mammal species during the Late Quaternary 'megafauna' extinctions: Life history and ecology, but not body size. Proc Roy Soc Lond B 269:2221–2227

Kay CE (2002) False gods, ecological myths, and biological reality. In: Kay CE, Simmons RT (eds) Wilderness and political ecology. University of Utah Press, Salt Lake City, pp 238–261

Kooyman B, Newman ME, Cluney C, Lobb M, Tolman S, McNeill P, Hills LV (2001) Identification of horse exploitation by Clovis hunters basesd on protein analysis. Am Ant 66:686–691

Linares O (1993) Mamíferos megaherbívoros en la Guayana venezolana: Indicios de aridez durante el Cuaternario tardío – Holoceno. V Reunión Anual del Proyecto PICG-281 "Climas cuaternarios de América del Sur", Santiago de Chile, Chile, pp 52

Lister AM (2004) The impact of Quaternary ice ages on mammalian evolution. Philosoph Trans Roy Soc Lond 359:221–241

Lyons SK, Smith FA, Brown JH (2004) Of mice, mastodons and men: Human-mediated extinctions on four continents. Evol Ecol Res 6:339–358

MacFadden B (2000) Middle Pleistocene climate change recorded in fossil mammal teeth from Tarija, Bolivia, and upper limit of the Ensenadan land-mammal age. Quat Res 54:121–131

MacNeish RS, Berger R, Protsch R (1970) Megafauna and man from Ayacucho, highland Peru. Science 168:975–977

Marchant R, Boom A, Behlin H, Hooghiemstra H, Melief B, Van Geel B, Van Der Hammen T, Wille M (2004) Colombian vegetation at the Last Glacial Maximum: A comparison of model- and pollen-based biome reconstructions. J Quat Sci 19:721–732

Marshall L, Berta A, Hoffstetter R, Pascual R, Reig OA, Bombin M, Mones A (1984) Mammals and stratigraphy: Geochronology of the continental mammal-bearing Quaternary of South America. Palaeovertebrata Mémoire Extraordinaire, pp 1–76

Martin PS, Klein RG (eds) (1984) Quaternary extinctions. A prehistoric revolution. University of Arizona Press, Tucson

Martínez GA (1997) A preliminary report on Paso Otero 5, a late Pleistocene site in the Pampean region. Curr Res Pleistocene 14:53–54

Mazzanti DL (1997) Excavaciones arqueológicas en el Sitio Cueva Tixi, Buenos Aires, Argentina. Latin Am Ant 8:55–62

Mazzanti DL, Quintana C (1997) Asociación cultural con fauna extinguida en el sitio arqueológico Cueva Tixi, provincia de Buenos Aires, Argentina. Revista Española de Antropología Americana 27:11–21

McCulloch RD, Bentley MJ, Purves RR, Hulton NJR, Sugden DE, Clapperton CM (2000) Climatic inferences from glacial and palaeoecological evidence at the last glacial termination, southern South America. J Quat Sci 15:409–417

Melo DJ de, Cássia Tardin Cassab R de, Ventura dos Passos F (2005) Coleção de mamíferos Pleistocênicos de Águas de Araxá, no Museu de Ciências da Terra. X Congresso de ABEQUA eletronic version without pagination

Messineo P, Politis GG (2006) The Campo Laborde site: New evidence for the Holocene survival of pleistocene megafauna in the Argentine pampas. International Council for Archaeozoology, 10th. Conference, abstracts, Mexico, pp 127–128.

Miotti L (1993) La ocupación humana en la Patagonia austral durante el Holoceno. In: Iriondo M (ed) El Holoceno en la Argentina. Cadinqua, Buenos Aires, Argentina, pp 94–130

Miotti L, Cattaneo R (1997) Bifacial/Unifacial technology c. 13,000 years ago in southern Patagonia. Curr Res Pleistocene 14:62–64

Montes C, De Lamo DA, Zavatti J (2000) Distribución de abundancias de guanacos (*Lama guanicoe*) en los distintos ambientes de Tierra del Fuego, Argentina. Mastozoología Neotropical 7:23–31

Morgan GS (1999) Late-Pleistocene climatic change and neotropical bat extinctions. Curr Res Pleistocene 16:133–133

Nabel P, Cione AL, Tonni EP (2000) Environmental changes in the pampean area of Argentina at the Matuyama–Brunhes (C1r-C1n) Chrons boundary. J South Am Earth Sci 162:403–412

Nami HH, Nakamura I (1995) Cronología radiocarbónica con AMS sobre muestras de hueso procedentes del sitio Cueva del Medio (Última Esperanza, Chile). Anales Instituto de la Patagonia 23:125–133

Neves W, Cornero S (1997) What did South American Paleoindians eat? Curr Res Pleistocene 14:93–96

Nowak RM (1999) Walker's mammals of the world, 6th ed., vols I & II. Johns Hopkins University Press, Baltimore

Nowak RM, Paradiso JL (1983) Walker's mammals of the world. Johns Hopkins University Press, New York

Núñez L, Varela J, Casamiquela R, Villagrán C (1994) Cuenca de Tagua-Tagua en Chile: el ambiente Pleistoceno superior y ocupaciones humanas. Revista Chilena de Historia Natural 67: 503–519

Owen-Smith N (1987) Pleistocene extinctions: The pivotal role of megaherbivores. Paleobiology 13:351–362

Pardiñas UFJ (1999) Fossil murids: Taxonomy, palaeoecology and biostratigraphy. In: Tonni EP, Cione AL (eds) Quaternary vertebrate palaeontology in South America. Quat South Am Antarct Peninsula 12:225–253

Pardiñas UFJ, Cione AL, San Cristóbal J, Verzi D, Tonni EP (2004) A new Last Interglacial continental vertebrate assemblage in central-eastern Argentina. Curr Res Pleistocene 21:111–112

Pardiñas UFJ, Gelfo J, San Cristóbal J, Cione AL, Tonni EP (1996) Una asociación de organismos marinos y continentales en el Pleistoceno superior en el sur de la provincia de Buenos Aires, Argentina. Actas XIII Congreso Geológico Argentino y III Congreso de Exploracion de Hidrocarburos 5:95–111

Pascual R, Ortega Hinojosa E, Gondar D, Tonni EP (1965) Las edades del Cenozoico mamalífero de la Argentina, con especial atención a aquellas del territorio Bonaerense. Anales Comision de Investigaciones Cientificas 6:165–193

Petit JR, Jouzel J, Raynaud D, Barkov NI, Barnola JM, Basile I, Bender M, Chappellaz J, Davis M, Delaygue G, Delmotte M, Kotlyakov V, Legrand M, Lipenkov V, Lorius C, Pepin L, Ritz C, Saltzman E, Stievenard M (1999) Climate and atmospheric history of the past 420,000 years from the Vostok ice core, Antarctica. Nature 399:429–436

Pielou EC (1992) After the Ice Age. The University of Chicago Press, Chicago

Pires Ferreira J, Pires Ferreira E, Kaulicke P (1976) Preceramic utilization in the central Peruvian Andes. Science 194:483–490

Politis G, Gutiérrez M (1998) Gliptodontes y cazadores-recolectores de la región pampeana (Argentina). Latin Am Ant 9:111–134

Politis GG, Johnson E, Gutiérrez M, Hartwell WT (2003) Survival of Pleistocene fauna: New radiocarbon dates on organic sediments from La Moderna (pampean region, Argentina). In: Miotti L, Salemme M, Flegenheimer N (eds) Where the south winds blow. Ancient evidence of Paleo South Americans. Center for the Study of the First Americans, pp 45–50

Politis G, Messineo PG, Kaufmann C (2004) El poblamiento temprano de las llanuras pampeanas de Argentina y Uruguay. Complutum 15:207–224

Politis G, Prado JL, Beukens RP (1995) The human impact in Pleistocene-Holocene extinctions in South America – The Pampean case. In: Johnson E (ed) Ancient peoples and landscapes. Museum of Texas Tech University, Lubbock, pp 187–205

Prado JL, Menegaz A, Tonni EP, Salemme M (1987) Los mamíferos de la fauna local Paso Otero (Pleistoceno tardío), provincia de Buenos Aires. Aspectos paleoambientales y bioestratigráficos. Ameghiniana 24:217–233

Prieto R (1996) Late Quaternary vegetational and climatic changes in the Pampa grassland of Argentina. Quat Res 45:73–88

Prevosti FJ, Soibelzon L, Prieto A, San Roman M, Morello F (2003) The southernmost record of Ursidae: The occurrence of *Pararctotherium* (Tremarctinae) in the latest Pleistocene of southern Patagonia (Chile). J Vert Paleontol 23:709–712

Pujos F, Salas R (2004) A new species of *Megatherium* (Mammalia: Xenarthra: Megatheriidae) from the Pleistocene of Sacaco and Tres Ventanas, Peru. Palaeontology 47:579–604

Quintana C (1996) Diversidad del roedor *Microcavia* (Caviomorpha, Caviidae) de América del Sur. Mastozoología Neotropical 3:63–86

Ramírez Rozzi FV, Errico FD, Zárate M (2001) Le site Paléo-Indien de Piedra Museo (Patagonie). Sa contribution au débat sur le premier peuplement du continent Américain. Comptes Rendus de l'Académie des Sciences – Series IIA – Earth Planet Sci 331:311–318

Raup DM (1992) Extinction. WW Norton, New York

Retallack GJ (1998) Fossil soils and completeness of the rock and fossil records. In: Donovan SK, Paul CR (eds) The adequacy of the fossil record. Wiley, Chichester

Roosevelt A, Lima da Costa M, Lopes Machado C, Michab M, Mercier N, Valladas H, Feathers J, Barnett W, Imazio da Silveira M, Henderson A, Sliva J, Chernoff B, Reese D, Holman J, Toth N, Schick K (1996) Paleoindian cave dwellers in the Amazon: The peopling of the Americas. Science 272:373–384

Rossello EA, Bor-Ming J, Tsung-Kwei L, Petrocelli JL (1999) New 4,300 YR 14C age of glyptodonts at Luján river (Buenos Aires, Argentina) and its implications. II Simposio Sudamericano de Geología Isotópica, Córdoba, Argentina, pp 105–110

Roy K (2001) Pleistocene extinctions. In: Briggs D, Crowther P (eds) Palaeobiology II. Blackwell Science, London

Sala E (2006) Top predators provide insurance against climatic change. Trends Ecol Evol 21:479–480

Sarno RJ, Franklin WL (1999) Maternal expenditure in the polygynous and monomorphic guanaco: Suckling behavior, reproductive effort, yearly variation, and influence on juvenile survival. Behav Ecol 10:41–47

Schmidt Dias A, Jacobus AL (2002) The antiquity of the peopling of southern Brazil. Curr Res Pleistocene 18:17–20

Scillato-Yané G, Carlini AA, Tonni EP, Noriega JI, (2005) Paleobiogeography of the late Pleistocene pampatheres of South America. J South Am Earth Sci 20:131–138

Scillato-Yané G, Carlini AA, Vizcaíno SF, Ortiz Jaureguizar E (1995) Los xenartros. In: Alberdi MT, Leone G, Tonni EP (eds) Evolución biológica y climática de la región pampeana durante los últimos cinco millones de años. Monografías del Museo Nacional de Cienicas Naturales 12:181–209

Sheldon PR (1998) Plus ça change – a model for stasis and evolution in different environments. Palaeogeogr Palaeocl Palaeoecol 127: 227–238

Simberloff D (1986) Are we on the verge of a mass extinction in tropical rain forests? In: Elliott DK (ed) Dynamics of extinction. Wiley Interscience, New York, pp 165–180

Simpson G (1980) Splendid isolation. Yale University Press, New Haven

Soibelzon LH (2002) Los Ursidae (Carnivora, Fissipedia) fósiles de la República Argentina. Aspectos sistemáticos y paleoecológicos. Ph.D. Dissertation, Universidad Nacional de La Plata

Soibelzon LH, Tonni EP, Bond M (2005) The fossil record of the South American short-faced bears (Ursidae, Tremarctinae). J South Am Earth Sci 20:105–113

Steadman DW, Martin PS, MacPhee RDE, Jull AJT, McDonald HG, Woods CA, Iturralde-Vinent M, Hodgins GWL (2005) Asynchronous extinction of late Quaternary sloths on continents and islands. PNAS 102:11763–11768

Steig EJ (2006) The south-north connection. Nature 444:152–153

Suárez R (2003) Paleoindian components of northern Uruguay: New data on early human occupations of the late Pleistocene and early Holocene. In: Miotti L, Salemme M, Flegenheimer N (eds) Where the south winds blow. Ancient evidence of Paleo South Americans. Center for the Study of the First Americans, pp 29–36.

Thompson LG (2000) Ice core evidence for climate change in the Tropics: Implications for our future. Quat Sci Rev 19:19–35

Tonni EP (1990) Mamíferos del Holoceno en la provincia de Buenos Aires. Paula-Coutiana 4:3–21

Tonni EP, Alberdi MT, Prado JL, Bargo MS, Cione AL (1992) Changes of mammal assemblages in the pampean region (Argentina) and their relation with the Plio-Pleistocene boundary. Palaeogeogr Palaeocl Palaeoecol 95:179–194

Tonni EP, Cione AL (1997) Did the Argentine pampean ecosystem exist in the Pleistocene? Curr Res Pleistocene 14:131–133

Tonni E.P., Cione A (eds). 1999. Quaternary vertebrate palaeontology in South America. Quaternary of South America and Antarctic Peninsula 12, 1–310.

Tonni EP, Cione AL, Figini A (1999a) Predominance of arid climates indicated by mammals in the pampas of Argentina during the late Pleistocene and Holocene. Palaeogeogr Palaeocl Palaeoecol 147:257–281

Tonni EP, Cione AL, Figini A (2001) Chronology of Holocene pedogenetic events in the pampean area of Argentina. Curr Res Pleistocene 18:124–127

Tonni EP, Fidalgo F (1978) Consideraciones sobre los cambios climáticos durante el Pleistoceno tardío – Reciente en la provincia de Buenos Aires. Aspectos ecológicos y zoogeográficos relacionados. Ameghiniana 15:235–253

Tonni EP, Huarte R, Carbonari JE, Figini AJ (2003) New radiocarbonic chronology for the Guerrero member of the Luján Formation (Buenos Aires, Argentina). Paleoclimatic significance. Quat Intl 109/110:45–48

Tonni EP, Nabel P, Cione AL, Etchichury M, Tófalo R, Scillato-Yané GJ, San Cristóbal J, Carlini A,Vargas D (1999b) The Ensenada and Buenos Aires formations (Pleistocene) in a quarry near La Plata, Argentina. J South Am Earth Sci 12:273–291

Tonni EP, Politis G (1980) La distribución del guanaco (Mammalia, Camelidae) en la provincia de Buenos Aires durante el Pleistoceno tardío y Holoceno. Los factores climáticos como causa de su retracción. Ameghiniana 17:53–66

Tonni EP, Scillato-Yané GJ (1997) Una nueva localidad con mamíferos Pleistocenos en el norte de Argentina. Aspectos paleozoogeográficos. Atas do VI Congreso da Associação Brasileira do Estudos do Quaternário, Curitiba, Brazil, pp 345–348

Ubilla M, Perea D (1999) Quaternary vertebrates of Uruguay: A biostratigraphic, biogeographic and climatic overview. In: Tonni EP, Cione AL (eds) Quaternary vertebrate palaeontology in South America. Quat South Am Antarct Peninsula 12:75–91

Van Valkenburgh B (2001) The dog-eat-dog world of carnivores. A review of past and present carnivore community dynamics. In: Stanford C, Bunn HT (eds) Meat-eating and human evolution. Oxford University Press, Oxford, pp 101–121

Verzi DH, Tonni EP, Scaglia OA. San Cristóbal J (2002) The fossil record of the desert adapted South American rodent *Tympanoctomys* (Rodentia, Octodontidae). Paleoenvironmental and biogeographic significance. Palaeogeogr Palaeocl Palaeoecol 179:149–158

Vialou AV (2003) Santa Elina rockshelter, Brazil: Evidence of the coexistence of man and Glossotherium. In: Miotti L, Salemme M, Flegenheimer N (eds) Where the south winds blow: Ancient evidence of Paleo South Americans. Center for the Study of the First Americans, Portland, pp 21–28

Vivo M, Carmignotto AP (2004) Holocene vegetation change and the mammal faunas of South America and Africa. J Biogeograp 31:943–957

Waguespack NM, Surovell TA (2003) Clovis hunting strategies, or how to make out on plentiful resources. Am Ant 68: 333–352

Ward P (1997) The call of distant mammoths. Springer, Berlin

Webb SD, Marshall LG (1982) Historical biogeography of recent South American land mammals. Spl Publ Pymatuning Lab Ecol 6:39–52

Whitlock C, Bartlein PJ (1997) Vegetation and climate change in northwest America during the past 125 kyr. Nature 388:57–61

Wille M, Hooghiemstra H, van Geel B, Behling H, de Jong A, van der Borg K (2003) Submillennium-scale migrations of the rainforest savanna boundary in Colombia: ^{14}C wiggle-matching and pollen analysis of core Las Margaritas. Palaeogeogr Palaeocl Palaeoecol 193:201–223

Wilmers CC, Getz WM (2005) Gray wolves as climate change buffers in Yellowstone. PLOS Biology 3:571–576

Wilmers CC, Post E (2006) Predicting the influence of wolf-provided carrion on community dynamics under climate change scenarios. Global Changes, Biol 12:403–409

Wilmers CC, Post ES, Peterson RO, Vucetich JA (2006) Predator disease out-break modulates topdown, bottom-up and climatic effects on herbivore population dynamics. Ecol Lett 9:383–389

Youngblood D (2002) From Beringia to Chile in less than 250 years? Note on the rapid traversal of continents. Curr Res Pleistocene 18:66–68

Zárate M, Espinosa M, Ferrero L (1995) La Horqueta II, río Quequén Grande: Ambientes sedimentarios de la transición Pleistoceno-Holoceno. IV Jornadas Geologicas y GeofísicasBonaerenses 1:195–204

Zetti J, Tonni EP, Fidalgo F (1972) Algunos rasgos de la geología superficial en las cabeceras del Arroyo del Azul (Provincia de Buenos Aires). Etnia 15:6–11

8. The Elusive Evidence: The Archeological Record of the South American Extinct Megafauna

Luis Alberto Borrero*
Instituto Multidisciplinario de Historia y Ciencias Humanas
CONICET
Saavedra 15 – Piso 5
(1083 ACA) Buenos Aires, Argentina
dipa.imhicihu@conicet.gov.ar

Keywords South America • Late Glacial archeology • megafaunal extinctions

Introduction

Here I consider the evidence for the interactions of humans and South American late Pleistocene megafauna, a subject not usually covered in much detail in general compilations dealing with Pleistocene extinctions (cf. Martin and Wright, 1967; Klein and Martin, 1984; Barnosky et al., 2004; Steadman et al., 2005; Koch and Barnosky, 2006). The apparent lack of interest is surprising, considering that South America, according to Martin and Steadman (1999:38), lost "over 50 genera of large mammals, more than any other continent", or at least 40 genera according to Cione and coauthors (2003:10). The Pampas alone were populated by 38 extinct herbivore genera in excess of 100 kg, 20 of which were megaherbivores, a fauna that has no living analog on the planet (Fariña, 1996; Prevosti and Vizcaíno, 2006). This situation alone must make us acknowledge South America to be an exceptional case, worthy of an intensive and detailed study.

The temporal focus of this chapter is on the period 14,000–8,000 BP,[1] the timespan in which most of the evidence and serious claims for the presence of humans in South America is concentrated. A few claims for the presence of humans up to 40,000 years ago or even more exist in the literature (Beltrao et al., 1986; Guidon and Delibrias, 1986; Lumley et al., 1988), but will not be dealt with in any detail here. None of these earlier sites offers solid evidence for the presence of humans, much less for association with Pleistocene megafauna.

As for the process of human colonization, all the available evidence for human entry area into South America points toward Panama, established as a bridge since around 3 million years ago (Coates and Obando, 1996). However, the information available from that region is not abundant. The main evidence consists of a fluted projectile point assemblage found at Cueva de los Vampiros, dated ca. 11,500–9,000 BP, unfortunately without good bone preservation (Pearson and Cooke, 2002; Pearson, 2004; Ranere, 2006).

In dealing with the association of humans and megafauna I will recognize two classes of associations: (1) physical associations, which simply refers to bones and tools found side by side, or in the same deposit; and (2) behavioral associations, which require the demonstration of human activities related to megafauna (see Haynes and Stanford, 1984). In order to make this distinction, the taphonomic history of the deposits must be taken into account. Taphonomic analyses allow for a greater degree of understanding with respect to vertical migration and other post-depositional site formation processes.

A human behavioral context implicates interaction, which can be measured directly by the presence of cutmarks on bones. Other measures such as the presence of burned bones or association with tools, hearths, and other human modified features, are indirect and should be qualified on a case by case basis. Burning, for example, tends to be equivocal, since it can also be the result of post-depositional processes. Less

* Address for correspondence: dipa.imhicihu@conicet.gov.ar

G. Haynes (ed.), *American Megafaunal Extinctions at the End of the Pleistocene*, 145–168.

compelling evidence of behavioral interaction is sometimes offered by bone disarticulation or breakage patterns. All these measures always require a taphonomically informed approach. In other words, it is necessary that the fossil samples are studied having in mind a multiplicity of alternative processes that can mimic human actions. It must be stated at the onset that the number of bone assemblages analyzed with a taphonomic perspective is minimal in South America, a limitation that should not deter its consideration.

It is not always easy to recognize cutmarks on bones. The bones of some megafauna, especially Edentata, are characterized by a particular texture that many times makes it difficult to recognize marks. Moreover, many times the bones are diagenetically altered, making even more difficult the use of cutmarks for the identification of human interaction with animals. Under these conditions many cases of association have to rely on less reliable markers, like those mentioned above, and the general context.

The fossil record being an imperfect image of the processes that occurred in the past, it becomes necessary to exert caution before accepting evidence that can be used to construct our interpretations. All this said, it can be surmised that the resolution of the extinctions debate –humans versus climate– is not near. I will try to assess the importance of the human factor, but I believe that if hunting by humans can not be demonstrated as an important cause, it does not follow that climate was the effective agent. The original extinction paper by Martin (1973) included an argument of invisibility, suggesting that if extinctions occurred then we should not find any evidence (Saxon, 1979; Haynes, 2006). The fossil record of North America can be used to argue that lack of visibility is not a problem, since even open-air sites were preserved and discovered (Haynes, 1991). Early humans in South America chose to use caves and rockshelters more regularly, which are locations that offer better preservation and discovery possibilities than open-air sites. For that reason it is possible to use the fossil record to discuss issues of coexistence, association, and behavior.

Once those cases that simply present physical associations are removed from the discussion, the emphasis will be on those sites for which a behavioral interpretation can be maintained. Regional variation is important in the evaluation of the human role in the extinction of megafauna in South America. This is due to the fact that the process of human colonizing was necessarily patchy, with some areas colonized later than others, including some that may have remained empty until late in the Holocene. This is one reason to concentrate the discussion on regional rather than continental patterns. Another reason is that at this scale there is better chronological control. One important chronological consideration involves the degree of overlap between humans and prey. The concept of First Contact considers "initial human encounters with endemic species native to continents or islands where humans were never previously resident" (MacPhee and Marx, 1997:173; Martin and Steadman, 1999:18). It basically refers to the evidence for the oldest presence of humans coexisting with megafauna. The concept of Last Dated Appearance (LDA) or Last Appearance Events (LAE) of megafauna (Alroy, 1999:119–120) refers to the youngest date for a particular taxon, independent of its archeological status. It can be directly measured with taxon-dates. My particular use of taxon-dates differs from that established by Grayson (1991). Instead of attributing the date of a deposit to all the bones that it contains, I count one taxon-date whenever bone, hair, or muscle of a particular taxon is directly dated (Borrero, 1997). This is a more taphonomically sensitive marker of temporal overlap.

Finally, there is the issue of the form the interaction took. An interpretation of hunting or scavenging can be invoked in certain cases, but this is usually a difficult distinction. The importance of recognizing hunting derives from the fact that this is the mode of exploitation that indicates a role for humans in the process of extinction. Anyway, scavenging probably is an opportunistic and complementary strategy.

The Record

The information about faunal distributions in South America at the end of the Pleistocene is very uneven (Pascual et al., 1966; Patterson and Pascual, 1972; Simpson, 1980). Importantly, most findings in South America are referred to the biostratigraphy of the Pampean area (Oliveira, 1999:64; Tonni et al., 1999:15). Not all countries in South America have a good record of the late Pleistocene. While extremely abundant data exist for Brazil and Chile, in the Guyanas or Paraguay this kind of information is extremely scarce. Some important regional biases also appear when the different biomes are considered. Almost none of the forested environments, including the Amazon basin or the Yungas belt that extends East of the Andes from Venezuela to NW Argentina, offer good evidence for megafauna in human contexts at the end of the Pleistocene. Having these limitations in mind, I will now introduce the main body of data organized by Regions and countries (Fig. 8.1). Given the emphasis on the human factor, the records of Paraguay and Guyanas are excluded from this review, since they lack late Pleistocene archeological evidence.

Intermediate Area

Colombia

Relatively little research was conducted on the initial peopling of this country, especially in the Northwest, the most probable entry area to South America for hunter-gatherers coming from the North. Three cordilleras with deep valleys in between constitute the backbone of Andean Colombia, with the famous Sabana de Bogotá located west of the eastern cordillera. Tropical rainforest is concentrated in the southwest of the country.

FIGURE 8.1. South America. Most important sites mentioned in the text.

Sites located in the savanna offer the best available information for the late Pleistocene. The open-air site of Tibitó, located near a marsh at ca. 2,500 m asl, must be mentioned. The evidence includes the presence of lithic artifacts physically associated with bones of *Cuvieronius hyodon*, *Haplomastodon* (probably *Stegomastodon waringi*; Alberdi and Prado, 1995:286), *Equus* (*Amerhippus*) *lasallei*, and modern animals (Correal Urrego, 1981; Alberdi and Prado, 2004). A single radiocarbon date of 11,740 ± 110 BP (GrN-9375) on bone is available. The published information for Tibitó is not detailed enough for a solid assessment of the behavioral status of the association. Other sites with association of megafauna and humans were recently mentioned in the literature, such as Tocaima, but we are still waiting for the presentation of detailed analysis of these assemblages (Correal Urrego, 1986, 1993).

Several sites with late Pleistocene radiocarbon dates are known at the Sabana de Bogotá, such as El Abra (Hurt et al., 1977) or Tequendama (Correal Urrego and Van der Hammen, 1977). However, they lack megafaunal remains. This is also the case at La Palestina, Antioquia, a site dated between 10,400 and 10,200 BP (López Castaño, 1995), and at the oldest occupations recorded on the Colombian mountain forests (Gnecco and Mora, 1997; Gnecco, 2000).

Summing up, the evidence available from Colombia is indicative only of possible coexistence of humans and megafauna.

Venezuela

Another region that may be relevant to the initial peopling of South America is located in the north of Venezuela. This country is characterized by cordilleras, sierras, and coastal Caribbean lowlands. The extensive Guyana highlands covered by tropical rainforests dominate the south of the country. It is in the coastal lowlands that the presence of late Pleistocene faunas is well documented. Several claims for association between humans and megafauna were made on the basis of evidence from the sites Rancho Peludo, Muaco, Cucuruchu, and Taima Taima. The artifacts found in physical association with the bones of megafauna are attributed to the El Jobo industry, characterized by thick bifacial projectile points (Cruxent, 1970; Oliver and Alexander, 2003). Most of these sites present indisputable evidence of mixing with recent materials, including modern glass (Lynch, 1974). Only at Taima Taima can it be suggested that a causal relationship may exist, even when not all authorities agree on this (Lynch, 1990:18–19). Bones of mastodont (originally classified as *Haplomastodon* sp., probably *Stegomastodon waringi*, according to Alberdi and Prado, 1995:286) and other species were recovered at a spring in Taima Taima (Cruxent, 1970; Ochsenius and Gruhn, 1979; Gruhn and Bryan, 1984; Oliver and Alexander, 2003). The mastodont bones displayed "butchering marks made with a sharp knife", as well as a fragment of a projectile point in the innominate (Gruhn and Bryan, 1984). However, the quality of the archeozoological work is very poor. In addition to the self-proclaimed inexperience of the author, the study was conducted without the benefit of either a comparative collection or a specialized bibliography (Casamiquela, 1979:64). These limitations, not to mention the absence of a taphonomic perspective, seriously reduce the significance of this finding. A number of radiocarbon dates were produced for this site, not all in perfect stratigraphic agreement. However, it may be argued that the mastodont is dated ca. 13,000 BP by three radiocarbon dates of 12,980 ± 85 (SI-3316), 13,000 ± 200 (Birm-802) and 13,860 ± 120 BP (USGS-247) on masticated twigs from its stomach contents (Bryan and Gruhn, 1979:57).

Even though the elements for a primary association may exist at Taima Taima, I do not think there is a basis for remarks like "the mastodont was killed, skinned, gutted, and stripped of meat – at least partially – in situ" (Casamiquela, 1979:70). At most, some kind of involvement with megafauna can be considered, on the basis that the mastodont is semi-articulated – which indicates lack of important perturbation in this sector of the site, giving credibility to the location of the projectile point. The presence of other species, such as *Macrauchenia* sp., *Glyptodon* sp., represented by three scutes, or three *Equus* sp. molars (Gruhn and Bryan, 1984) is difficult to explain within a human behavioral context. Summing up, the case for Taima Taima qualifies as interesting, but far from clear.

The site El Vano, located in the Sierra Barbacoas at 1,250 m asl, is characterized by a physical association between *Eremotherium rusconi* and El Jobo artifacts. A bone date of 10,710 ± 60 BP (B-95602) is available (Jaimes, 2003). Credible cutmarks were identified on some of the bones, supporting exploitation but not necessarily hunting. The presence of fractures was also interpreted as resulting from human activities, but this evidence is more equivocal than cutmarks.

The general impression is that the indications offered by the sites from Venezuela can not be confidently used to support human predation on megafauna.

Ecuador

Two main cordilleras characterize the Andean portion of this country, with an extensive Amazonian rainforest to the east. The numerous findings of *Cuvieronius hyodon* in the highlands and *Stegomastodon* sp. in the lowlands (Prado et al., 2005) are well known (Hoffstetter, 1986). However, the physical association between humans and fauna was never maintained with a convincing behavioral argument (Stothert et al., 2003). A few sites dated at the Pleistocene-Holocene Transition, including Chobshi cave (Lynch and Pollock, 1981) and Cubilan (Temme, 1982), do not contain megafanual remains. Classic findings of human bones attributed to the Pleistocene like the Otavalo man or the Punin calvarium (Sullivan et al., 1925) proved to be much younger when submitted to radiocarbon dating (Davies, 1978; Hoffstetter, 1986: 233).

Recent research identified the recurrent presence of arthritis in metapodials of *Equus* (*A.*) *andium* from several sites at the Ecuadorian Andes, a rare pathology related to recessive alleles and suggestive of stressful conditions (Alberdi and Prado, 2004:194). This infirmity should have made horses an interesting target for early human hunters, but no behavioral association confirms this expectation.

Central Andes

Peru

Peru is an environmentally diverse country, with the Pacific coastal expanse, the Andean cordillera, and the eastern rainforests. The Pacific coasts of Peru are among the driest environments on earth. However, a variety of habitats exist, some of them highly productive during the end of the Pleistocene. For that reason the coastal tar pits located in the area known as Pampa de los Fósiles accumulated the remains of megafauna such as *Equus* sp., *Scelidodon* sp., *Holmesina* sp., *Cuvieronius hyodon*, several carnivores including *Smilodon populator*, among others (Hoffstetter, 1986; Alberdi and Prado, 1995), and human artifacts on the nearby surfaces (Table 8.1). The

TABLE 8.1. Radiocarbon dates for sites in Perú.

Site	Date(BP)	Lab	Material	Source
Talara	11,200 ± 115	SI-1415	Shells at "campsite"	Richardson, 1978:282
Talara	8,125 ± 80	SI-1414	Shells at "campsite"	Richardson, 1978:282
Pampa de los Fósiles	10,380 ± 170	GIF-5160	-	Chauchat, 1988
Pampa de los Fósiles	10,200 ± 180	GIF-3781	-	Chauchat, 1988
Pampa de los Fósiles	9,810 ± 180	GIF-4161	-	Chauchat, 1988
Pampa de los Fósiles	9,600 ± 170	GIF-5162	-	Chauchat, 1988
Pampa de los Fósiles	9,490 ± 170	GIF-4914	-	Chauchat, 1988
Pampa de los Fósiles	9,300 ± 160	GIF-4915	-	Chauchat, 1988
Pampa de los Fósiles	9,360 ± 170	GIF-5161	-	Chauchat, 1988
La Cumbre	10,535 ± 280	GX-2019	Mastodont bone	Ossa and Moseley, 1972
La Cumbre	12,360 ± 700	GX-2494	Mastodont bone	Ossa and Moseley, 1972

regional archeology includes Paijan tools, and at least one fishtail projectile point (Richardson, 1978). The chronology is not well established. Radiocarbon dates on the apatite fraction of mastodont bone of 10,535 ± 280 (GX-2019) and 12,360 ± 700 BP (GX-2494) are known, as are older ages using the Uranium series dating method (Ossa and Moseley, 1972; Falguères et al., 1994). However, as remarked by Chauchat, the bones and tools are in different geological contexts, and there is "no trace of human agency either on their bones or as a cause of their death" (Chauchat, 1988:58).

The Andean cordillera separates the coast from the rainforests. Several sites – especially caves – are located between ca. 4,200 and 3,800 m asl, and provided vertebrate faunas. Caverna Huargo, near La Unión at 4,050 m asl, produced bones of *Paleolama*, *Scelidotherium s.l.*, and *Parahipparion* (Hoffstetter, 1986: 230). Radiocarbon dates of 13,460 ± 700 (BVA-8) and 21,070 ± 90 BP (Beta-134596) were derived from bone. Scratches observed on one of the bones were considered cutmarks (Cardich, 1973), but can not be differentiated from trampling marks and thus do not constitute good evidence of the action of humans. The presence of a diversity of Camelidae is inferred from morphometric studies (Cardich and Izeta, 1999–2000).

The horse remains of Uchkumachay, located at 4,050 m asl, were found together with eight lithic flakes (Pires-Ferreira et al., 1976). No cutmarks were found on the bones and there are no radiocarbon dates. With so small a sample there will be always questions of vertical migration, complicating the interpretation. However, even if the flakes are in primary position, they do not constitute sufficient evidence to infer hunting or even exploitation of megafauna.

The remains recovered at Pikimachay cave in the valley of Ayacucho, at 2,800 m asl, are usually mentioned in relation with the peopling of America (MacNeish, 1976, 1979). Several species of megafauna including *Megatherium tarijense*, *Equus andium,* and fossil camelids (Hoffstetter, 1986: 230) were found at different depths in the sediments of this cave, in physical proximity to so called lithic tools. The radiocarbon dates obtained on bones range from more than 20,000 to ca. 11,000 BP (MacNeish, 1976, 1979). Skepticism about the association as well as the character of the tools, many of which are made on volcanic tuff (of which the cave is made), is widespread (Lynch, 1974, 1990; Rick, 1988; Dillehay et al., 1992). The abundant remains of ground sloth, include bones and dung, suggest that a sloth denning site is at least partially implicated in the accumulation of the bones. None of the basic criteria needed to defend a behavioral association between humans and megafauna are present in the published reports.

There are many sites with Transition dates that do not present megafauna, including Guitarrero (Lynch, 1980), Tres Ventanas (Engel, 1970), Quirihuac (Ossa, 1978), Pachamachay (Rick, 1980), Telarmachay (Lavallée et al., 1985), and Lauricocha (Cardich, 1964). Either cervids, vicuñas (*Vicugna vicugna*), or guanacos (*Lama guanicoe*) are the dominant vertebrates at these sites. Megafauna is present basically in paleontological contexts. Potential prey animals such as mastodonts (*Cuvieronius hyodon*) are known in several open-air localities, reaching altitudes over 2,500 m asl near Cusco or Huancayo (Hoffstetter, 1986:227), but never in association with humans. The general impression is that the coexistence between humans and megafauna was short and not very interactive.

Bolivia

A vast altiplano at an altitude of about 4,000 m asl surrounded by high cordilleras characterizes Bolivia. However, the better known Middle to late Pleistocene faunal locality in Bolivia – Tarija – is located ca 2,000 m asl (McFadden and Wolff, 1981). The remains of *Megatherium* sp., *Macrauchenia patachonica*, *Equus (A.) insulatus*, *Hippidion devillei*, *Cuvieronius hyodon,* and *Smilodon populator* are present, but Glyptodontinae are the most abundant (Hoffstetter, 1986; Bond, 1999; Alberdi and Prado, 2004). The chronology for the late Pleistocene is very poor (MacFadden and Wolff, 1981), and there is no indication of association with humans.

At the locality of Ñuapua there are human bones physically associated with megafauna *Macrauchenia patachonica* or *Toxodon ensenadensis* (McFadden and Wolff, 1981; Hoffstetter, 1986; Bond, 1999), but this evidence is difficult to accept as proof of contemporaneity. Dates ranging between > 21,000 and 6,600 ± 370 BP are available for the same human skeleton (Lynch, 1990:16), and there are indications that Ñuapua is a palimpsest in a lacustrine environment.

Lowlands and Pampas

Brazil

This vast country is dominated by the extensive Amazonian tropical lowlands and the extremely dry N.E. The southern extreme of Brazil with highlands characterized by *Araucaria* forests is part of the Pampas biogeographic province. The evidence for late Pleistocene faunas is relatively abundant (Paula Couto, 1979), but fossil contexts with megafauna together with humans are scarce.

Santa Elina Rockshelter is a site located in a savanna environment, in Mato Grosso. There are claims for human presence during the late Pleistocene (Vialou et al., 1995) with radiocarbon dates ranging from 10,120 ± 60 BP (Gif-8954) to 23,320 ± 1,000 (Gif-9365). Older ages were provided by the Uranium-Thorium dating method. The evidence at Santa Elina consists of stone tools in physical association with osteoderms, a fragment of maxilla and skull, one molar, and vertebrae of *Glossotherium* aff. *G. lettsomi*. According to Vialou (2003:23), two osteoderms are pierced and partially abraded, which in itself does not constitute a proof of contemporaneity with humans. There are many problems with the interpretation of this site. A list of 31 radiocarbon ages has numerous reversals (Vialou et al., 1995:659) indicating the possibility of vertical migration. The wide vertical spread of thousands of osteoderms throughout most of the sequence may be one result. It is also clear that the identification of cutmarks can not be reliably done because the preservation of bone is very bad (Vialou, 2003:23). Without a careful taphonomic analysis it is not possible to understand the formation of the deposits recovered at Santa Elina.

The Lagoa Santa district, well known thanks to Lund's findings in several caves during the 19th century, produced a number of late Pleistocene assemblages characterized by the remains of *Eremotherium* sp., *Smilodon populator*, and *Toxodon platensis,* among others (Simpson and Paula Couto, 1957; Paula Couto, 1979). This district is also famous for the presence of an undated human skeleton in the late Pleistocene deposits of Lapa Vermelha IV. Several radiocarbon dates on charcoal are physically associated with the bones of *Scelidotherium* sp. and with quartz tools (Laming-Emperaire et al., 1975; Delibrias et al., 1986; Prous, 1986) (Table 8.2). However, the evidence for behavioral association or even for coexistence is not clear (Lynch, 1990:20; Piló et al., 2005; Neves and Hubbe, 2005).

Faunal remains and tools found at Formation Touro Passo, Rio Grande do Sul, were among the initial late Pleistocene claims for association between humans and *Glossotherium* sp. in South Brazil (Bombin and Bryan, 1978; Miller, 1987). The original chronology is based on a radiocarbon date of 12,770 ± 220 BP (SI-801) on a skull of *Glossotherium myloides* from Lajeado dos Fósseis (Schmidt Días and Jacobus, 2003:42). However, the context is unreliable, with unclear stratigraphy and description of the findings (Oliveira, 1999:64–65; Prous and Fogaca, 1999:25; Schmidt Días and Jacobus, 2003). As Oliveira puts it, "Only fragmentary and isolated bones occur near the dated levels. No dates directly associated with fossil vertebrates were reported" (Oliveira, 1999:65). The age of the Touro Passo formation was based on its correlation with the Sopas formation, which is now firmly dated as older than 40,000 radiocarbon years (Ubilla et al., 1994; Ubilla and Perea, 1999). Thus, there is no evidence of cultural association between tools and faunal remains in the Touro Passo Formation. A similar lack of context applies to several dates recently published by Meggers and Miller (2002).

TABLE 8.2. Radiocarbon dates for sites in Brazil.

Site	Date(BP)	Lab	Material	Source
Lapa Vermelha	10,200 ± 220	Gif-3727	Charcoal	Delibrias et al, 1986; Prous, 1986:174
Lapa Vermelha IV	11,680 ± 500	Gif-3726	Charcoal	Schmitz, 1987; Prous, 1986:174
Lapa Vermelha IV	12,960 ± 300	Gif-3906	Charcoal	Delibrias et al, 1986
Lapa Vérmelha IV	15,300 ± 400	Gif-3905	Charcoal	Delibrias et al, 1986
Lapa Vérmelha	9,580 ± 200	Gif-3208	Charcoal	Delibrias et al, 1986
Lapa Vérmelha IV	9,330 ± 60	Beta-84439	Charcoal	Prous, 1986:176–177
Lajeado dos Fosseis	12,770 ± 220	SI-801	*Glossotherium myloides* bone	Schmidt Días and Jacobus, 2003:42

The presence of late Pleistocene faunas in the different regions of Brazil is well attested (Paula Couto, 1979; Cartelle and Bohorquez, 1986; Oliveira, 1999). Claims for isolated indications of human modification of megafaunal bones exist (Cartelle and Fonseca, 1981; Prous and Fogaca, 1999:25), but most Brazilian sites with Pleistocene-Holocene Transition dates do not contain megafaunal remains (Prous, 1986, 1991a, b; Schmitz, 1987; Guérin, 1991; Roosevelt et al., 1996; Prous and Fogaca, 1999; Kipnis, 2002). Other sites, such as Toca do sitio do Meio (Guidon, 1986), present late Pleistocene dates in contexts whose resolution is very difficult to evaluate with the published information. In sum, the Brazilian evidence displays very little evidence concerning behavioral interaction between Pleistocene fauna and humans.

Uruguay

Uruguay is part of the biogeographic province of the Pampas and of the La Plata river basin. Archeological sites with late Pleistocene radiocarbon ages are known, like those at Isla de Arriba and Arroyo del Tigre in the Uruguay river (Suárez and López, 2003). However, these materials were never studied in any depth, and are almost useless for the discussion about contemporaneity between humans and megafauna.

Recent research at the locality of Pay Paso 1 in the Cuareim basin produced evidence for human occupations with dates between 9,500 and 9,000 BP (Suárez, 2003; Suárez and López, 2003) (Table 8.3). Late Pleistocene fauna, including one *Glyptodon* sp. scute, was found in association with bifacial lithic tools, red ocher, and fragments of carbonized plants (Suárez, 2003:31). An association with a scute is not easy to explain behaviorally, as it is hard to relate with hunting or even exploitation of megafauna. Even contemporaneity is difficult to infer. Further work is needed to understand the faunal context of the inhabitants of Pay Paso 1. At Pay Paso 4 some bones of *Stegomastodon* sp. were found in physical proximity

TABLE 8.3. Radiocarbon dates for sites in Uruguay.

Site	Date(BP)	Lab	Material	Source
Y-58	11,200 ± 500	Gif-4412	Charcoal	Suárez and López, 2003
K87	10,420 ± 90	Kn-2531	Charcoal	Suárez and López, 2003
Pay Paso 1	9890 ± 75	Rt-1445	Charcoal	Suárez, 2003
Pay Paso 1	9280 ± 200	Uru-248	Charcoal	Suárez, 2003
Pay Paso 1	9120 ± 40	Beta-156973	Charcoal	Suárez, 2003
Pay Paso 1	8570 ± 150	Uru-246	Charcoal	Suárez, 2003

to three bifaces, but no relationship can be defended since no cutmarks or impact fractures were described.

The diversity of late Pleistocene faunas in Uruguay is well reported (Mones and Francis, 1973; Ubilla and Perea, 1999), including *Mylodon darwinii*, *Macrauchenia patachonica*, *Doedicurus clavicaudatus,* and *Glyptodon clavipes*. They belong to the Dolores formation, and the available chronology is older than the dates obtained at the archeological sites. Therefore, no behavioral interaction has been demonstrated in Uruguay.

Chile

Chile is a long and narrow country, located west of the Andes. In addition to a diversity of coastal Pacific habitats, it is characterized by the absolute desert of Atacama in the North, a system of valleys between two ranges in the center, and the archipelagos of Cabo de Hornos in the South.

In the northern extreme of the country there are many archeological sites with late Pleistocene dates, but only at Tuina, located in the Atacama desert, is there extinct fauna, specifically a fragment of an innominate of *Equus* sp. (Nuñez et al., 2001) (Table 8.4). The rest of the early archeological faunal assemblages consists only of modern species.

The remains of mastodonts are abundant in Central Chile, but only rarely are they in archeological contexts. Not too far away from Santiago the lagoon of Taguatagua (also spelled Tagua Tagua) is located, well known at least since Charles Darwin's visit in 1833 (Darwin, 1860). At this site at least two *loci* with artifacts and mastodont bones were found: Taguatagua 1, dated ca. 11,400–11,000 BP (Casamiquela et al., 1967; Montané, 1968), and Taguatagua 2 (Nuñez et al., 1994) dated ca. 10,190–9,700 BP. At Taguatagua the bones of one adult mastodont (*Cuvieronius hyodon*) were found, together with the bones of *Equus* (*A.*) sp. (Alberdi and Prado, 2004:192), *Antifer niemeyeri* (Casamiquela, 1999), and modern fauna, including small animals such as fish, frogs, etc. The small fauna looks very much like the taphonomic background noise expected in a lagoon location. At Taguatagua 2 at least two Fell Cave projectile points were recovered, as well as remains of several mastodonts, including a fragment of tusk with geometric designs. The disarticulated remains are dominated by pelves, skulls, and vertebrae. The presence of cutmarks is recorded on 7.6% of the elements (C. García, 2005). The evidence from Taguatagua ranks among the best in existence in South America. Even when it is not easy to defend that hunting took place, it still remains a plausible interpretation of the data.

At the locality of Los Vilos, in the central semiarid coastal area, there is evidence of physical association between humans and the megafaunal taxa *Hippidion* sp, and *Mylodon* sp. or *Macrauchenia* sp. (Jackson, 2003; Jackson et al., 2003; Jackson et al., 2005). The dates obtained at these sites, including one of 13,500 ± 65 BP (NSRL-11081) on a *Mylodon* sp. vertebra, are slightly older than those obtained at nearby coastal shell middens (Jackson et al., 2003). The preservation of bones is bad, hindering any analysis of cutmarks. It must be stated that the shallow character of some of the deposits containing the materials, ca. 10 cm depth, requires caution in the interpretation of association. Sandy deposits are known for their dynamic nature, and the bones of Pleistocene mam-

TABLE 8.4. Radiocarbon dates for sites in North and Central Chile.

Site	Date(BP)	Lab	Material	Source
Tuina 5 EIV	10,820 ± 630	SI-3112	*Equus* sp. sacrum	Cartajena et al., 2005
Taguatagua 1	11,000 ± 170	-	-	Montané, 1968
Taguatagua 1	11,320 ± 300	-	-	Montané, 1968
Taguatagua 1	11,380 ± 320	GX-1205	Charcoal	Nuñez et al., 1994; Montané, 1968
Taguatagua 2	9,900 ± 100	Beta-45519	Charcoal	Nuñez et al., 1994
Taguatagua 2	10,190 ± 130	Beta-45520	Charcoal	
Taguatagua 2	9,710 ± 90	Beta-45518	Charcoal	Nuñez et al., 1994
Quereo Layer Q I-II	11,100 ± 150	N-2963	Wood	Nuñez et al., 1983
Quereo Layer Q II	11,100 ± 150	N-2962	Wood	Nuñez et al., 1983
Quereo Layer Q I	11,400 ± 155	N-2964	Wood	Nuñez et al., 1983
Quereo Layer Q	10,925 ± 85	SI-3391	Wood	Nuñez et al., 1983
Quereo Layer Q I	11,400 ± 145	N-2966	Wood	Nuñez et al., 1983
Quereo Layer Q I	11,600 ± 190	N-2965	Wood	Nuñez et al., 1983
Quereo (Philips' Mastodont)	9,100 ± 300	GaK-2984	*Cuvieronius hyodon* bone	Paskoff, 1971
El Membrillo	13,500 ± 65	NSRL-11081	-	Jackson, 2003
Monte Verde MV-7	12,740 ± 440	TX-5375	Wood	Dillehay and Pino, 1997
Monte Verde MV-7	12,650 ± 130	TX-4437	Wood	Dillehay and Pino, 1997
Monte Verde MV-6	12,450 ± 150	OXA-381	Wood	Dillehay and Pino, 1997
Monte Verde MV-6	12,230 ± 140	Beta-6755	Wood	Dillehay and Pino, 1997
Monte Verde MV-6	12,000 ± 250	OXA-105	Ivory	Dillehay and Pino, 1997
Monte Verde MV-6	11,990 ± 200	TX-3760	Bone	Dillehay and Pino, 1997
Monte Verde	13,565 ± 250	TX-3208	Charcoal	Dillehay and Pino, 1997
Monte Verde MV-6	11,790 ± 200	TX-5374	Wood	Dillehay and Pino, 1997
Monte Verde MV-6	11,920 ± 120	TX-5376	Wood	Dillehay and Pino, 1997
Monte Verde MV-7	12,780 ± 240	Beta-59082	Burned artifact	Dillehay and Pino, 1997
Monte Verde MV-7	12,420 ± 130	Beta-65842	Wood	Dillehay and Pino, 1997

mals, lithics, and ceramics were also found on the surface (Jackson, 2003).

Not too far from the locality of Los Vilos is located the site Quereo in a paleolacustrine setting (Nuñez et al., 1983, 1994; Labarca, 2003). Two layers were identified, Quereo 1 and 2. Quereo 1 is dated between 11,600 ± 190 (N-2965) and 10,925 ± 85 (SI-3391) BP (Table 8.4), and presents remains of *Equus (A.) neogeus*, *Cuvieronius hyodon.*, *Antifer* sp., *Mylodon* sp., and *Palaeolama* sp. (Casamiquela, 1984; Nuñez et al., 1994; Labarca and López, 2006). So called expedient human artifacts are described, whose cultural status is difficult to accept, since no effort was made to assess the impact of water abrasion on the bones. No cutmarks were found on the bones. This case is difficult to defend as proof of human activities. Nuñez and coauthors (1983) described one horse skull as "collapsed" as a result of throwing a rock over the head. Not only it is an implausible hunting tactic, but it also makes no taphonomic sense in terms of the distribution of the horse bones.

The layer Quereo 2 is radiocarbon dated 11,100 ± 150 BP (N-2962). Bones of two individuals of *Cuvieronius hyodon* were recovered, some of which present presumed cutmarks (Nuñez et al., 1983:25). The presence of *Equus (A.) neogeus* and *Antifer* sp. is also recorded (Casamiquela, 1984; Labarca and López, 2006). The study of the mastodont bones from the Phillips collection (Paskoff, 1971), collected at the end of the 19th century and perhaps attributable to Quereo 2, showed the existence of impairing paleopathologies (Labarca, 2003:160).

With the available data it is difficult to accept the evidence from Quereo as more than a case of naturally deposited bones. Water movement might have been implicated in the distribution of bones (Labarca, 2003), and the presence of amphibians and molluscs indicates a lacustrine habitat. The status of the bone artifacts is equivocal. Only some cutmarks can be used for a serious discussion of human involvement. The hunting scenario stated by Nuñez and coauthors (1983) is highly implausible, and the alternative of scavenging offered by Labarca (2003:171) makes more sense of the data.

In south central Chile there are several locations presenting bone remains of mastodont. Monte Verde is the better known of these locations (Dillehay, 1997). This is a controversial archeological site dated ca. 12,500 BP, where most of the evidence related to human activities is made of wood or plants in general. Mastodont bones and hide are present and dated 11,990 ± 200 BP (TX-3760) and 12,000 ± 250 BP (OXA-105) (Dillehay and Pino, 1997:46). However, the presence of dirt embedded in the bones (Karathanasis, 1997) indicates that at least some of the bones were collected from a defleshed carcass. The mastodont bones should be attributed to *Cuvieronius hyodon* (Alberdi and Prado, 1995; Labarca and López, 2006). A single *Palaeolama* bone was also found (Casamiquela and Dillehay, 1989). The number of mastodont bone fragments is 414. Human activities at the site do not indicate exploitation of either mastodont or *Palaeolama* for food. There are other sites that record the presence of mastodonts in South Central Chile, such as one found at Nochaco and dated ca. 16,000 BP (Labarca and López, 2006:96), and many other undated findings. However, none have any signal of human activities.

Further south, at the Baño Nuevo site, several human burials with early Holocene dates were found (Mena et al., 1999, 2003) (Table 8.5). There are also lithic artifacts and bones of guanaco and fox in the early Holocene layers. *Mylodon* sp. osteoderms were recovered in the same layers, four of which

TABLE 8.5. Radiocarbon dates for sites in South Chile.

Site	Date(BP)	Lab	Material	Source
Baño Nuevo	8,530 ± 160	Beta-90892	Charcoal	Velásquez and Mena, 2006
Baño Nuevo	8,695 ± 25	UCIAMS-10099	Charcoal	Velásquez and Mena, 2006
Baño Nuevo	8,850 ± 50	CAMS-36633	Human bone	Velásquez and Mena, 2006
Baño Nuevo	8,880 ± 50	CAMS-36634	Human bone	Velásquez and Mena, 2006
Baño Nuevo	8,890 ± 90	Beta-90889	Charcoal	Velásquez and Mena, 2006
Baño Nuevo	8,945 ± 40	CAMS-101894	Human bone	Mena et al., 2003
Baño Nuevo	8,950 ± 60	CAMS-101893	Human bone	Mena et al., 2003
Baño Nuevo	8,950 ± 50	CAMS-79933	Human bone	Mena et al., 2003
Baño Nuevo	8,975 ± 20	UCIAMS-10095	Coprolite	Velásquez and Mena, 2006
Baño Nuevo	8,990 ± 30	UCIAMS-10098	Human bone	Velásquez and Mena, 2006
Baño Nuevo	9070 ± 25	UCIAMS-10091	Charcoal	Velásquez and Mena, 2006
Baño Nuevo	9,070 ± 50	CAMS-80532	Plant remains	Velásquez and Mena, 2006
Baño Nuevo	9,155 ± 25	CAMS-10087	Charcoal	Velásquez and Mena, 2006
Baño Nuevo	9,200 ± 80	Beta-90888	Charcoal	Velásquez and Mena, 2006
Baño Nuevo	9,245 ± 25	UCIAMS-10093	Charcoal	Velásquez and Mena, 2006
Baño Nuevo	9,260 ± 25	UCIAMS-10093	*Ctenomys* sp. bone	Velásquez and Mena, 2006
Baño Nuevo	9,435 ± 25	UCIAMS-10097	Charcoal	Velásquez and Mena, 2006
Baño Nuevo	9,530 ± 25	UCIAMS-10094	Charcoal	Velásquez and Mena, 2006
Baño Nuevo	11,240 ± 40	CAMS-71702	Bone	Velásquez and Mena, 2006
Baño Nuevo	11,250 ± 50	CAMS-72356	Bone	Velásquez and Mena, 2006
Baño Nuevo	11,255 ± 30	UCIAMS-10105	*Mylodon* osteoderm	Velásquez and Mena, 2006
Baño Nuevo	11,265 ± 35	UCIAMS-10106	*Mylodon* osteoderm	Velásquez and Mena, 2006

(continued)

TABLE 8.5. (continued)

Site	Date(BP)	Lab	Material	Source
Baño Nuevo	11,480 ± 50	CAMS-32685	*Mylodon* osteoderm	Mena et al., 2003
Baño Nuevo	11,665 ± 50	UCIAMS-19491	*Macrauchenia* sp. bone	Velásquez and Mena, 2006
Baño Nuevo	12,000 ± 35	UCIAMS-10110	Ungulate bone	Velásquez and Mena, 2006
Baño Nuevo	12,320 ± 30	UCIAMS-10109	Ungulate bone	Velásquez and Mena, 2006
Baño Nuevo	12,325 ± 30	UCIAMS-10101	Wood	Velásquez and Mena, 2006
Baño Nuevo	12,400 ± 30	UCIAMS-10111	Ungulate bone	Velásquez and Mena, 2006
Baño Nuevo	12,510 ± 30	UCIAMS-10107	*Mylodon* osteoderm	Velásquez and Mena, 2006
Baño Nuevo	13,480 ± 3	UCIAMS-10100	Wood	Velásquez and Mena, 2006
Cueva Sofía 1	11,570 ± 60	PITT-0684	Charcoal	Prieto, 1991
Cueva Sofía 1	10,710 ± 70	OxA-8635	*Lama guanicoe* bone	Massone and Prieto, 2004
Cueva Sofía 1	10,780 ± 60	OxA-9319	*Hippidion saldiasi* bone	Massone and Prieto, 2004
Cueva Sofía 1	10,310 ± 160	OxA-9504	*Hippidion saldiasi* bone	Massone and Prieto, 2004
Cueva Sofía 1	10,140 ± 120	OxA-9505	*Pseudalopex culpaeus* bone	Massone and Prieto, 2004
Cueva Sofía 1	12,250 ± 110	OxA-9506	*Mylodon darwinii* bone	Massone and Prieto, 2004
Cueva Sofía 1	12,990± 490	Pitt-0399	*Mylodon darwinii* bone	Prieto, 1991
Cueva del Medio	10,930 ± 230	Beta-39081	Charcoal	Nami and Menegaz, 1991
Cueva del Medio	10,310 ± 70	GrN-14913	Charcoal	Nami and Nakamura, 1995
Cueva del Medio	9,770 ± 70	Beta-40281	Bone	Nami and Nakamura, 1995
Cueva del Medio	9,595 ± 115	PITT-0344	Charcoal	Nami, 1987
Cueva del Medio	10,350 ± 130	Beta-58105	Burnt bone	Nami and Nakamura, 1995
Cueva del Medio	10,430 ± 80	Beta-52522	Charcoal	Nami and Nakamura, 1995
Cueva del Medio	10,550 ± 120	GrN-14911	Burnt bone	Nami and Nakamura, 1995
Cueva del Medio	10,710 ± 100	NUTA1811	*Hippidion saldiasi* bone	Nami and Nakamura, 1995
Cueva del Medio	10,930 ± 230	Beta-39081	Charcoal	Nami and Nakamura, 1995
Cueva del Medio	12,390 ± 180	PITT-0343	Charcoal	Nami, 1987
Cueva del Medio	10,860 ± 160	NUTA-2331	*Hippidion saldiasi* bone	Nami and Nakamura, 1995
Cueva del Medio	11,040 ± 250	NUTA-2197	*Lama* cf. *owenii* bone	Nami and Nakamura, 1995
Cueva del Medio	10,430 ± 100	NUTA-1734	*Lama* cf. *owenii* bone	Nami and Nakamura, 1995
Cueva del Medio	11,120 ± 130	NUTA-1737	*Lama* cf. *owenii* bone	Nami and Nakamura, 1995
Cueva del Medio	10,960 ± 150	NUTA-2330	*Lama* cf. *owenii* bone	Nami and Nakamura, 1995
Cueva del Medio	10,450 ± 100	NUTA-1735	*Lama guanicoe* bone	Nami and Nakamura, 1995
Cueva del Medio	10,710 ± 190	NUTA-2332	*Lama guanicoe* bone	Nami and Nakamura, 1995
Cueva del Medio	10,850 ± 130	NUTA-1812	*Lama guanicoe* bone	Nami and Nakamura, 1995
Cueva Fell	10,720 ± 300	W-915	Charcoal	Bird, 1988
Cueva Fell	11,000 ± 170	I-3988	Charcoal	Bird, 1988
Cueva Fell	10,080 ± 160	I-5146	Charcoal	Bird, 1988
Cueva de los Chingues	11,210± 50	Beta-147744	*Hippidion saldiasi* bone	San Román et al., 2000
Tres Arroyos Va-b	11,880 ± 250	Beta-20219	Bones	Massone, 2004
Tres Arroyos Vb	10,420 ± 100	Dic-2733	Bones	Massone, 2004
Tres Arroyos Va	10,280 ± 110	Dic-2732	Bones	Massone, 2004
Tres Arroyos Va	10,600 ± 90	Beta-101023	Charcoal	Massone, 2004
Tres Arroyos Va	10,580 ± 50	Beta-113171	Charcoal	Massone, 2004
Tres Arroyos Va	10,575 ± 65	OxA-9245	*Dusicyon avus* bone	Massone, 2004
Tres Arroyos Va	10,630 ± 70	OxA-9246	*Vicugna* sp. bone	Massone, 2004; Massone and Prieto, 2004
Tres Arroyos Va	10,685 ± 70	OxA-9247	*Hippidion* bone	Massone, 2004; Massone and Prieto, 2004
Tres Arroyos Va	11,085 ± 70	OxA-9248	*Panthera onca mesembrina* bone	Massone, 2004
Tres Arroyos Va	10,130 ± 210	OxA-9666	Charcoal	Massone, 2004

were dated as late Pleistocene (Table 8.5). Mena and coauthors interpret this as a case of secondary association (Mena et al., 2003). A late Pleistocene radiocarbon date was also obtained for *Macrauchenia* sp. at Baño Nuevo (Velásquez and Mena, 2006).

In the southern extreme of Chile, in Patagonia, some of the best evidence of association between humans and megafauna has been discovered, including a close association of broken and burnt bones with several hearths and lithic and bone tools (Bird, 1988). Several caves recorded important information in the southern extreme of the Pali Aike Volcanic Field. Cueva Fell (Bird, 1938, 1988) is the type site used to understand the sequence of human occupations of Patagonia. The remains of *Hippidion saldiasi*, *Mylodon darwinii*, and guanaco were found in association with hearths and lithic artifacts, including Fell Cave projectile points, also known as "fishtails" (Bird, 1988). Three radiocarbon dates between 11,000 ± 170 (I-3988) and 10,080 ± 160 BP (I-5146) are available (Table 8.5). For years it was accepted that humans were the exclusive bone accumulators at Cueva Fell. However, re-study of the bone collection from the lower layers indicated the presence of carnivore marks particularly on the horse bones, which were attributed to an extinct feline (Borrero and Martin, 1996). These marks were taken as an indication of alternating use of the cave by humans and carnivores (Borrero, 2005). An analysis of the abundant canid remains suggested the presence of *Canis familiaris* at least since the early Holocene (Clutton-Brock, 1977), although other specialists have attributed these bones to *Dusicyon avus* (Caviglia, 1986).

Other sites present more equivocal evidence. For example the cave Pali Aike (Bird, 1988) contained abundant bones of at least seven individuals of *Mylodon darwinii* in contact with human artifacts. A radiocarbon date on undetermined bones of 8,639 ± 450 BP (C-485) is widely cited, but can only be used as a minimal age. The remains of *Hippidion saldiasi* and modern fauna were also recovered. The presence of human burials at the rear of the cave, where most of the megafaunal remains were recovered, complicates the interpretation of the stratigraphy (Borrero, 2005). No cutmarks were found on the bones of Pleistocene fauna from Pali Aike.

At the Cerro Sota site a collective human burial physically associated with megafaunal bones and hairs was discovered. The hairs were attributed to Pleistocene horse and ground sloth (Whittford, in Bird, 1988), and a recent reanalysis of the horse bones attributed them to *Hippidion saldiasi*, the only species known in Patagonia (Alberdi and Prieto, 2000). Bones of *Mylodon darwinii* and guanaco were also found. The human bones were dated to ca. 3,900 BP (Hedges et al., 1992), and on that basis it is possible to infer that no behavioral relationship between humans and megafauna is implicated at this site, where humans arrived at the site thousands of years after the megafauna.

Two paleontological sites that contribute to our understanding of the faunas available at the time of human arrival to the region were recently discovered and studied. These are Cueva de los Chingues (San Román et al., 2000) and Cueva del Puma (Martin et al., 2004).

The remains of *Hippidion saldiasi*, *Mylodon* sp., *Arctotherium tarijense*, a large size felid, and guanaco were found at Cueva de los Chingues (Prevosti et al., 2003; Soibelzon, 2004). A radiocarbon date of 11,210 ± 50 BP (Beta-147744) on a horse phalange II is available. This paleontological deposit was found below a mixed paleontological and archeological Holocene deposit.

At Cueva del Puma, the unweathered bones of extinct animals were found on the surface of a hidden dark chamber, including a complete femur of *Arctotherium* sp. which was dated 10,345 ± 75 BP (Ua-21033). Bones of *Hippidion saldiasi, Panthera onca mesembrina, Dusicyon avus, Mylodon* sp., and Camelidae were found both on the surface and in stratigraphy. A *Lama* sp. radio-ulna recovered below the surface was dated 11,575 ± 80 BP (Ua-21035). Their location at a dark chamber that can only be accessed through a small passage, together with the presence of carnivore marks, suggests that it was a carnivore den (Martin et al., 2004). *Panthera onca mesembrina* is a candidate to be the main agent of bone accumulation. Tonni and coauthors (2003a:611) think that *Mylodon* was not a potential prey for *Panthera onca*, but it was not possible for the ground sloths to find their way into the cave through the narrow lava tubes that constitute its access. It is clear that only parts of the sloth carcass were transported to the interior. A similar situation applies to Cueva Lago Sofía 4 (see below). Moreover, carnivore feces from Cueva del Milodon stored at the British Museum of Natural History contain Mylodontidae osteoderms, which constitutes further evidence of interaction between both animals.

Important archeological sites were found in the region of Ultima Esperanza, on the Pacific side of the Andean cordillera. At Cueva Lago Sofía 1, bones of *Hippidion saldiasi*, *Mylodon darwinii*, *Dusicyon avus*, and modern species were found together with hearths and bone and lithic artifacts (Prieto, 1991; Alberdi and Prieto, 2000; Jackson and Prieto, 2005). Several radiocarbon dates including 11,570 ± 60 and 10,140 ± 120 BP are available (Massone and Prieto, 2004) (Table 8.5). Under the archeological layer ground sloth remains were dated 12,990 ± 490 BP (Pitt-0399). The archeological layer appears to be consistent in terms of the presence of cutmarks on the horse and ground sloth bones and the limited vertical distribution of bones and tools.

An important association was also found at Cueva del Medio (Nami, 1987; Nami and Nakamura, 1995). There are bones of *Hippidion saldiasi*, *Lama* cf. *owenii,* and *Mylodon* (?) *listai* (Nami and Menegaz, 1991; Alberdi and Prieto, 2000), as well as basin shaped hearths and lithic artifacts, including two Fell Cave projectile points. Seventeen radiocarbon dates locate the human occupation between 11,120 ± 130 BP (NUTA-1737) and 9,595 ± 115 (Pitt-0344) Ground sloths were dated 11,570 ± 100 BP (AA-12578) and 11,990 ± 100 (AA-12577). Cutmarks were recognized on the horse bones (Fig. 8.2). Bones of *Panthera onca mesembrina* were found below the human occupations (Fig. 8.3).

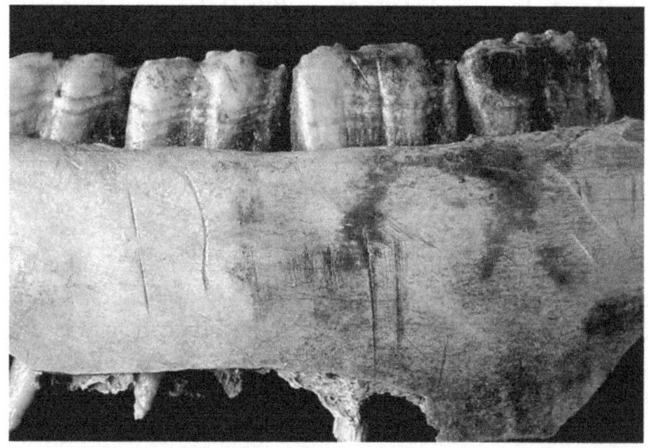

FIGURE 8.2. Mandible of *Hippidion saldiasi* with cutmarks, Cueva del Medio, Ultima Esperanza, Chile (See Nami and Menegaz, 1991:124).

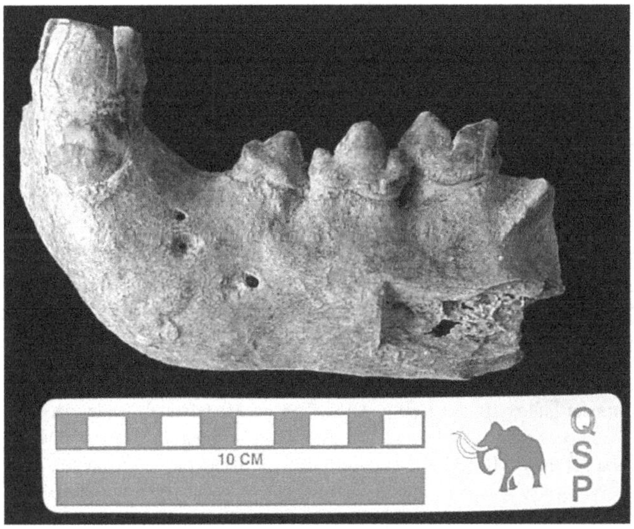

FIGURE 8.3. Mandible of *Panthera onca mesembrina*, Cueva del Medio, Ultima Esperanza, Chile.

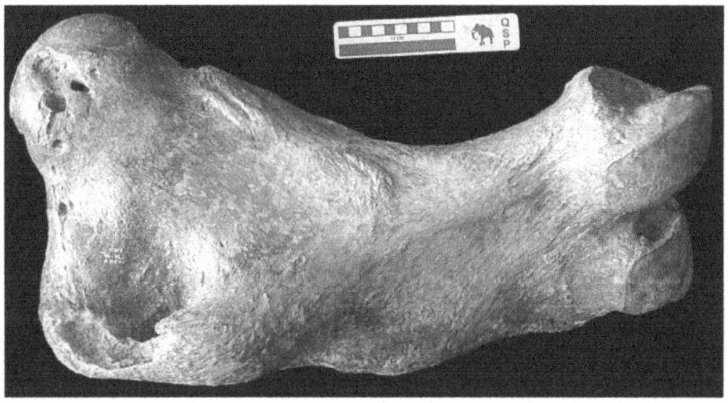

FIGURE 8.4. Femur of *Mylodon darwinii*, Alero Dos Herraduras, Ultima Esperanza, Chile. Notice two punctures at the neck of the head.

In Ultima Esperanza there are also some paleontological sites of approximately the same age which are relevant for our understanding of the late Pleistocene faunas.

Cueva del Milodon is world famous due to the excellent preservation of skin, hair, muscular tissue, and bone (Borrero et al., 1991; Borrero, 1999). The most abundant vertebrate at the cave is *Mylodon darwinii*, whose bones, hair, skin, and dung were found in several locations within this large cave. Recovery of DNA from sloth and *Smilodon populator* bones was successful (Höss et al., 1996; Barnett et al., 2005). There are 19 radiocarbon dates on dung between 12,200 ± 400 (Sa-49) and 13,560 ± 180 BP (A-1390), three on skin between 10,400 ± 410 (R-4299) and 13,040 ± 300 BP (W-2998), and six on bone between 10,300 ± 480 BP (LP-49) and 13,260 ± 115 BP (LU-794). The *Mylodon* remains include a fragment of skull of a newborn or fetal individual (Tonni et al., 2003a). The remains of *Hippidion saldiasi*, *Panthera onca mesembrina*, *Smilodon populator*, and extinct and modern camelids were also recovered (Hauthal, 1899; Nordenskjold, 1900; Saxon, 1979; Borrero et al., 1991; Mol et al., 2003; Barnett et al., 2005). A fragment of a bear femur was described by Smith-Woodward, probably attributable to *Arctotherium s.* (Smith-Woodward, 1900; Soibelzon, 2004). A phalanx of *Macrauchenia patachonica* was found by Nordenskjold (Bond, 1999: 188) and remains of *D. avus* were described by Roth (1902).

Cueva Lago Sofía 4 is a dark cave in the vicinity of Cueva Lago Sofía 1 (Borrero et al., 1997; Alberdi and Prieto, 2000). An assemblage of bones sealed in calcium carbonate was recovered from the inner chambers. The higher number of bone specimens is for camelids, followed by abundant remains of *Mylodon* sp. (from at least four juvenile individuals) and *Hippidion saldiasi*. Bones of *Smilodon* sp. were also found (Canto, 1991). The site was interpreted as a carnivore den. One radiocarbon date on a *Mylodon* sp. vertebra of 11,590 ± 100 BP (Pitt-0940) and two on *Mylodon* sp. osteoderms of 11,050 ± 60 (NSRL-3341) and 13,400 ± 90 BP (AA-11498) are available.

At the Dos Herraduras rockshelter the bones of *Mylodon* sp. were found within a tephra layer fingerprinted to the Reclus volcano (Borrero and Massone, 1994), which had an explosive eruption ca. 12,600 BP (Stern, 1992; McCulloch et al., 2005). A standard radiocarbon date of 11,380 ± 50 BP (LP-421) on a *Mylodon* rib and an AMS radiocarbon date of 12,825 ± 110 BP (AA-12574) on an osteoderm are available. The second date makes more sense in terms of the age of the tephra layer that covers the bones. A ground sloth femur presents impressive carnivore punctures (Fig. 8.4). These marks are large enough to be considered the result of an extinct large carnivore. The evidence adds to the body of data that can be used to infer a predator-prey interaction between extinct carnivores and ground sloths.

At Cueva Ventana, not too far away from Dos Herraduras, horse remains were recovered (Alberdi and Prieto, 2000). Both sites present Holocene human occupations, but no claim for association with megafauna was made.

The site Tres Arroyos, located on what today is the island of Tierra del Fuego, offers solid evidence for human occupation. Five basin shaped hearths are dated by charcoal between 10,130 ± 210 BP (OxA-9666) and 10,600 ± 90 BP (Beta-101023) (Latorre, 1998; Borrero, 2003; Massone, 2004). The hearths and numerous lithic tools were found in association with broken, cutmarked, and burnt bones of camelids, *Hippidion saldiasi*, and *Mylodon* sp. (Latorre, 1998; Alberdi and Prieto, 2000). One date on *Hippidion saldiasi* is 10,685 ± 70 BP (OxA-9247), while a *Lama* sp. bone was dated 10,630 ±70 BP (OxA-9246). A *Canis* (*Dusycion*) *avus* mandible was found at late Holocene Layer III (Borrero, 2003). A radiocarbon date made on the bone itself produced a result of 10,575 ± 65 (OxA-9245). In this particular case the action of rabbits (*Oryctolagus cuniculus*), introduced to Tierra del Fuego in the 1930s and 1950s, may explain the upward migration of this mandible. Effectively, abundant rabbit bones were recovered in Layer V, in all cases closely associated with tunnels. These were identified as rabbit burrows by comparison with the size, content, and other properties of several excavated burrows in North Tierra del Fuego (Borrero and Martin, 1999).

Lynchailurus colocolo was recovered in Layer IV (Prevosti, 2006). The only taxon-date for *Panthera onca mesembrina* was obtained at Tres Arroyos; a metapodial found below the human occupation was dated 11,085 ± 70 BP (OxA-9248).

Another late Pleistocene faunal assemblage, TA 14(30), was discovered at an open air location in front of the Tres Arroyos rockshelter. There is one radiocarbon date of 12,280 ± 110 BP (Beta-101056). A few lithic flakes were found in physical contact with the bones of *Mylodon* sp. and *Hippidion saldiasi*, but this is attributed to vertical migration from an upper human occupation dated ca. 2,200 BP (Borrero, 2003; Massone, 2004).

The review of the Chilean data indicates evidence of behavioral interaction in Taguatagua, Cueva Lago Sofía 1, Cueva del Medio, Cueva Fell, and Tres Arroyos. The evidence from other sites indicates only the potential for relevance.

Argentina

This is a highly varied country that extends between 23° and 55° S latitude. In the high altitude Puna environments of the Northwest of the country, there are several archeological sites with Transition ages but lacking megafauna. The sites include Inca Cueva 4 (Aschero, 1979), Huachichocana III (Yacobaccio and Madero, 1992), Quebrada Seca 3 (Aschero et al., 1991) and Pintoscayoc (Hernández Llosas, 1999; Elkin and Rosenfeld, 2001). At least two sets of evidence indicate the presence of megafauna in the area immediately before the arrival of humans. Dates between 13,300 and 12,500 BP for megamammal dung attributed to Megatheriinae were obtained at Peñas de las Trampas 1.1 and Cueva Cacao 1A (Powell et al., 2001; Martinez, 2003), while horse remains attributed to *Hippidion devillei* dated between 12,500 and 10,200 BP were recovered at Barro Negro, Jujuy (Alberdi et al., 1986).

South of the Puna there is evidence of late Pleistocene human presence at site Agua de la Cueva (Table 8.6), on the eastern face of the Andes, at 2,900 m asl in central Argentina (García, 2003a). The faunal remains include the possible presence of an extinct camelid, *Lama gracilis* (Neme et al., 1998), but the dominant prey are modern camelids.

The remains of *Mylodon* sp., *Hippidion* sp., and Megatheriidae, plus a few lithics, were found at Gruta del Indio, at the eastern foot of the Andes. More than 30 radiocarbon dates indicate the presence of megafauna between > 31,000 and ca. 9,000 BP (García, 2003b; Long et al., 1998) (Table 8.6). The indications of human presence include at least 11 flakes found within a ground sloth dung layer (Lagiglia, 1979; García and Lagiglia, 1998–1999) plus two hearths dated ca. 10,500–10,900 BP. The "possible time of… coexistence of man and extinct Pleistocene megafauna" is calculated to be about 1,400 radiocarbon years,

Table 8.6. Radiocarbon dates for sites in Central Argentina.

Site	Date(BP)	Lab	Material	Source
Agua de la Cueva 2b	9,840 ± 90	Beta-26781	Charcoal	García, 2003a, c
Agua de la Cueva 2a	10,350 ± 220	Beta-26250	Charcoal	García, 2003c
Agua de la Cueva 2a	10,950 ± 190	Beta-61409	Charcoal	García, 2003c
Agua de la Cueva 2b	9,210 ± 70	Beta-64539	Charcoal	García, 2003c
Agua de la Cueva 2b	10,240 ± 60	Beta-61408	Charcoal	García, 2003c
Agua de la Cueva t2b	9,760 ± 160	Beta-61410	Charcoal	García, 2003c
Gruta del Indio	10,950 ± 60	GrN-5558	Megafaunal dung	García, 2003b
Gruta del Indio	10,530 ± 140	A-1638	Charcoal	García, 2003b
Gruta del Indio	10,135 ± 95	A-9486	Charcoal	García, 2003b
Gruta del Indio	9,905 ± 140	A-9489	Charcoal	García, 2003b
Gruta del Indio	10,195 ± 80	A-9497	Charcoal	García, 2003b
Gruta del Indio	10,170 ± 70	A-9498	Charcoal	García, 2003b
Gruta del Indio	8,045 ± 55	GrN-5394	Charcoal	García, 2003a
Gruta del Indio	9,770 ± 85	A-9491	Charcoal	García, 2003b
Gruta del Indio	9,825 ± 95/90	A-9492	Charcoal	García, 2003b
Gruta del Indio	9,890 ± 75	A-9495	Charcoal	García, 2003b
Gruta del Indio	9,990 ± 75	A-9496	Charcoal	García, 2003b
Gruta del Indio	10,610 ± 210	A-1351	Megafaunal dung	García, 2003b
Gruta del Indio	11,820 ± 180	A-1371	Megafaunal dung	García, 2003b
Gruta del Indio	12,375 ± 115	A-9571	Megafaunal dung	García, 2003b
Gruta del Indio	11,040 ± 130	A-9570	Megafaunal dung	García, 2003b
Gruta del Indio	10,900 ± 185	A-9493	Megafaunal dung	García, 2003b
Gruta del Indio	9,560 ± 90	GrN-5772	Bone	García, 2003b
Gruta del Indio	9,590 ± 120	LP-860	Charcoal	García, 2003b
Gruta del Indio	9,580 ± 100	LP-941	Charcoal	García, 2003b
Gruta del Indio	9,510 ± 90	LP-991	Charcoal	García, 2003b
Gruta del Indio	9,160 ± 90	LP-986	Charcoal	García, 2003b
Gruta del Indio	9,700 ± 110	LP-876	Charcoal	García, 2003b
Gruta del Indio	8,920 ± 110	LP-854	Charcoal	García, 2003b
Gruta del Indio	7,430 ± 90	LP-873	Charcoal	García, 2003b
Gruta del Indio	7,860 ± 90	LP-845	Charcoal	García, 2003b

(continued)

TABLE 8.6. (continued)

Site	Date(BP)	Lab	Material	Source
Gruta del Indio	8,990 ± 90	LP-925	Megafaunal dung	García, 2003b
Gruta del Indio	30,200 ± 800	LP-929	Megafaunal dung	García, 2003b
Gruta del Indio	30,800 ± 700	LP-918	Megafaunal dung	García, 2003b
Gruta del Indio	28,670 ± 720	LP-1072	Megafaunal dung	García, 2003b
Gruta del Indio	24,140 ± 510	LP-1075	Megafaunal dung	García, 2003b
Gruta del Indio	>37,610	Beta-152587	Megafaunal dung	García, 2003b
Gruta del Indio	29,530 ± 540	Beta-152588	Megafaunal dung	García, 2003b
Gruta del Indio	36,400 ± 200	Beta-152589	Megafaunal dung	García, 2003b

but no evidence for interaction was found (García, 2003b:33). The activities of burrowing animals were identified, and the wide vertical distribution of the flakes suggests secondary contexts.

The Pampas is the area that presents the most complete sequence of Quaternary mammals in South America and it constitutes the basis on which the faunas from other areas are interpreted. The framework provided by Ameghino was reorganized and refined by generations of paleontologists, providing schemes of Mammal Ages (Pascual et al., 1966) and biozones (Cione and Tonni, 1999) which are widely used to organize the paleontological information from several regions in South America.

At the site Arroyo Seco 2 the bones of *Megatherium americanum*, *Hippidion principale*, *Equus* (*Amerhippus*) *neogeus*, and guanaco, found in association with lithic artifacts, present cutmarks (Fidalgo et al., 1986; Politis et al., 1995). The chronology based on several radiocarbon dates suggests a late Pleistocene exploitation of these animals, since some early Holocene dates could not be replicated when the same bones were redated (Table 8.7) (Politis et al., 2003). The additional presence of *Eutatus*, cf. *Mylodon*, *Hemiauchenia paradoxa*, *Glossotherium robustum*, *Toxodon platensis*, *Macrauchenia patachonica*, and *Glyptodon* sp. (Fidalgo et al., 1986; Bond, 1999; Miotti and Salemme, 1999; Cione et al., 2003) is also recorded, but without any evidence of behavioral association with humans. Thus, this complex locality contains three fossil and several modern taxa that were exploited by humans and evidence of contemporaneity of several other extinct species whose presence in the site is not yet well understood (Gutierrez et al., 2000; Politis et al., 1995).

At the site Paso Otero 5, dated 10,190 ± 120 BP (AA-19291) and 10,440 ± 100 BP (AA-39363), there is an association between human artifacts, including a Fell Cave projectile point, and *Megatherium americanum*, *Glossotherium* sp., *Glyptodon* sp., *Toxodon* sp., *Hemiauchenia* sp., *Equus* (*A.*) *neogeus*, and guanaco (Martinez, 2001). A taphonomic study of this site suggests that megafaunal bones were used as fuel (Gutiérrez et al., 2001; Joly et al., 2005). Unfortunately the state of preservation of the bones precludes any analysis of cutmarks.

Cerro La China 1 is a small site where a scute of *Eutatus* sp. and Fell Cave projectile points were found coexisting in an archeological deposit dated between 10,520 ± 75 and 10,804 ± 75 BP (Flegenheimer and Zárate, 1997) (Table 8.7). It is difficult to infer the existence of behavioral interaction on this basis, since it is equally possible that the association resulted from economic, symbolic, or post-depositional causes. Other archeological sites in the Cerro La China locality have late Pleistocene dates, but no bone preservation.

Cueva Tixi produced remains of *Eutatus seguini* and *Dusicyon avus*, as well as a long list of modern species (Mazzanti and Quintana, 2001). Notably, cutmarks were found only on the bones of the modern species (Valverde, 2001). Cueva Tixi presents dates of ca. 10,000 BP. Other sites such as Los Pinos, Amalia, Burucuya, or La Brava provided indications of human utilization with late Pleistocene radiocarbon dates. However, megafaunal bones are not present. Problems of bone preservation might be implicated in some cases, but the presence of modern fauna suggests this is not always the case.

In addition to the evidence from Arroyo Seco 2, Paso Otero 5, and Cueva Tixi, there are radiocarbon dates from paleontological sites. A rib of Mylodontidae from Salto de Piedra was dated 11,940 ± 80 BP (LP-1193), bones of *Megatherium americanum* from Campo del Arce were dated 13,070 ± 120 (LP-174), and bones of *Equus* (*Amerhippus*) *neogeus* from Zanjón Seco were dated 10,290 ± 130 (LP-1235) (Tonni et al., 2003b). Thus, there is chronological support for overlap between humans and megafauna in the Pampas. The evidence indicates the presence of some loci of megafaunal utilization, but no clear indication of the mode of their exploitation. Martinez and Gutierrez (2004) suggested that these animals were obtained by scavenging.

The archeology of the Pampas is also characterized by the presence of two sites with evidence for megafauna with Holocene dates.

The site La Moderna produced three radiocarbon dates ca. 7,000–7,500 BP on bones of *Doedicurus clavicaudatus* in association with human artifacts (Politis and Gutierrez, 1998). Bones of *Mylodon* sp., *Sclerocalyptus* sp., and *Tolypeutes* sp. are also present. Radiocarbon dates were also made on the soil humates fraction of sediment samples from the archeological layers, with results of 8,356 ± 65 (DRI-3012) and 7,448 ± 109 BP (DRI-3013) (Politis et al., 2003: 46). As for the existence of Holocene survival of Pleistocene megafauna, the evidence from site Campo Laborde also provided early Holocene radiocarbon dates for bones of *Megatherium americanum* found associated with human artifacts (Messineo et al., 2004).

The site El Trebol in Northern Patagonia produced bones of *Canis* (*D.*) *avus*, several Mylodontinae osteoderms and bones of modern fish and mammals which were found

TABLE 8.7. Radiocarbon dates for sites in the Pampas, Argentina.

Site	Date(BP)	Lab	Material	Source
Arroyo Seco 2	10,500 ± 90	AA-9049	*Glossotherium robustus* bone	Politis et al., 2003
Arroyo Seco 2	12,240 ± 110	-	-	Politis et al., 2003
Arroyo Seco 2	12,200 ± 170	CAMS-58182	*Megatherium americanum* bone	Politis et al., 2003
Arroyo Seco 2	11,590 ± 90	AA-7965	*Equus neogeus* bone	Politis and Madrid, 2001
Arroyo Seco 2	11,250 ± 100	AA-7964	*Toxodon platensis* bone	Politis and Madrid, 2001
Arroyo Seco 2	8,890 ± 90	TO-1504	*Equus* sp. bone	Politis and Madrid, 2001
Arroyo Seco 2	8,390 ± 240	LP-53	*Megatherium americanum* bone	Politis and Madrid, 2001
Arroyo Seco 2	12,240 ± 110	OxA-4591, same as AA-9049	*Glossotherium robustus* bone	Politis and Madrid, 2001
Arroyo Seco 2	7,320 ± 50	TO-1506	*Megatherium americanum* bone	Politis and Madrid, 2001
Arroyo Seco 2	11,000 ± 100	OxA-4590	*Equus neogeus* bone	Politis and Madrid, 2001
Arroyo Seco 2	12,200 ± 170	CAMS-58182	*Megatherium americanum* bone (same as TO-1506)	Politis and Madrid, 2001
Arroyo Seco 2	11,750 ± 70	CAMS-16389	*Toxodon platensis* bone (same as AA-7964)	Politis and Madrid, 2001
Paso Otero 5	10,190 ± 120	AA-19291	Megamammal bone	Holliday et al., 2003
Paso Otero 5	10,440 ± 100	AA-39363	*Megatherium americanum* bone	Holliday et al., 2003
La Moderna	12,330 ± 370	TO-1507	*Doedicurus clavicaudatus* bone	Politis and Madrid, 2001
La Moderna	7,010 ± 100	TO-1507–1	*Doedicurus clavicaudatus* bone	Politis and Gutiérrez, 1998
La Moderna	7,510 ± 370	TO-1507–2	*Doedicurus clavicaudatus* bone	Politis and Gutiérrez, 1998
La Moderna	7,460 ± 80	TO-2610	*Doedicurus clavicaudatus* bone	Politis and Gutiérrez, 1998
La Moderna	6,555 ± 160	Beta-7824	*Doedicurus clavicaudatus* bone	Politis and Gutiérrez, 1998
Cerro La China Site 1	10,804 ± 75	AA-8953	Charcoal	Flegenheimer and Zárate, 1997
Cerro La China Site 1	10,520 ± 75	AA-8954	Charcoal	Flegenheimer and Zárate, 1997
Cerro La China Site 1	10,745 ±75	AA-8952	Charcoal	Flegenheimer and Zárate, 1997
Cerro La China Site1	10,790 ± 120	A-1327	Charcoal	Flegenheimer and Zárate, 1997
Cerro La China 1	10,720 ± 150	I-12741	Charcoal	Flegenheimer and Zárate, 1997
Cerro La China Site 3	10,610 ± 180	AA-1328	Charcoal	Flegenheimer and Zárate, 1991
Cerro La China Site 2	11,150 ± 135	AA-8955	Charcoal	Flegenheimer and Zárate, 1997
Cerro La China Site 2	10,560 ± 75	AA-8956	Charcoal	Flegenheimer and Zárate, 1997
Cerro Sombrero Abrigo 1	10,725 ± 90	AA-4765	Charcoal	Flegenheimer, 2003
Cerro Sombrero Abrigo 1	10,270 ± 85	AA-4766	Charcoal	Flegenheimer, 2003
Cerro Sombrero Abrigo 1	10,675 ± 110	AA-4767	Charcoal	Flegenheimer, 2003
Cerro Sombrero Abrigo 1	10,480 ± 70	AA-5220	Charcoal	Flegenheimer, 2003
Cerro Sombrero Abrigo 1	8,060 ± 140	AA-5221	Charcoal	Flegenheimer, 2003
El Abra Cave	9,834 ± 65	AA-38098		Mazzanti, 2003
Los Pinos Shelter	9,570 ± 150	LP-630	Charcoal	Mazzanti, 2003
Los Pinos Shelter	10,465 ± 65	AA-24045	Charcoal	Mazzanti, 2003
Los Pinos Shelter	10,415 ± 70	AA-24046	Charcoal	Mazzanti, 2003
Tixi Cave	10,045 ± 95	AA-12131	Charcoal	Mazzanti and Quintana, 2001
Tixi Cave	10,375 ± 90	AA-12130	Charcoal	Mazzanti and Quintana, 2001
La Brava Cave	9,670 ± 120	LP-550	Charcoal	Mazzanti, 2003
Burucuyá Cave	10,000 ± 120	LP-863	Charcoal	Mazzanti, 2003
Amalia Site 2	10,425 ± 75	AA-35499	Charcoal	Mazzanti, 2003

together with human artifacts (Hajduk et al., 2004). There is mention of the presence of cutmarks on some osteoderms. No radiocarbon dates are available. At Cuchillo Cura an intriguing discovery was made, since DNA extracted from dung belongs to a small ground sloth not yet represented by skeletal material. Dung was dated 13,750 ± 230 BP (GX 21149) and 14,665 ± 150 BP (Ua 13871) (Hofreiter et al., 2003).

Several sites from Southern Patagonia offer important information for our understanding of the relationships between humans and extinct mammals (Table 8.8). At the site Los Toldos 3, bones of *Hippidion saldiasi*, *Lama* (*Vicugna*) *gracilis*, and modern camelids were found together with charcoal and lithic artifacts. A single date is 12,600 ± 650 BP (published with no lab number), but the context of the sample is not clearly defined (Cardich et al., 1973). The finding of *Hippidion saldiasi* in Layer 10, which has a date on charcoal of 8,750 ± 480 BP (no lab number), was used to defend a Holocene survival of horse in Patagonia. However, the lack of a taphonomic evaluation of the faunas and sediments conspire against the credibility of this affirmation. At the nearby site Los Toldos 2, the remains of *Hippidion saldiasi* and *Lama* (*V.*) *gracilis* were also found in archeological context.

Not too far from Los Toldos is located the site El Ceibo, where bones of *Lama* (*V.*) *gracilis* and modern camelids were found in association with human tools (Miotti et al., 1999).

TABLE 8.8. Radiocarbon dates for sites in South Patagonia, Argentina.

Site	Date(BP)	Lab	Material	Source
Cerro Casa de Piedra 7	10,530 ± 620	UGA-7385	Megafaunal dung	Civalero and Aschero, 2003
Cerro Casa de Piedra 7 Layer 17	9,640 ± 190	UGA-7384	Wood	Civalero and Aschero, 2003
Cerro Casa de Piedra 7 Layer 17	9,100 15	LP-364	Charcoal	Civalero and Aschero, 200
Los Toldos 3 Level 11	12,600 ± 600	-	Charcoal	Cardich et al., 1973
Los Toldos 3 Level 10	8,750 ± 480	-.	Charcoal	Cardich et al., 1973
El Ceibo	Ca. 9,500	-	-	Miotti and Salemme, 2003
Piedra Museo	9230 ± 105	LP-949	*Lama guanicoe* bone	Miotti et al., 2003
Piedra Museo	9710 ± 105	LP-859	*Lama guanicoe* bone	Miotti et al., 2003
Piedra Museo	10,400 ± 80	AA-8428	Camelidae bone	Miotti et al., 2003
Piedra Museo	10,470 ± 60	OxA-9249	Charcoal	Miotti et al., 2003
Piedra Museo	10,470 ± 65	GRA-9837	Charcoal	Miotti et al., 2003
Piedra Museo	10,390 ± 70	OxA-8527	*Lama guanicoe* bone	Miotti et al., 2003
Piedra Museo	10,925 ± 65	OxA-8528	*Hippidion saldiasi* bone	Miotti et al., 2003
Piedra Museo	11,000 ± 65	AA-27950	Charcoal	Miotti et al., 2003
Piedra Museo	12,890 ± 90	AA-20125	Charcoal	Miotti et al., 2003; Miotti et al., 1999
Piedra Museo	9,950 ± 75	OxA-509	Charcoal	Miotti et al., 2003
Piedra Museo	9,350 ± 130	OxA-9508	Charcoal	Miotti et al., 2003
Piedra Museo	10,100 ± 110	OxA-9507	*Lama* sp. bone	Miotti et al., 2003
Casa del Minero 1	10,999 ± 55	AA-37207	Charcoal	Paunero, 2003
Casa del Minero 1	10,967 ± 55	AA-37208	Charcoal	Paunero, 2003
Cueva de la Mesada	9,090 ± 40	Beta-135963	Charcoal	Paunero, 2003

The lack of chronological, stratigraphic, or taphonomic information makes an evaluation difficult in this case.

Two important sites are located south of Los Toldos, both in the central plateau of the province of Santa Cruz: Casa del Minero 1 and Piedra Museo. At the site Casa del Minero 1 flaked bones, hearths, and lithic artifacts were found in association with bones of *Hemiauchenia* cf. *paradoxa*, *Lama (Vicugna) gracilis*, guanaco, and probably *D. avus* (Paunero, 2003; Paunero et al., 2004). There are two radiocarbon dates of 10,999 ± 55 BP (AA-37207) and 10,967 ± 55 BP (AA-37208). Thus two extinct camelids were found at this site in clear association with human artifacts.

At the Piedra Museo site, hearths and lithic artifacts were found in association with bones of *Hippidion saldiasi*, *Lama (V.) gracili*, *Mylodon* sp., and guanaco (Miotti, 1996; Alberdi et al., 2001; Miotti et al., 2003). There are twelve radiocarbon dates between 9,230 ± 105 BP (LP-949) and 11,000 ± 65 BP (AA-27950), plus one isolated date of 12,890 ± 90 BP (AA-20125) on charcoal (Table 8.8). This outlier does not overlap statistically with the other dates.

At the site Las Buitreras the remains of *Mylodon listai* (an adult and a juvenile), *Dusicyon avus,* and two teeth of *Hippidon saldiasi* were found (Sanguinetti, 1976; Scillato Yané, 1976; Alberdi and Prado, 2004). A claim for behavioral association of the ground sloth remains and several flakes (Sanguinetti and Borrero, 1977) is another case of mere physical association. A case for the operation of a process of vertical migration of guanaco bones and flakes is held responsible for this association (Borrero and Martin, 2006).

Other sites with Transition dates do not present bones of megafauna, such as Cerro Tres Tetas (Paunero, 1993–1994) and Chorrillo Malo 2 (Borrero and Franco, 1997). Also, layers with Pleistocene fauna were discovered below the initial human occupation of Cerro Casa de Piedra 7 (Civalero and Aschero, 2003), Traful 1 (Crivelli Montero et al., 1993), and Baño Nuevo (Mena et al., 2003). All this evidence indicates either late arrival of humans or lack of earlier interaction with megafauna.

The evidence from at least two regions in Argentina, the Pampas and southern Patagonia, indicates a certain degree of interaction between humans and megafauna. The latter is confirmed by equally compelling evidence produced by four sites in southern Chile.

Discussion

It is difficult to say much at a continental scale on the basis of the evidence of taxon-dates, since these are very rare and are strictly concentrated in the Pampas and Patagonia (Borrero, 1997; Alberdi and Prado, 2004). Accordingly, differences between Last Occurrences and First Contact dates are as of yet difficult to assess. Our grip on the length of the coexistence of humans and megafauna is equivocal, since we have good and reasonably abundant dates only for only a very few species at a limited number of sites. The conclusion is that most of the megafauna remains very poorly dated.

Some of the temporal implications derived from the few regions for which reliable data exist are extremely important. For example, there is no evidence for the Holocene survival of any of the large mammals in Ultima Esperanza (Borrero, 1999), or extra-Andean Patagonia (Borrero, 1997), but there is support for the survival of *Doedicurus clavicaudatus,*

Megatherium americanum, and other species into the early Holocene in the Pampas (Politis et al., 1995: 196–197; Politis and Gutierrez, 1998; Messineo et al., 2004) and *Mylodon* sp. in Cuyo (García and Lagiglia, 1998–1999). Available chronological data indicate that extinctions occurred in Ultima Esperanza or the central plateau of south Patagonia within 1,000 years after the first human contact (Miotti, 1996; Borrero, 1999; Paunero et al., 2004), within 1,400 years at Cuyo and within 3,000 years in the Pampas (Politis and Gutierrez, 1998).

If it is true that Pleistocene mammals survived the arrival of humans for more than 1,000 years in the Southern part of South America (11,000 to 10,000 BP), then the timing of their extinction could have been later than is usually accepted. There is no way to use the chronological record at a finer scale, since the end of the Pleistocene is characterized by the radiocarbon plateaus that tend to concentrate events of different age (Becker, 1993).

Even in the absence of good chronological resolution there are patterns that deserve further exploration in the future. For example, *Stegomastodon platensis*, the mastodont from the Pampas, is not present in the dozens of terminal Pleistocene archeological and paleontological sites of that region (Alberdi and Prado, 1995:286). Considering that some of the best evidence for behavioral association in South America concerns mastodonts, this pattern is interesting and might suggest extinction before human arrival in the Pampas.

A medium-sized fox, *Dusicyon avus*, is the only representative of the Pleistocene fauna that apparently became extinct during the late Holocene in more than one region, including Southern extra-Andean Patagonia (Miotti and Berman, 1988) and Pampa (Tonni and Politis, 1981). However, it must be emphasized that it is not yet clear if this is a true extinction or a case of evolution of the species (Berman and Tonni, 1987). On the other hand, in only one case the remains of this fox were directly dated, at the Tres Arroyos site in Tierra del Fuego, where a mandible produced a result of 10,575 ± 65 (OxA-9245).

There are also more recent but not better understood extinctions, such as that of the Malvinas fox *Dusicyon australis*, which was last seen in 1876 (see MacPhee and Flemming, 1999). But it is probable that this canid, world famous due to Darwin's description (1860), "was either a relic of a domesticated form of *Dusicyon* or a feral hybrid evolved from a cross between a domestic *Canis* species and a *Dusicyon* species" (Clutton-Brock, 1977:1341).

If we take a close look at the late Pleistocene faunal associations, it is clear that extant species dominate most of the archeological assemblages and that megafaunal taxa are rarely abundant (i.e., Politis et al., 1995; Miotti et al., 1999). Horses are usually treated by archeologists as a hunted animal, whose presence is widespread in South America, many times in clear association with humans, and in some cases considered the dominant prey. At Piedra Museo the count of horse bones on which specialization was predicated is only nine (Alberdi et al., 2001). Cueva del Medio is another site for which it was maintained that a horse-hunting specialization existed, but the remains attributed to camelids are numerically dominant and the Minimum Number of Individuals for horse is only two (Nami and Menegaz, 1991). Horse remains are also scarce at many other sites. Existing evidence suggests that it was hunted, but not that it was a preferred prey. Opportunistic exploitation of Pleistocene mammals is the best available explanation for the recorded archeological patterns. Either scavenging or hunting is indicated in different cases. The former could have been specifically the case for ground sloth and other megaherbivores. Only the mastodont remains from Taguatagua fit comfortably in a discussion of selective hunting tactics. This is probably a limitation derived from the fact that it is one of the few open-air sites, together perhaps with La Moderna, with the potential to be a killsite. The coexistence of projectile points and megafauna at sites like Cueva Fell or Cueva del Medio also can be used for this discussion, but clearly these are locations that were positioned away from eventual kill areas. When the hypothesis of extinction of megafauna by Overkill is examined in this light, the result is that there is not much evidence in support. It is not the lack of a fossil record of the extinction, or the "invisible evidence" to which Martin refers. Rather than that, it is what the fossil record appears to tell us, namely that there is coexistence between humans and megafauna, and there is interaction, but with the exception of a few cases active organized hunting is very difficult to substantiate.

It may not be possible for humans to take mammal-hunting to extinction levels, since that would require the implementation of sub-optimal strategies (Smith, 1983; but also see Surovell and Waguespack, 2008, Chapter 5). This situation highlights the problem of explaining the disappearance of the last members of a population, since "harvesting becomes economically inefficient once the resource gets scarce" (Owen-Smith, 1999:60). However, if other values are built into the model, such as the existence of a "show-off" strategy among hunters, as suggested by O'Connell (2000), then the possibility exists that humans pursued large and rare animals because of its social payoffs: "Recent work on traditional foragers… shows that hunters routinely ignore commonly encountered, easily captured prey that would ensure the subsistence security of mates and offspring in favor of other targets, larger in size but less often seen and much less regularly taken" (O'Connell, 2000:174; see also Bliege Bird and Smith, 2005). In conclusion, there is no impediment against an active human participation in the taking and exploitation of rare or difficult to obtain prey.

In order to evaluate hunting, some considerations about behavior and survival of prey are also needed. Lack of defenses based on the naiveté of the animals was a basic component of Martin's classic argument for overkill (see Diamond, 1984). Inspired by his recent work on re-introduction of carnivores into ecosystems, J. Berger and coauthors (2001, also Pyare and Berger, 2003) found that anti-predator responses are not learned very quickly by mammals exposed to new carnivores in their habitats. This supports the notion that: "...a first tier of highly vulnerable species succumb to predation or other ecological traumas to which they cannot adjust in time to escape extinction" (Martin and Steadman, 1999:47).

We should look into this matter in more detail. Perhaps the survivors – for example, bison in North America and guanaco in South America – are just quick learners, as opposed to slow learners like horse or ground sloth (see Berger et al., 2001:1039). Martin and Steadman listed some required properties: "To survive late Quaternary Extinctions in South America it paid to be relatively small, cryptic in habitat, fleet, and fecund. It also helped to favor dense cover, or, conversely, wide open spaces in the high dry inter-Andean plateau (...) South American capybara, tapir, two camelids, various deer, and the spectacled bear possess at least some of the above attributes" (Martin and Steadman, 1999:38). Living in rough terrain (Alroy, 1999:139) and being an aquatic mammal are other invoked properties (Stanley, 1984). However, the case may not be so simple. Using uniformitarianism as a guiding principle, Gary Haynes suggested that proboscideans, which clearly did not survive into the Holocene in continental North America, were quick learners (Haynes, 1991:107–108). Moreover, the glyptodont *Doedicurus clavicaudatus* which lived in the Pampas up to ca. 8,000 BP cannot be considered small, fecund, or rapid, nor was it an inhabitant of difficult terrain. Furthermore, it was one of those armored animals that must have needed highly productive environments where travel and search time for forage was low, in other words where it could easily find food (Morse, 1980:127). Therefore, it seems that for the time being there is not much in the survival attributes or the ability to learn antipredator tactics that help us to understand the list of survivors.

In sum, we still do not know what caused the demise of so many species at the end of the Pleistocene, but together with a taphonomic assessment of the evidence, a growing chronological framework, and revised systematics, we do have better and more focused questions to ask in the near future.

Conclusions

Extinction caused by humans – sometimes combined with environmental or climatic factors – is one of the recurrent explanations favored by some South American authors (Cione et al., 2003; Alberdi and Prado, 2004; Prous and Fogaca, 1999). However, rarely anything beyond the impressive chronological overlap between the time of the extinctions and the time of the arrival of humans is invoked and the interpretation of the archeological evidence is just as rarely aided by taphonomic insight. In dealing with the factors behind the extinctions, it is probably safe to affirm that there is no single factor producing the extinction of large mammals (Barnosky, 1990; Minta et al., 1999; Gittleman and Gompper, 2001; Brenchley and Harper, 1998:336; MacPhee and Flemming, 1999:36; Owen-Smith, 1999:67). Even though at this time I am little inclined to discuss the specific causes of the extinctions, I have here concentrated on an evaluation of the human factor, which appears not to be substantiated by the archeological record of South America. The available information about behavioral interaction indicates that megafauna was at most a complementary resource. The degree of association with lithic tools is varied, but very few cases can support a behavioral argument of synchronic interaction. In the cases with a defensible association, the number of involved megafaunal individuals is low in comparison with modern species. One conclusion then is that humans and Pleistocene faunas coexisted in many regions of South America, and that in some situations that coexistence led to a moderate degree of interaction. It is difficult to tell if that interaction took the form of hunting or scavenging. Other forms of interaction such as avoidance may also explain at least part of the available evidence, but will be difficult to discuss with the fossil record. Anyway, the result was that by the end of the Pleistocene –and apparently early Holocene in some regions– the megafauna were gone. After that, South American human hunters were feeding on vicuñas, guanacos, deer, or a variety of smaller resources, a pattern that was basically established some time before the disappearance of the megafauna. This conclusion in no way falsifies the hypothesis of Overkill, but it does show that the existing empirical evidence can not be used in its support.

Summing Up

1. The age difference between Last Dated Appearance of Pleistocene megafauna and First Contact with humans is around 1,000 radiocarbon years at some regions and was perhaps longer at places where Holocene survival is suggested. Thus, the length of coexistence can be measured in thousands of years. The extinctions of South American megafauna are spread over a long time period.
2. Some places, including the south of the Intermediate Area, most of the Central Andes, and the regions closer to the Andes in Patagonia, were colonized by humans after the disappearance of the megafauna.
3. The information for interaction between humans and megafauna is very uneven. It is non-existent for the rainforests, it is available but not really compelling for the Andes, and it is present but minimal for the Pacific coast. It is important and well documented only for the pampas and plateaus of the east and south of South America.
4. There are very few sites with integrity that present behavioral associations between humans and megafauna, in spite of the relatively abundant evidence for the presence of megafaunal bones, and the existence of many archeological

sites with Transition radiocarbon dates (Hoffstetter, 1986; Alberdi and Prado, 1992).

In the future we will need more evidence from open-air sites, and perhaps we should work guided by a combination of geoarcheological and taphonomic perspectives. This is the only way to surmount the present bias toward samples from caves and rockshelters.

Acknowledgments. Thanks to Gary Haynes for his invitation to contribute this chapter. Fabiana Maria Martin discussed in detail the evidence from southern Patagonia and contributed the bone photographs. Marcia Bianchi Villelli and Ramiro Barberena helped me with the map.

Note

1 Radiometric dates in this chapter are noted as radiocarbon years Before Present (that is, uncalibrated), with 1950 CE/AD considered to be the "present."

References

Alberdi MT, Prado JL (1992) El registro de *Hippidion* Owen, 1869 y *Equus (Amerhippus)* Hoffstetter, 1950 (Mammalia, Perissodactyla) en América del Sur. Ameghiniana 29:265–284

Alberdi MT, Prado JL (1995) Los mastodontes de América del Sur. In: Alberdi MT, Leone G, Tonni EP (eds) Evolución biológica y climática de la región pampeana durante los últimos cinco millones de años. Un ensayo de correlación con el Mediterráneo occidental. Consejo Superior de Investigaciones Científicas, Madrid, pp. 277–291

Alberdi MT, Prado JL (2004) Caballos fósiles de América del Sur. Una historia de tres millones de años. INCUAPA, Olavarría

Alberdi MT, Prieto A (2000) Restos de Hippidion saldiasi en las cuevas de la Patagonia chilena. Anales del Instituto de la Patagonia 28:147–171

Alberdi MT, Fernández J, Menegaz AN, Prado JL (1986) Hippidion Owen 1869 (Mammalia. Perissodactyla) en sedimentos del Pleistoceno tardío de la localidad Barro Negro (Jujuy, Argentina). Estudios Geológicos 42:487–493

Alberdi MT, Miotti L, Prado JL (2001) *Hippidion saldiasi* Roth, 1899 (Equidae, Perissodactyla), at the Piedra Museo Site (prov. of Santa Cruz, Argentina): Its implications for the regional economy and environmental reconstructions. J Archaeol Sci 28:411–419

Alroy J (1999) Putting North America's end-Pleistocene megafaunal extinction in context: Large-scale analyses of spatial patterns, extinction rates, and size distributions. In: MacPhee RDE (ed) Extinctions in near time: Causes, contexts and consequences. Kluwer/Plenum, New York, pp 105–144

Aschero CA (1979) Un asentamiento acerámico en la Quebrada Inca Cueva. Informe preliminar sobre el sitio IC4. In: Actas de las Jornadas de Arqueología del Noroeste argentino, pp 159–163

Aschero CA, Elkin D, Pintar EL (1991) Aprovechamiento de recursos faunísticos y producción lítica en el precerámico tardío. Un caso de estudio: Quebrada Seca 3 (Puna meridional Argentina). In: Actas del VI Congreso Nacional de Arqueología Chilena, pp 147–162

Barnett R, Barnes I, Phillips MJ, Martin LD, Harington CR, Leonard JA, Cooper A (2005) Evolution of the extinct sabretooths and the American cheetah-like cat. Curr Biol 15:R589–590

Barnosky A (1990) The late Pleistocene event as a paradigm for widespread mammal extinction. In: Donovan SK (ed) Mass extinctions. Columbia University Press, New York, pp 235–254

Barnosky A, Koch PL, Feranec RS, Wing SL, Shabel AB (2004) Assessing the causes of late Pleistocene extinctions on the continents. Science 306:70–75

Becker B (1993) An 11,000 year German Oaks and Pine dendrochronology for radiocarbon calibration. Radiocarbon 35:201–213

Beltrao MC de MC, Enriquez CR, Danon J, Zuleta E, Poupeau G (1986) Thermoluminescence dating of burnt cherts from the Alice Boer site (Brazil). In: Bryan AL (ed) New evidence for the Pleistocene peopling of the Americas, Center for the Study of Early Man, Orono, pp 203–213

Berger J, Swenson JE, Persson IL (2001) Recolonizing carnivores and naive prey: Conservation lessons from Pleistocene extinctions. Science 291:1036–1039

Berman WD, Tonni EP (1987) *Canis (Dusicyon) avus* Burmeister, 1864 (Carnivora, Canidae) en el Pleistoceno tardío y Holoceno de la provincia de Buenos Aires. Aspectos sistemáticos y bioestratigráficos relacionados. Ameghiniana 24:245–250

Bird J (1938) Antiquity and migrations of the early inhabitants of Patagonia. Geograph Rev 28:250–275

Bird J (1988) Travels and archaeology in South Chile. University of Iowa Press, Iowa City

Bliege Bird R, Smith EA (2005) Signaling theory, stratigraphic interaction, and symbolic capital. Curr Anthropol 46 (2):221–248

Bombin M, Bryan AL (1978) New perspectives on Early Man in southwestern Rio Grande do Sul, Brazil. In: Bryan AL (ed) Early Man in America from a circum-Pacific perspective. Occasional Papers 1, University of Alberta, pp 301–302

Bond M (1999) Quaternary native ungulates of southern South America. A synthesis. Quat South Am Antarct Peninsula 12:177–205

Borrero LA (1977) La extinción de la megafauna: Su explicación por factores concurrentes. La situación en Patagonia austral. Anales del Instituto de la Patagonia 8:81–93

Borrero LA (1997) La extinción de la megafauna en la Patagonia. Anales del Instituto de la Patagonia 25:89–102

Borrero LA (1999) The faunas of the Pleistocene/Holocene boundary in the Seno de la Ultima Esperanza, Chile. In: Driver J (ed) Zooarchaeology of the Pleistocene/Holocene boundary, BAR International Series 800, Oxford, pp 59–62

Borrero LA (2003) Taphonomy of the Tres Arroyos 1 rockshelter. Quat Intl 109–11:87–93

Borrero LA (2005) Taphonomy of late Pleistocene faunas at Fuego-Patagonia. J South Am Earth Sci 20:115–120

Borrero LA, Franco NV (1997) Early Patagonian hunter-gatherers: Subsistence and technology. J Anthropol Res 53:219–239

Borrero LA, Martin FM (1996) Tafonomía de carnívoros: un enfoque regional. In: Gómez Otero J (ed) Arqueología, Sólo Patagonia, CENPAT, Puerto Madryn, pp 189–198

Borrero LA, Martin FM (1999) The taphonomy of small mammals at Tres Arroyos Rockshelter. Paper presented at the 64th Annual Meeting of the Society for American Archaeology, Mar. 24–28 Chicago, IL

Borrero LA, Martin FM (2006) Revisiting the association of Pleistocene faunas and humans at cave Las Buitreras, Santa Cruz, Argentina. Manuscript

Borrero LA, Massone M (1994) Arqueología de los aleros Dos Herraduras, Ultima Esperanza, Chile. In: Actas y Memorias XI Congreso Nacional de Arqueología Argentina 2:288–289

Borrero LA, Lanata JL Cárdenas C (1991) Reestudiando cuevas:Nuevas excavaciones en Ultima Esperanza. Anales del Instituto de la Patagonia 20:101–122

Borrero LA, Martin FM, Prieto A (1997) La Cueva Lago Sofía 4, Ultima Esperanza, Chile: U madriguera felina del Pleistoceno tardío. Anales del Instituto de la Patagonia 25:103–122

Brenchley PJ, Harper DAT (1998) Palaeoecology: Ecosystems, environments and evolution. Chapman & Hall, London

Bryan AL, Gruhn R (1979) The radiocarbon dates of Taima Taima. In: Ochsenius C, Gruhn R (eds) Taima Taima: A late Pleistocene Paleoindian kill site in northernmost South America. Final Report of 1976 Excavations. Programa CIPICS, Federal Republic of Germany, pp 53–58

Canto J (1991) Posible presencia de una variedad de *Smilodon* en el Pleistoceno tardío. Anales del Instituto de la Patagonia 20:96–99

Cardich A (1964) Lauricocha. Fundamentos para una prehistoria de los Andes centrales. Studia Praehistorica 3, Centro Argentino de Estudios Prehistóricos, Buenos Aires

Cardich A (1973) Excavaciones en la caverna de Huargo, Perú. Revista del Museo Nacional 39:11–29

Cardich A, Cardich L, Hajduk A (1973) Secuencia arqueológica y cronología radiocarbónica de la cueva 3 de Los Toldos (provincia de Santa Cruz, Argentina). Relaciones de la Sociedad Argentina de Antropología 7:85–123

Cardich A, Izeta A (1999–2000) Revisitando Huargo (Perú). Análisis cuantitativos aplicados a restos de Camelidae del Pleistoceno tardío. Anales de Arqueología y Etnología 54–55:29–46

Cartajena I, Nuñez L, Grosjean M (2005) Las arqueofaunas del Arcaico temprano en la vertiente occidental de la Puna de Atacama. In: Actas del XVI Congreso Nacional de Arqueología Chilena, pp 507–517

Cartelle C, Bohorquez GA (1986) Descricao das pre-maxilas de *Nothrotherium maquinense* (Lund) Lydekker, 1899 (Edentata, Megalonychidae) e de *Eremotherium laurillardi* (Lund) Cartelle and Bohorquez, 1982 (Edentata, Megatheridae). Iheringia 11:9–14

Cartelle C, Fonseca JS (1981) Espécies do genero *Glossotherium* no Brasil. In: Anais do II Congresso Latino-Americano de Paleontología 2:805–818

Casamiquela RM (1979) An interpretation of the fossil vertebrates of the Taima-Taima site. In: Ochsenius C, Gruhn R (eds) Taima-Taima. A late Pleistocene Paleo Indian kill site in northernmost South America. Final report of 1976 excavations. CIPICS, Federal Republic of Germany, pp 59–76

Casamiquela RM (1984) Critical catalogue of some Chilean fossil vertebrates. I. The deers: Complementary considerations on *Antifer* (*Antifer niemeyeri* n. Sp.), the Pleistocene giant deer. Quat South Am Antarct Peninsula 2:41–50

Casamiquela RM (1999) The vertebrate record of the Pleistocene of Chile. Quat South Am Antarct Peninsula 12:91–107

Casamiquela RM, Dillehay T (1989) Vertebrates and invertebrates faunal analysis. In: Dillehay T (ed) Monte Verde. A late Pleistocene settlement in Chile, vol 1. Paleoenvironment and site context. Smithsonian Institution Press, Washington DC, pp 205–210

Casamiquela R, Montané J, Santana R (1967) Convivencia del hombre con el mastodonte en Chile central. Noticiario Mensual del Museo Nacional de Historia Natural 11(132):1–6

Caviglia SE (1986) Nuevos restos de cánidos tempranos en sitios arqueológicos de Fuego-Patagonia. Anales del Instituto de la Patagonia 16:85–93

Chauchat C (1988) Early hunter-gatherers on the Peruvian coast. In: Keatinge RW (ed) Peruvian prehistory. Cambridge University Press, Cambridge, pp 41–66

Cione A, Tonni EP (1999) Biostratigraphy and chronological scale of uppermost Cenozoic in the pampean area, Argentina. Quat South Am Antarct Peninsula 12:23–51

Cione A, Tonni EP, Soibelzon L (2003) The broken zig zag: Late Cenozoic large mammal and tortoise extinctions in South America. Revista Museo de Historia Natural Bernardino Rivadavia 5(1):1–19

Civalero MT, Aschero CA (2003) Early human occupations at Cerro Casa de Piedra 7, Santa Cruz Province, Patagonia, Argentina. In: Miotti L, Salemme M, Flegenheimer N (eds) Where the south winds blow: Ancient evidence of Paleo South Americans. Center for the Study of the First Americans, Texas A&M, pp 141–147

Clutton-Brock J (1977) Man-made dogs. Science 197:1340–1342

Coates AG, Obando JA (1996) The geological evolution of the Central America isthmus. In: Jackson JBC, Budd AF, Coates AG (eds) Evolution and environment in tropical America. University of Chicago Press, Chicago, pp 21–56

Correal Urrego G (1981) Evidencias culturales y megafauna Pleistocénica en Colombia. Fundación de Investigaciones Arqueológicas Nacionales, Banco de la República, Bogotá

Correal Urrego G (1986) Apuntes sobre el medio ambiente Pleistocénico y el hombre prehistórico en Colombia, In: Bryan AL (ed) New evidence for the Pleistocene peopling of the Americas. Center for the Study of Early Man, Orono, pp 115–131

Correal Urrego G (1993) Nuevas evidencias culturales Pleistocenicas y megafauna en Colombia. Boletín de Arqueología 8(1):3–12

Correal Urrego G, Van der Hammen T (1977) Investigaciones arqueologicas en los abrigos rocosos de Tequendama. Banco Popular, Bogotá

Crivelli Montero EA, Curzio D, Silveira MJ (1993) La estratigrafía de la cueva Traful 1 (Pcia del Neuquén). Praehistoria 1:9–160

Cruxent JM (1970) Projectile points with Pleistocene mammals in Venezuela. Am Ant 44:223–225

Darwin C (1860) Journal of researches into the natural history and geology of the countries visited during the voyage of H.M.S. Beagle Round the World, John Murray, London

Davies DM (1978) Some observations on the Otavalo skeleton from Imbabura Province, Ecuador. In Bryan AL (ed) Early Man in America from a circum-Pacific perspective, University of Alberta, Edmonton, p 273

Delibrias G, Guillier MT, Labeyrie J (1986) Gif Natural Radiocarbon Measurements X. Radiocarbon 28:9–68

Diamond J (1984) Historic extinctions: A rosetta stone for understanding prehistoric extinctions. In: Martin PS, Klein R (eds) Quaternary extinctions: A prehistoric revolution. University of Arizona Press, Tucson, pp 824–862

Dillehay TD (ed) (1997) Monte Verde: A late Pleistocene settlement in Chile. Vol. 2. The archaeological context and interpretation. Smithsonian Institution Press, Washington DC

Dillehay TD, Pino M (1997) Radiocarbon chronology. In: Dillehay T (ed) Monte Verde: A late Pleistocene settlement in Chile, vol 2:

The archaeological context and interpretation. Smithsonian Institution Press, Washington DC, pp 41–52
Dillehay TD, Ardila Calderón G, Politis G, de Moraes Coutinho Beltrao M da C (1992) Earliest hunters and gatherers of South America. J World Prehist 6:145–204
Elkin D, Rosenfeld S (2001) La explotación de camélidos en Pintoscayoc 1 (Jujuy). In: Mengoni Goñalons GL, Olivera DE, Yacobaccio HD (eds) El uso de los camélidos a través del tiempo. Ediciones del Tridente, Buenos Aires, pp 29–64
Engel F (1970) Explorations of the Chilca canyon, Peru. Curr Anthropol 11:55–58
Falguères C, Fontugne M, Chauchat C, Guadelli JL (1994) Datacions radiométriques de l'extinction des grandes faunes Pléistocenes au Perou. Comptes Rendus de L'Academie des Sciences, Serie II 319(2):261–266
Fariña R (1996) Trophic relationships among Lujanian mammals. Evolut Theor II (2):125–134
Fidalgo F, Meo Guzmán L, Politis G, Salemme S, Tonni EP (1986) Investigaciones arqueológicas en el sitio 2 de Arroyo Seco (partido de Tres Arroyos, provincia de Buenos Aires, República Argentina). In: Bryan AL (ed) New evidence for the Pleistocene peopling of the Americas. Center for the Study of Early Man, Orono, pp 221–269
Flegenheimer, N (2003) Cerro El Sombrero: A locality with a view. In: Miotti L, Salemme M, Flegenheimer N (eds) Where the south winds blow: Ancient evidence of Paleo South Americans. Center for the Study of the First Americans, Texas A&M pp 51–56
Flegenheimer N, Zárate M (1997) Considerations on radiocarbon and calibrated dates from Cerro La China and Cerro El Sombrero, Argentina. Curr Res Pleistocene 14:27–28
García A (2003a) Exploitation territory at Agua de la Cueva (Southern area) site (11,000–9000 RCYBP). In: Miotti L, Salemme M, Flegenheimer N (eds) Where the south winds blow: Ancient evidence of Paleo South Americans. Center for the Study of the First Americans, Texas A&M, pp 83–86
García A (2003b) On the coexistence of men and extinct Pleistocene megafauna at Gruta del Indio (Argentina). Radiocarbon 45:33–39
García A (2003c) Los primeros pobladores de los Andes centrales argentinos. Zeta Editor, Mendoza
García A, Lagiglia H (1998–1999) Avances en el estudio del registro Pleistocénico tardío de la Gruta del Indio (Mendoza). Cuadernos del INAPL 18:167–174
García C (2005) Análisis zooarqueológicos de restos óseos de mastodontes (*Cuvieronius hyodon*) procedentes de Taguatagua 2, Chile central. In: Actas del XVI Congreso Nacional de Arqueología Chilena, pp 529–536
Gittleman JL, Gompper ME (2001) The risk of extinction: What you don't know will hurt you. Science 291:997–999
Gnecco C (2000) Ocupación temprana de bosques tropicales de montaña. Universidad de Cauca, Colombia
Gnecco C, Mora S (1997) Late Pleistocene/early Holocene tropical forest occupation at San Isidro and Peña Roja. Antiquity 71: 683–690
Grayson D (1991) Late Pleistocene mammalian extinctions in North America: Taxonomy, chronology, and explanations. J World Prehist 5:193–232
Gruhn R, Bryan AL (1984) The record of Pleistocene extinctions at Taima-Taima, northern Venezuela. In: Martin PS, Klein R (eds) Quaternary extinctions: A prehistoric revolution. University of Arizona Press, Tucson, pp 128–137
Guérin C (1991) La faune de vertébrés du Pleistocéne supérieur de Sao Raimundo Nonato. Comptes Rendus de L'Académie des Sciences, Serie II 312:567–572
Guidon N (1986) Las unidades culturales de Sao Raimindo Nonato –S.E. del Estado de Piauí – Brasil. In: Bryan AL (ed) New perspectives on the Pleistocene peopling of the Americas. Center for the Study of Early Man, Orono, pp 157–171
Guidon N, Delibrias G (1986) Carbon 14 dates point to man in the Americas before 32,000 years ago. Nature 320:769–771
Guthrie RD (1984) Mosaics, allelochemics, and nutrients: An ecological theory of late Pleistocene megafaunal extinctions. In: Klein RG, Martin PS (eds) Quaternary extinctions: A prehistoric revolution. University of Arizona Press, Tucson, pp 259–298
Gutierrez M, Politis G Johnson E (2000) La ocupacion más temprana en Arroyo Seco 2. Paper presented at the II Congreso de Arqueología de la Región Pampeana Argentina, Mar del Plata
Gutiérrez M, Martinez G, Nielsen-Marsh C (2001) Alteraciones diagenéticas y preservación diferencial de los conjuntos óseos de la localidad arqueológica Paso Otero. Estudios de Geología 56:291–299
Hajduk A, Albornoz A, Lezcano MJ (2004) El *Mylodon* en el patio de atrás. Informe preliminar sobre los trabajos en el sitio El Trébol, ejido rubano de San Carlos de Bariloche, Provincia de Rio Negro. In: Civalero MT, Fernandez PM, Guraieb AG (eds) Contra viento y marea. Arqueologia de Patagonia, INAPL, Buenos Aires, pp 715–731
Hauthal R (1899) Reseña de los hallazgos en las cavernas de Ultima Esperanza. Revista del Museo de La Plata 9:409–420
Haynes G (1991) Mammoths, mastodonts, and elephants: Biology, behavior, and the fossil record. Cambridge University Press, Cambridge
Haynes G (2006) A review of some attacks on the Overkill hypothesis, with special attention to misrepresentations and doubletalk. Quat Intl 169–170:84–94
Haynes G, Stanford D (1984) On the possible utilization of *Camelops* by Early Man in North America. Quat Res 22:216–230
Hedges REM, Housley RA, Bronk CR, von Klinken GJ (1992) Radiocarbon dates from the Oxford AMS system: Archaeometry datelist. Archaeometry 34:337–357
Hernández Llosas MI (1999) Pintoscayoc. Arqueologia de quebradas altas en Humahuaca. Ph.D. Dissertation, University of Buenos Aires, Buenos Aires
Hoffstetter R (1986) High Andean mammalian faunas during the Plio-Pleistocene. In: Vuilleumier F, Monasterio N, (eds) High altitude tropical biogeography. Oxford University Press, Oxford, pp 218–245
Hofreiter M, Betancourt JL, Pelliza Sbriller A, Markgraf V, McDonald HG (2003) Phylogeny, diet, and habitat of an extinct ground sloth from Cuchillo Cura, Neuquen. Quat Res 59:364–378
Holliday V, Martinez G, Johnson E, Buchanan B (2003) Geoarchaeology of Paso Otero 5 (pampas of Argentina). In: Miotti L, Salemme M, Flegenheimer N (eds), Where the south winds blow: Ancient evidence of Paleo South Americans. Center for the Study of the First Americans, Texas A&M, pp 37–43
Höss M, Dilling A, Currant A, Pääbo S (1996) Molecular phylogeny of the extinct ground sloth *Mylodon darwinii*. Proc Natl Acad Sci 93:181–185
Hurt W, Van der Hammen T, Correal G (1977) El Abra rockshelters, sabana de Bogotá, Colombia, South America. Occasional Papers Indiana University Museum, 2

Jackson D (2003) Evaluating evidence of cultural association of *Mylodon* in the semiarid region of Chile. In: Miotti L, Salemme M, Flegenheimer N (eds) Where the south winds blow: Ancient evidence of Paleo South Americans. Center for the Study of the First Americans, Texas A&M, pp 77–81

Jackson D, Prieto A (2005) Estrategias tecnologicas y conjunto litico del contexto Paleoindio de Cueva Lago Sofia 1, Ultima Esperanza, Magallanes. Magallania 33(1):115–120

Jackson D, Méndez C, Seguel R (2003) Late Pleistocene human occupation in the semiarid coast of Chile: A comment. Curr Res Pleistocene 20:35–37

Jackson D, Méndez C, López P, Jackson D, Seguel R (2005) Evaluación de un asentamiento arqueológico en el semiárido de Chile: Procesos de formación, fauna extinta y comportamiento cultural. Interacciones en Antropología 6:139–151

Jaimes A (2003) Una nueva localidad Paleo-India en el noroccidente de Venezuela. Maguare 17:46–64

Joly D, March RJ, Martinez G (2005) Les os brûlés de Paso Otero 5: un témoignage posible de l'utilisation de l'os comme combustible par des chasseurs-cueilleurs de la fin due Pléistocéne, Archescience Revue d'Archèometrie 29:83–93

Karathanasis AD (1997) X-ray diffraction, X-ray fluorescence, and differential scanning calorimetry analyses of sediments from selected features and bones. In: Dillehay T (ed) Monte Verde. A late Pleistocene settlement in Chile, vol 2. The archaeological context. Smithsonian Institution Press, Washington DC, pp 817–824

Kipnis R (2002) Long-term land tenure systems in Central Brazil. Evolutionary ecology, risk-management, and social geography. In: Fitzhugh B, Habu J (eds) Beyond foraging and collecting: Evolutionary change in hunter-gatherer settlement systems. Kluwer/Plenum, New York, pp 181–230

Klein R, Martin PS (eds) (1984) Quaternary extinctions: A prehistoric revolution. Arizona University Press, Tucson

Koch P, Barnosky A (2006) Late Quaternary extinctions: State of the debate. Ann Rev Ecol Evolut Syst 37:215–250

Labarca R (2003) Relación hombre-mastodonte en el semiárido chileno: El caso de Quebrada Quereo (IV Region, Coquimbo). Boletín Chileno del Museo Nacional de Historia Natural 52:151–175

Labarca R, López PG (2006) Los mamíferos finipleistocénicos de la Formación Quebrada Quereo (IV Region, Chile). Biogeografía, bioestratigrafía e inferencias paleoambientales. Mastozoología Neotropical 13(1):89–101

Lagiglia H (1979) Dinámica cultural en el centro oeste y sus relaciones con áreas aledañas Argentinas y Chilenas. In: Actas del VII Congreso de Arqueología de Chile II:531–560

Laming-Emperaire A, Prous A, Vilhena de Moraes A, Beltrao M (1975) Grottes et abris de la région de Lagoa Santa, Minas Gerais, Brésil. EPHE, París

Latorre C (1998) Paleontología de mamíferos del alero Tres Arroyos 1, Tierra del Fuego. Anales del Instituto de la Patagonia 26:77–90

Lavallée D, Julien M, Wheeler JC, Karlin C (1985) Telarmachay. Chasseurs et pasteurs préhistoriques des Andes. ADPS-ERC, Paris

Long A, Martin PS. Lagiglia HA (1998) Ground sloth extinction and human occupation at Gruta del Indio, Argentina. Radiocarbon 40(2):693–700

López Castaño C (1995) Evidencias Paleoindias en el valle medio del río Magdalena. Boletín de Arqueología 2:3–23

Lopez Castaño, Pino J, Realpe A (2003) Stratigraphic, chronometric and lithic technological evidence for late Pleistocene and early Holocene occupations in the middle Magdalena river, Colombia, South America. Curr Res Pleistocene 20:46–49

Lumley H, Lumley M, Beltrao M, Yokoyama Y, Labeyrie J, Danon J, Delibrias G, Falguieres C, Bischoff J (1988) Présence d'outils taillès associes a une faune datee au Pléistocène moyen dans la Toca da Esperanca, Région de Central, Etat de Bahía, Brésil. L'Anthropologie 91:917–942

Lynch TF (1974) The antiquity of Early Man in South America. Quat Res 4:356–377

Lynch TF (1980) Guitarrero cave. Early Man in the Andes. Academic Press, New York

Lynch TF (1990) Glacial-age man in South America? A critical review. Am Ant 55:12–36

Lynch TF, Pollock S (1981) Chobshi cave and its place in Andean and Ecuadorian archaeology. In: Harten LB, Warren CN, Tuohy DR (eds) Anthropological papers in memory of Earl H. Swanson. Idaho State Museum of Natural History, Pocatello, pp 19–40

MacNeish R (1976) Early Man in the New World. Am Scientist 63(3):316–327

MacNeish R (1979) The Early Man remains from Pikimachay cave, Ayacucho basin, Highgland Peru. In: Humphrey RL, Stanford D (eds) Pre-Llano cultures of the Americas: Paradoxes and possibilities. Anthropological Society of Washington, Washington DC, pp 1–48

MacPhee RDE, Marx PA (1997) The 40,000 year plague: Humans, hyperdisease and first-contact extinctions. In: Goodman SM, Patterson BD (eds) Natural change and human impact in Madagascar. Smithsonian Institution Press, Washington DC, pp 169–217

MacPhee RDE, Flemming C (1999) *Requiem aeternam*: The last five hundred years of mammalian species extinctions. In: MacPhee RDE (ed) Extinctions in near time. causes, contexts, and consequences. Kluwer/Plenum, New York, pp 333–371

Martin FM, Prieto A, San Román M, Morello F, Prevosti F, Cárdenas P, Borrero LA (2004) Late-Pleistocene megafauna at Cueva del Puma, Chile. Curr Res Pleistocene 21:101–103

Martin PS, Wright HEJ (eds) (1967) Pleistocene extinctions: The search for a cause. Yale University Press, New Haven

Martin PS (1973) The discovery of America. Science 179:969–974

Martin PS, Steadman DW (1999) Prehistoric extinctions on islands and continents. In: MacPhee RDE (ed) Extinctions in near time: Causes, contexts, and consequences. Kluwer/Plenum, New York, pp 17–55

Martinez G (2001) 'Fish-tail' projectile point and megafauna: New evidence from Paso Otero 5 (Argentina). Antiquity 75:523–528

Martinez G, Gutiérrez M (2004) Tendencias en la explotación humana de la fauna durante el Pleistoceno final y Holoceno en la región pampeana (Argentina). In: Mengoni Goñalons GL (ed) Zooarchaeology of South America. BAR International Series 1298, pp 81–98

Martinez JG (2003) Ocupaciones humanas tempranas y tecnología de caza en la microregión de Antofagasta de la Sierra (10000–7000 AP). Ph.D. Dissertation, Universidad Nacional de Tucumán

Massone M (2004) Los cazadores después del hielo. Centro de Investigaciones Barros Arana, Santiago

Massone M, Prieto A (2004) Evaluacion de la modalidad cultural Fell 1 en Magallanes. Chungara, Vol. Esp. 1:303–315

Mazzanti D (2003) Human settlements in caves and rockshelters during the Pleistocene-Holocene transition in the eastern Tandilia Range, pampean region, Argentina. In: Miotti L, Salemme M, Flegenheimer N (eds) Where the south winds blow: Ancient evidence of Paleo South Americans. Center for the Study of the First Americans, Texas A&M, pp 57–61

Mazzanti D, Quintana C (eds) (2001) Cueva Tixi: Cazadores y recolectores de las sierras de Tandilia oriental. 1. Geología, Paleontología y Zooarqueología. Universidad Nacional de Mar del Plata, Mar del Plata

McCulloch RD, Fowgill CJ, Sugden DE, Bentley MJ, Kubik PW (2005) Chronology of the last glaciation in central Strait of Magellan and Bahia Inutil, southernmost South America. Geografiska Annaler 87A(2):289–312

McFadden BJ, Wolff RG (1981) Geological investigations of late Cenozoic vertebrate-bearing deposits in southern Bolivia. In: Anais do II Congresso Latino-Americano de Paleontología 2:765–778

Meggers B, Miller E (2002) Hunter-gatherers in Amazonia during the Pleistocene-Holocene transition. Under the canopy. In: Mercader J (ed) The archaeology of tropical rain forests. Rutgers University Press, New Brunswick, pp 291–316

Mena FJ, Stafford TW Jr, Southon J (1999) Direct AMS radiocarbon dating on human bones for Baño Nuevo, central Patagonian Andes, Chile. Curr Res Pleistocene 16:71–73

Mena FJ, Reyes O, Stafford TW Jr, Southon J (2003) Early human remains for Baño Nuevo-1 Cave, central Patagonian Andes, Chile. Quat Intl 109–110:113–121

Menegaz A, Nami HG (1994) Late Pleistocene faunal diversity in Ultima Esperanza (Chile): Further data from Cueva del Medio. Curr Res Pleistocene 11:93–95

Meltzer D, Adovasio J, Dillehay T (1994) On the Pleistocene human occupation at Pedra Furada, Brazil. Antiquity 68:695–714

Messineo P, Politis G, Rivas MI (2004) Cazadores tempranos y megamamíferos tardìos en la región pampeana: el sitio Campo Laborde. Paper presented at the XX Congreso Nacional de Arqueología Argentina, Río Cuarto

Miller E (1987) Pesquisas arqueológicas Paleoindígenas no Brasil occidental. Estudios Atacameños 8:37–61

Minta SC, Kareiva PM, Peyton Curlee A (1999) Carnivore research and conservation: Learning from history and theory. In: Clark TW, Peyton Curlee A, Minta SC, Kareiva PM (eds) Carnivores in ecosystems. The Yellowstone experience. Yale University Press, New Haven/London, pp 323–404

Miotti L (1996) Piedra Museo. Nuevos datos para la ocupación Pleistocena en Patagonia. In: Gómez Otero J (ed) Arqueologia, Sólo Patagonia, CENPAT, Puerto Madryn, pp 27–38

Miotti L, Vazquez M, Hermo D (1999) Piedra Museo. Un Yamnagoo Pleistocénico de los colonizadores de la meseta de Santa Cruz. El estudio de la arqueofauna. Soplando en el Viento, Neuquén-Buenos Aires, pp 113–136

Miotti L, Salemme M (1999) Biodiversity, taxonomic richness and specialists-generalists during late Pleistocene/early Holocene times in Pampa and Patagonia (Argentina, southern South America). Quat Intl 53/54:53–68

Miotti L, Salemme M, Rabassa J (2003) Radiocarbon chronology at Piedra Museo locality. In: Miotti L, Salemme M, Flegenheimer N (eds) Where the south winds blow: Ancient evidence of Paleo South Americans. Center for the Study of the First Americans, Texas A&M, pp 99–104

Miotti L, Berman D (1988) Mamíferos del Holoceno tardío de Punta Bustamante. Resúmenes de las V Jornadas de Paleontología de Vertebrados, pp 68–69, La Plata

Miotti L, Salemme M (2003) When Patagonia was colonized: People's mobility at high latitudes during Pleistocene/Holocene transition. Quat Intl 109–110:95–111

Mol D, van Bree PJH, McDonald GH (2003) De Amsterdamse collectie fossielen uit de Grot van Ultima Esperanza (Patagonië, Zuid-Chili). Grondboor & Hamer 2:26–31

Mones A, Francis JC (1973) Lista de los vertebrados fósiles del Uruguay, II. Comunicaciones Paleontológicas del Museo de Historia Natural de Montevideo 1(4):39–97

Montané J (1968) Paleoindian remains from Laguna de Tagua Tagua, central Chile. Science 161:1137–1138

Morse DH (1980) Behavioral mechanisms in ecology. Harvard University Press, Cambridge/London

Nami HG (1987) Cueva del Medio: Perspectivas arqueológicas para la Patagonia austral. Anales del Instituto de la Patagonia 16:103–109

Nami HG, Menegaz A (1991) Cueva del Medio: Aportes para el conocimiento de la diversidad faunística hacia el Pleistoceno/Holoceno en Patagonia austral. Anales del Instituto de la Patagonia 20:117–132

Nami HG, Nakamura T (1995) Cronología radiocarbónica con AMS sobre muestras de hueso procedentes del sitio Cueva del Medio (Ultima Esperanza, Chile). Anales del Instituto de la Patagonia 23:125–134

Neme G, Gil A, García A (1998) Preliminary zooarchaeological results at Agua de la Cueva, Sector Sur Shelter. Curr Res Pleistocene 15:139–140

Neves WA, Hubbe M (2005) Cranial morphology of early Americans from Lagoa Santa, Brazil: Implications for the settlement of the New World. Proc Natl Acad Sci 102:18309–18314

Nordenskjold E (1900) Jackttagelser och fynd i grottor vid Ultima Esperanza i sydvestra Patagonien. Konglinga Svenska Vetenskaps Handlingar 33(3):1–24

Nuñez L, Varela J, Casamiquela R (1983) Ocupación Paleoindio en Quereo. Reconstrucción multidisciplinaria en el territorio semiárido de Chile. Universidad del Norte, Antofagasta

Nuñez L, Casamiquela RM, Schippacasse V, Niemeyer H, Villagrán C (1994) Cuenca de Taguatagua en Chile: El ambiente del Pleistoceno y ocupación humana. Revista Chilena de Historia Natural 67(4):503–519

Nuñez L, Grosjean M, Cartajena I (2001) Human occupation and climate change in the Puna de Atacama. Science 298:821–824

Ochsenius C, Gruhn R (1979) Taima-Taima: A late Pleistocene Paleo-Indian kill site, northernmost South America. Final report of the 1976 excavation. Programa CIPICS, Federal Republic of Germany

Oliver JR, Alexander CS (2003) Ocupaciones humanas del Pleistoceno temprano en el occidente de Venezuela. Maguera 17:83–245

Oliveira ÉV (1999) Quaternary vertebrates and climates of South Brazil. Quat South Am Antarct Peninsula 12:61–73

O'Connell JF (2000) An emu hunt. In: Anderson A, Murray T (eds) Australian archaeologist. Collected papers in honour of Jim Allen. The Australian National University, Canberra, pp 172–181

Ossa P (1978) Paiján in early Andean prehistory: The Moche valley evidence. In: Bryan AL (ed) New evidence for the Pleistocene peopling of the Americas. University of Alberta, Edmonton, pp 290–295

Ossa P, Moseley M (1972) La Cumbre: A preliminary report on the early lithic occupation of the Moche Valley. Ñaupa Pacha 9:1–16

Owen-Smith RN (1999) The interaction of humans, megaherbivores, and habitats in the late Pleistocene extinction event. In: MacPhee RDE (ed) Extinctions in near time: Causes, contexts, and consequences. Kluwer/Plenum, New York, pp 57–69

Pascual R, Ortega Hinojosa E, Gondar D, Tonni EP (1966) IV. Vertebrata. In: Borrello A (ed) Paleontografía Bonaerense, Comisión de Investigaciones Científicas de la Provincia de Buenos Aires, La Plata

Paskoff R (1971) Edad radiométrica del mastodonte de Los Vilos: 9100 ± 300 años B.P. Noticiario Mensual 15:11

Patterson B Pascual R (1972) The fossil mammal fauna of South America. In: Keast A, Erk FC, Glass B (eds) Evolution, mammals and southern continents. State University of New York Press, Albany, pp 247–309

Paula Couto C (1979) Tratado de paleomastozoología. Academia Brasileira de Ciencias, Río de Janeiro

Paunero R (1993–1994) El sitio Cueva 1 de la localidad Cerro Tres Tetas (Ea. San Rafael, Pcia de Santa Cruz, Argentina). Anales de Arqueología y Etnología 48–49:73–90

Paunero R (2003) The presence of a Pleistocene colonizing culture in La Maria archaeological locality, Casa del Minero 1. In: Miotti L, Salemme M, Flegenheimer N (eds) Where the south winds blow: Ancient evidence of Paleo South Americans. Center for the Study of the First Americans, Texas A&M, pp 127–132

Paunero R, Cueto M, Frank A, Ghidini G, Rosales G, Skarbun F (2004) Comunicación sobre campaña arqueológica 2002 en la localidad La María, Santa Cruz. In: Civalero MT, Fernandez PM, Guraieb AG (eds) Contra viento y marea. Arqueologia de Patagonia, INAPL, Buenos Aires, pp 797–808

Pearson GA (2004) Pan-American Paleoindian dispersals and the origins of fishtail projectile points as seen through the lithic raw-material reduction strategies and tool-manufacturing techniques at the Guardia site, Turrialba, Costa Rica. In: Barton CM, Clark G, Yesner D, Pearson G (eds) The settlement of the American continents: A multidisciplinary approach to human biogeography. University of Arizona Press, Tucson, pp 85–102

Pearson GA, Cooke R (2002) The role of the Panamanian landbridge during the initial colonization of the Americas. Antiquity 76:931–932

Piló LB, Auler AS, Neves WA, Wang X, Cheng H, Edwards RL (2005) Geochronology, sediment provenance, and fossil emplacement at Sumidouro Cave, a classic late Pleistocene/early Holocene paleoanthropological site in eastern Brazil. Geoarchaeology 20:751–764

Pires-Ferreira J W, Pires-Ferreira E, Kaulicke P (1976) Prehistoric animal utilization in the Central Peruvian Andes. Science 194:483

Politis G, Gutierrez M (1998) Gliptodontes y cazadores-recolectores en la region pampeana de Argentina. Latin Am Ant 9(2): 111–134

Politis G, Madrid P (2001) Arqueología pampeana: Estado actual. In: Berberián E, Nielsen A (eds) Historia Argentina regional. Editorial Brujas, Córdoba, pp 737–814

Politis G, Prado JL, Beukens R (1995) The human impact in Pleistocene-Holocene extinctions in South America. In: Johnson E (ed) Ancient peoples and landscapes. Museum of Texas Tech University, Lubbock, pp 187–205

Politis G, Johnson E, Gutiérrez M, Hartwell WT (2003) Survival of Pleistocene fauna: New radiocarbon dates on organic sediments from La Moderna (pampean region, Argentina). In: Miotti L, Salemme M, Flegenheimer N (eds) Where the south winds blow: Ancient evidence of Paleo South Americans. Center for the Study of the First Americans, Texas A&M, pp 45–50

Powell JE, Aschero CA, Martinez J (2001) Megaterinos finipleistocénicos en sitios arqueológicos de la Puna meridional Argentina. Quintas Jornadas de Comunicaciones, Universidad Nacional de Tucumán, p 18

Prado JL, Alberdi MT, Azanza B, Sánchez B, Frasinetti D (2005) The Pleistocene Gompotheridae (Proboscidea) from South America. Quat Intl 126–128:21–30

Prevosti FJ (2006) New material of Pleistocene cats (Carnivora, Felidae) from southern South America, with comments on bioogeography and the fossil record. Geobios 39:679–694

Prevosti FJ, Vizcaíno S (2006) Paleoecology of the large carnivore guild from the late Pleistocene of Argentina. Acta Paleontologica Polonica 51(3):407–422

Prevosti FJ, Soibelzon LH, Prieto A, San Román M, Morello F (2003) The southernmost bear: *Paractotherium* (Carnivora, Ursidae, Tremarctinae) in the latest Pleistocene of south Patagonia, Chile. J Vert Paleontol 23:709–712

Prieto A (1991) Cazadores tempranos y tardíos en la Cueva Lago Sofía 1. Anales del Instituto de la Patagonia 20:75–100

Prous A (1986) Os mais antigos vestigios arqueológicos no Brasil central. In: Bryan AL (ed) New evidence for the Pleistocene peopling of the Americas. University of Alberta, Edmonton, pp 173–182

Prous A (1991a) Santana do Riacho – Tomo I. Arquivos de Museu de Historia Natural da Universidade Federal de Minas Gerais 12:3–384

Prous A (1991b) Fouilles de l'abri du Boquete, Minas Gerais, Brésil. Journal de la Société des Américanistes 77:77–109

Prous A, Fogaca E (1999) Archaeology of the Pleistocene/Holocene boundary in Brazil. Quat Intl 53/54:21–41

Pyare S, Berger J (2003) Beyond demography and delisting: Ecological recovery for Yellowstone's grizzly bears and wolves. Biol Conserv 113:63–73

Ranere T (2006) The Clovis colonization of central America. In: Morrow J, Gnecco C (eds) Paleoindian archaeology, a hemispheric perspective. University Press of Florida, Gainesville, pp 69–85

Richardson JB III (1978) Early Man on the Peruvian North coast, early maritime exploration and the Pleistocene and Holocene environment. In: Bryan AL (ed) Early Man in America. University of Alberta, Edmonton, pp 274–289

Rick J (1980) Prehistoric hunters of the high Andes. Academic Press, New York

Rick J (1988) The character and context of highland preceramic societies. In: Keatinge RW (ed) Peruvian prehistory. Cambridge University Press, Cambridge, pp 3–40

Roth S (1902) Nuevos restos de mamíferos de la caverna Eberhardt en Ultima Esperanza. Revista del Museo de La Plata 11:37–54

Roosevelt A, Costa L, Machado L, Michar N, Mercier N, Valdas H, Feathers J, Barnett W, Silveira M, Henderson H, Silva J, Chernoff B, Rease D, Homan J, Toth N, Schick R (1996) Paleoindian cave dwellers in the Amazon: The peopling of the Americas. Science 272:373–384

San Román M, Morello F, Prieto A (2000) Cueva de los Chingues (Parque Nacional Pali Aike), Magallanes, Chile. Historia natural y cultural I. Anales del Instituto de la Patagonia 28:125–145

Sanguinetti AC (1976) Excavaciones prehistóricas en la cueva Las Buitreras (Pcia de Santa Cruz). Relaciones de la Sociedad Argentina de Antropología 10:271–292

Sanguinetti AC, Borrero LA (1977) Los niveles con fauna extinta de la cueva Las Buitreras (Río Gallegos, Santa Cruz). Relaciones de la Sociedad Argentina de Antropología 11:167–178

Saxon EC (1976) La prehistoria de Fuego-Patagonia: Colonización de un habitat marginal. Anales del Instituto de la Patagonia 7:63–73

Saxon EC (1979) Natural prehistory: The archaeology of Fuego-Patagonian ecology. Quaternaria 21:329–356

Schmidt Días A, Jacobus AL (2003) Quão antico é o povoamento do sul do Brasil? Revista do CEPA 27(38):39–67

Schmitz PI (1987) Prehistoric hunter-gatherers of Brazil. J World Prehist 1:1–26

Scillato Yané G (1976) Sobre algunos restos de *Mylodon ¿ listai* (Edentata, Tardigrada) procedentes de la cueva "Las Buitreras" (Pcia de Santa Cruz, Argentina). Relaciones de la Sociedad Argentina de Antropología 10:309–312

Simpson GG (1980) Splendid isolation, the curious history of South American mammals. Yale University Press, New Haven

Simpson GG, Paula Couto C (1957) The mastodons of Brazil. Bull Am Mus Nat Hist 112(2):125–190

Smith EA (1983) Anthropological applications of Optimal Foraging Theory: A critical review. Current Anthropology 24:625–652

Smith-Woodward A (1900) On some remains of *Grypotherium* (*Neomylodon*) Lista and associated mammals from a cavern near Consuelo Cove, Last Hope Sound. Proc Zool Soc London 56:64–79

Soibelzon LH (2004) Revisión sistemática de los Tremarctinae (Carnivora, Ursidae) fósiles de América del Sur. Revista del Museo de Ciencias Naturales 6:107–133

Stanley SM (1984) Marine mass extinctions: A dominant role for temperatures. In: Nitecki MH (ed) Extinctions. University of Chicago Press, Chicago, pp 69–118

Steadman DW, Martin PS, MacPhee RDE, Jull AJT, McDonald HG (2005) Asynchronous extinction of late Quaternary sloths on continents and islands. Proc Natl Acad Sci USA 102:11763–11768

Stern C (1992) Implications of tephrochronology for the age of the human occupation of Tierra del Fuego. Res Explor 8:239–240

Stothert KE, Piperno DR, Andres TC (2003) Terminal Pleistocene/early Holocene adaptation in coastal Ecuador: The Las Vegas evidence. Quat Intl 109–110:23–43

Suárez R (2003) Paleoindian components of north Uruguay: New data on early human occupation of the late Pleistocene and early Holocene. In: Miotti L, Salemme M, Flegenheimer N (eds) Where the south winds blow. Ancient evidence of Paleo South Americans. Center for the Study of the First Americans, Texas A&M, pp 29–36

Suárez R, López JM (2003) Archaeology of the Pleistocene-Holocene transition in Uruguay: An overview. Quat Intl 109–110:65–76

Sullivan LR, Hellman M, Anthony HE (1925) The Punin calvarium. Anthropological papers of the American Museum of Natural History 23:309–337

Temme M (1982) Excavaciones en el sitio precerámico de Cubilán (Ecuador). Miscelánea Antropológica Ecuatoriana, Museo Banco Central, Guayaquil and Quito, pp 135–164

Tonni EP, Politis G (1981) Un gran cánido del Holoceno de la Provincia de Buenos Aires y el registro prehistórico de *Canis* (*Canis*) *familiaris* en las áreas Pampeana y Patagónica. Ameghiniana 18:251–265

Tonni EP, Carlini AA, Scillato Yané GJ, Figini AJ (2003a) Cronología radiocarbónica y condiciones climáticas en la Cueva del Milodón (sur de Chile) durante el Pleistoceno tardío. Ameghiniana 40(4):609–615

Tonni EP, Huarte RA, Carbonari JE, Figini AJ (2003b). New radiocarbon chronology for the Guerrero Member of the Lijan Formation (Buenos Aires, Argentina): Paleoclimatic significance. Quat Intl 109–110:45–48

Tonni EP, Cione A, Bond, M (1999) Quaternary vertebrate Paleontology in Argentina. Now and then. Quat South Am Antarct Peninsula 12:5–22

Ubilla M, Perea D (1999) Quaternary vertebrates of Uruguay: A biostratigraphic, biogeographic and climatic overview. Quat South Am Antarct Peninsula 12:75–90

Ubilla M, Perea D, Martinez S (1994) Paleofauna del Cuaternario tardío continental del Uruguay (Fm. Sopas y Fm. Dolores). Acta Geológica Leopoldensia 39(1):441–458

Valverde F (2001) Huellas y marcas sobre huesos. In: Mazzanti D, Quintana C (eds) Cueva Tixi: Cazadores y recolectores de las sierras de Tandilia oriental. 1. Geología, paleontología y zooarqueología. Universidad Nacional de Mar del Plata, Mar del Plata, pp 137–155

Van der Hammen T, Correal G (2001) Mastodontes en un humedal Pleistocenico en el valle del Magdalena (Colombia) con evidencias de la presencia del hombre en el periglacial. Boletín de Arqueología 16(1):4–36

Velásquez H, Mena F (2006) Distribuciones óseas de ungulados en la cueva Baño Nuevo 1 (XI Región, Chile): un primer acercamiento. Magallania 34(2):91–105

Vialou AV (2003) Santa Elina Rockshelter, Brazil: Evidence of the coexistence of man and *Glossotherium*. In: Miotti L, Salemme M, Flegenheimer N (eds) Where the south winds blow. Ancient evidence of Paleo South Americans. Center for the Study of the First Americans, Texas A&M, pp 21–28

Vialou AV, Aubry T, Benebdelmadi M, Cartelle C, Figuti L Fontugne M, Solari ME Vialou D (1995) Découverte de Mylodontinae dans un habitat préhistorique daté du Mato Grosso (Brésil): L'Abri rupestre de Santa Elina. Comptes Rendús Academie Sciences Paris 320, IIa:655–661

Yacobaccio HD, Madero C (1992) Zooarqueología de Huachichocana III (Jujuy, Argentina). Arqueología 2:149–188

9. *Insulae infortunatae*: Establishing a Chronology for Late Quaternary Mammal Extinctions in the West Indies

R.D.E. MacPhee*
Division of Vertebrate Zoology
American Museum of Natural History
Central Park West at 79th St
New York NY 10024, USA
macphee@amnh.org

Keywords Quaternary extinctions • West Indies • radiocarbon dating • extinction causation

"Islands are an enormously important source of information and an unparalleled testing ground for various scientific theories."

Ernst Mayr (1967)

Introduction

This volume is devoted to recent advances in understanding megafaunal extinctions in the New World during the LQ (for this and all other abbreviations used in this paper, see Table 9.1). A chapter on endemic land-mammal extinctions on the islands comprising the West Indies may therefore seem out of place, because (1) few of the species that once lived on the islands would have qualified as megafaunal, even under a generous definition of that term; and (2) while the WI non-volant mammal fauna eventually suffered near-total collapse, this happened quite recently, long after losses subsided on the mainlands. Yet to ignore the vicissitudes of the WI biota would be to overlook the only non-mainland context in the Western Hemisphere to suffer major LQ extinctions – one that might in principle help us to critically evaluate what we know, or think we know, about cause-effect relationships which forced dramatic losses elsewhere in the New World. For example, because of the proximity of these islands to the adjacent continents, any large-scale climatic changes sufficient to drive extinctions on the mainlands should also have had a determinable and coeval effect in the West Indies. If no appreciable effect can be detected given the data at hand – and, as we shall see, none has been – we need to ask why end-Pleistocene climate change should continue to be considered as the competent mechanism behind New World extinctions. As to the view that human impacts have been the prime movers in causing near-time extinctions, because *Homo sapiens* has occupied both islands and continents comparatively recently, losses due to direct anthropogenic effects should exhibit comparable features in both theaters. If features are not comparable, or seem to have worked on vastly different time scales, we need to ask why in this case as well. Finally, it is of great biological interest to assess, to the degree possible, how factors of physiology (e.g., body size) or macroecology (e.g., range collapse) that are putatively correlated with mainland extinctions might have affected the island fauna. Although cataloguing losses among Antillean birds and herps (including the often-overlooked chelonians) would also be pertinent to the development of these themes, in my view the story of faunal collapse in the West Indies is best told from the perspective of the group most affected, the mammals (for information on other vertebrate losses, see Pregill, 1981; Pregill and Olson, 1981; http://www.iucnredlist.org).

Dating, Scheduling, and Fuses

One of the few examples of consensus in modern studies of "near time" vertebrate extinctions (i.e., losses occurring in the past ~40,000 years) concerns how islands lose species. When people arrive, so it is widely agreed, the animals go

* Address for correspondence: macphee@amnh.org

G. Haynes (ed.), *American Megafaunal Extinctions at the End of the Pleistocene*, 169–193.

TABLE 9.1. Abbreviations used in text and Table 9.2.

1. Time conventions	
BCE	Before common era
CE	Common era (equivalent to AD dates)
Ma	Millions of years ago
ka	Thousands of years ago
BP	Radiocarbon years Before Present (1950 CE)
cal bp	Calibrated radiocarbon years Before Present (1950 CE)
2. Named intervals	
NU	Neogene unspecified (within last 23 Ma)
LQ or LQU	Late Quaternary unspecified (within last 250,000 years)
HU	Holocene unspecified (within last 10,000 years)
PI	Post-LGM interval (18,000–6,000 BP)
AI	Amerindian interval (~6,000–500 BP)
ME	Modern era (<500 BP)
3. Methods of age determination	
a	^{14}C age estimate, based directly on elements belonging to the target taxon (bones or teeth)
b	^{14}C age estimate, based on associated material (e.g., charcoal, unidentified bone scrap)
c	Uranium-series disequilibrium age estimate, based on associated $CaCO_3$ speleothem
d	Amino acid racemization age estimate, based on associated land-snail shell
e	$^{40}K/^{40}A$ (potassium/argon) dating method
f	General faunal attribution (i.e., target has been found in undated circumstances in association with LQ faunal elements)
g	Stratigraphic/geomorphological data
h	Observational evidence (i.e., LAD based on last recorded direct observation), date in yr CE
j	Association with *Rattus/Mus* in owl pellets (entry of OW murines into the West Indies assumed to be 1500 CE, regardless of island)
k	Amerindian archeological association (e.g., kitchen midden)
4. Other abbreviations	
AP	Antero-posterior
LAD	Last appearance date or datum
LGM	Last glacial maximum, ca. 20 ka (18,000 BP)
OW	Old World (in reference to alien murines)
WI	West Indian
*	Species of doubtful validity
und.	Undescribed (species)

– apparently very quickly (e.g., MacPhee and Flemming, 1999; Martin and Steadman, 1999). This inference is said to derive from basic considerations of island ecology: on most islands, resident species must bear a heavy burden in the form of intrinsic limiting factors such as small potential ranges, population size, and variable food supply (e.g. Dunn, 1994; Rosenzweig, 1995; Caughley and Gunn, 1996; Whittaker, 1998). Under such circumstances, so it is said, even rather mild perturbations of the local environment – let alone major events like the introduction of novel predators – may force drastic and immediate crashes.

Yet "immediate" turns out to be a rather elastic concept when it comes to actual cases, even in island contexts where the train of causation seems unambiguous. Consistent with the idea of speedy loss, on some Pacific islands extinctions of large endemic birds appear to have occurred within a few decades of human colonization (Holdaway, 1999; Worthy and Holdaway, 2002). However, on some other islands, the evidence is mounting for the contrary observation that extinction did not resemble "lightning war" so much as a sort of rolling thunder. In Madagascar, for example, according to new radiometric evidence, some of the giant lemurs and other "subfossil" species were still in existence in isolated places > 1,500 years after the projected arrival of humans (MacPhee and Burney, 1991; Alcover et al., 1998; Perez et al., 2005). Do these results imply that some groups are simply prone to more rapid disappearance than others, or does species' resilience to forms of endangerment vary with particular ecological roles? Is island size or habitat heterogeneity a decisive factor? Is there any obvious way to test these or any similar explanations in a fossil record? The picture is likely to remain puzzlingly complex.

Extinction theories tend to be severely underdetermined by reliable facts, a situation that is unlikely to change in the near term (cf. MacLeod, 2003). In this regard no aspect of data-gathering is more crucial than facts about time, as cause-effect interpretations depend on establishing the timing or "scheduling" of specific losses. (I prefer the latter term because it conveys the sense that the process may take a certain amount of time to reach completion.) Determination of such schedules is the main emphasis of this chapter, and basic to this undertaking is the critical assessment of so-called "last" appearance dates (LADs), or the point at which evidence of the continuing presence of a given species terminates (see also Fiedel, Chapter 2). For obvious reasons, a question of general importance is how to judge the likelihood that a given LAD is reasonably close to the actual disappearance date of the species in question. Generally speaking, LADs should be based on solid empirical evidence, such as radiometric age determinations or last recorded observations of a species in the wild. However, even for modern-era (ME) extinctions (i.e., losses within the last 500 years), the evidential basis for LAD determination is often quite meager (MacPhee and Flemming, 1999).

With still-living species one can sidestep the problem of being able to accurately predict when something will finally disappear by relying instead on census (or similar) evidence as a proxy for "ecological extinction" (i.e., the point at which a species is so reduced in numbers that the few individuals remaining have no more ecological relevance or hope of recovery; Purvis et al., 2000). How ecological extinction would look in an empirical fossil record remains uncertain, if only because of obdurate difficulties in inferring population sizes in such cases (e.g., Signor-Lipps effect), but in theory at least, the concept is of fundamental importance to understanding how extinctions took place in the West Indies. The basic issue is how to measure "fuse" length, or the amount of time between the theoretical onset of a factor having a magnitude sufficient to cause extinction, and the actual disappearance of the last members of the species thus impacted. As the following example from a different context shows, defining a "true" extinction date may be difficult even for the best-substantiated cases, where records are comparatively robust.

Current latest dates for the presence of woolly mammoth (*Mammuthus primigenius*) in various parts of continental Eurasia fall within a fairly narrow interval, 9,000–10,000 BP (MacPhee et al., 2002; Stuart et al., 2002, 2004; Kuzmin and Orlova, 2004). Since there are many hundreds of radiocarbon estimates for this species, it is not unreasonable to infer that woolly mammoths must have disappeared from what are now the continental parts of Eurasia no later than the beginning of the 10th millennium BP. Accepted latest dates for woolly mammoths in North America are earlier, by 500–750 radiocarbon years (cf. Meltzer and Mead, 1985); identical logic would suggest that North American populations disappeared no later than, and possibly somewhat before, Eurasian populations of the same species. The fact that no younger dates (or, at least, no dates considered to be reliable) have been reported for mainland Eurasian mammoths, despite much activity in this arena during recent years (e.g., Sulerzhitsky and Romanenko, 1999; Stuart et al., 2002; MacPhee et al., 2002), suggests that we should have considerable, if not absolute, confidence in the view that the last continental populations of *M. primigenius* collapsed just after the close of the Pleistocene. Furthermore, lack of numerous early Holocene dates might logically imply that by 10,000 BP woolly mammoths were ecologically extinct, if not completely extirpated, on the mainlands where they previously occurred.

Nevertheless, as we now know, woolly mammoths survived much longer than this on certain arctic and subarctic shelf islands (Wrangel Island and islands in the Bering Sea; Vartanyan et al., 1993; Guthrie, 2004; Yesner et al., 2005). Wrangel mammoths in particular lasted well into the late Holocene (current LAD, ~3,700 BP), a point that could never have been derived from first principles given the continental record for this species. Although it remains obscure how these elephants managed to persist on these relatively tiny landmasses, which only became isolated from adjacent mainlands after postglacial drowning of the Beringian shelf, the empirical pattern of late insular survival has held up to close scrutiny (cf. Vartanyan et al., 1995), and is, of course, tested in some sense each time a new mammoth specimen is dated. Yet we must never lose sight of the fact that all such "tests" are merely inductions, subject to rejection by even a single verifiable counter-example, no matter how many dates one can point to that are consistent with some previously-defined extinction schedule. Although all LADs are thus provisional, one is certainly justified in believing that LADs based on multiple, self-consistent criteria are more likely to be confirmed than denied by the next datum.

These considerations throw the deficiencies as well as the realities of the WI dating record into sharp relief. Recently, McFarlane and Lundberg (2004) estimated that approximately 60 radiometric age determinations with any bearing on the extinction of the WI fauna had appeared in the published literature. This amounts to an average of ~1 date *per species* of extinct WI land mammal; although a few more age estimates have been collected in recent years, the total is still < 100. This is an extremely thin foundation on which to base an extinction chronology, especially when many of the disappeared have little or nothing in the way of a dating framework to evaluate (see next section). For this reason, associational dating, which provides a "date" relative to something else of assumed or known age, continues to play a significant role in treatments of extinction scheduling of the WI fauna. Associations involving Old World (OW) murines (*Rattus* and *Mus*) are a good example: since these rodents did not live on these islands prior to early European times, one can reasonably assume that any valid association of extinct endemics with rats and mice must date to 1500 CE or later. (Obviously, to be meaningful this method requires a specific context, such as well-stratified sediments or owl pellets.) Another example of associational dating is to indirectly date a target taxon using ^{14}C-dated objects from the same layer or context, such as charcoal. However, charcoal is not inherently "better" than bone when it comes to interpreting what an age estimate means. For example, charcoal produced by wildfires obviously has an interpretative significance different from that found in a definite hearth. Further, the result of the radiocarbon analysis will be an estimate of the actual age of the wood from which the charcoal was derived, not the date at which it was burnt, which might be much later.

Quality of preservation is, of course, an unreliable guide to age, but it has played a conspicuous role historically, if only because of the paucity of firm LADs for the majority of extinct WI species. Thus Miller (1930) believed that the fresh appearance of owl-pellet material containing *Nesophontes*, unidentified mammal hair, and *Rattus* bones from the rock shelter of Monte Culo de Maco in southern Hispaniola indicated that *Nesophontes* might have survived into the 20th century. However, the most recent age determination on *Nesophontes* from this site is 680 ± 50 BP, not greatly different from the youngest nesophont material so far identified (skulls from Cueva Martin, Cuba, dated to 590 ± 50 BP). There is thus (still) no decisive evidence that *Nesophontes* persisted for centuries after the introduction of exotic vertebrates associated with European colonization (MacPhee et al., 1999).

Lack of radiometric dates for extinct WI mammals may be partly due to the historically small number of investigators who are interested in this fauna (and who have the money to pay for ^{14}C dating), but it also stems from the nature of fossil collecting on these islands. Cave deposits are practically the only places in which bones of extinct vertebrates are encountered; here they are continually exposed to hot, humid conditions, with the result that bone proteins are rapidly degraded and leached. Indeed, it is not uncommon to recover apparently well-preserved elements that, on testing, prove to retain so little native organic material that reliable dating is not possible (MacPhee and Flemming, 1999; Turvey et al., 2007).

As already noted, various additional factors may affect the trustworthiness of LAD determinations. Rarity of collection is certainly one: a significant number of WI taxa have been found in only one or a few localities (e.g., *Rhizoplagiodontia lemkei*, *Xenothrix mcgregori*), but this need not imply that these species were also rare in nature. Given the taphonomic conditions that prevail in the West Indies, species that were too large to be taken by the local predator guild, or never frequented caves, or lived in nonlimestone regions are simply less likely to be recovered paleontologically.

Roll Call of Extinction

The body of this chapter is devoted to distilling and interpreting chronometric data on WI extinctions and related subjects. To avoid breaking up the text by recurrently introducing issues affecting individual groups (including humans), I have relegated much of this material to the Appendix where it may be consulted as needed.

Scientific interest in the collapse of the land mammal fauna of the West Indies is long-standing (e.g., Miller, 1916, 1929a, b; Allen, 1942; Varona, 1974), but Morgan and Woods (1986) were the first authors to treat species losses in the West Indies in a comprehensive and integrated fashion, giving due attention to the quality of the evidence, systematics, and possible modes of causation. Since the appearance of their paper, taxonomic changes have taken some species off the extinction list, while others have been added as the result of new discoveries. Although the dating picture for WI extinctions remains woefully inadequate, it has been substantially, if unequally, improved for major taxa in the past 20 years. As discussed in detail in the following sections, Fig. 9.1 and Table 9.2 are intended to present these changes as efficiently as possible within the limits of this short paper. Together with the information presented in the Appendix, these treatments should provide up-to-date documentation on the extinction schedules and status of all valid, endemic, non-volant taxa of land mammals from the West Indies thought to be of late Quaternary age.

Although it has long been accepted that most LQ extinctions occurred after human arrival in the West Indies, thinking is changing as to how recent these losses might be (e.g., Steadman et al., 2005; MacPhee et al., 2007; Turvey et al., 2007). Figure 9.1 illustrates how, for a reasonably representative selection of taxa (27 species), new dates and interpretations are altering previously accepted models. In the case of *Amblyrhiza*, for example, earlier concepts of its extinction schedule allowed for its survival until ca.1600 CE (e.g., Anderson, 1984; Cole et al., 1994); more recently, U-series disequilibrium dates and other considerations (McFarlane et al., 1998a) have shown that it is considerably more likely that *Amblyrhiza* disappeared much earlier, in the later phases of Marine Isotope Stage (MIS) 5, ca. 125–75 ka. By contrast, it has also been demonstrated that some taxa (e.g., *Nesophontes edithae*) lasted longer than previously believed. In still other cases, debate continues. For example, radiometric, associational, and vague observational evidence all support the conclusion that *Isolobodon portoricensis* was still extant ca. 1500 CE. This species may well have survived until much later, although there is as yet no empirical evidence for Woods and Kilpatrick's (2005:1598) contention that it "survived...until the last few decades, and may still survive in certain remote areas" (see Fig. 9.1). As to *Neocnus* (="*Acratocnus*") *comes*, the alleged association that implies that this sloth survived into the ME is exclusively based on Miller's (1929a) statement that he found its remains in a cave deposit that also yielded domestic pig. The likelihood that this association was real has always seemed doubtful, and still does. However, in a surprising development, new chronometric studies have shown that some megalonychids persisted long after the estimated time of initial human arrival in the Greater Antilles (Steadman et al., 2005; MacPhee et al., 2007; see Appendix). It may be anticipated that such reinterpretations will continue to be made as the dating record continues to improve.

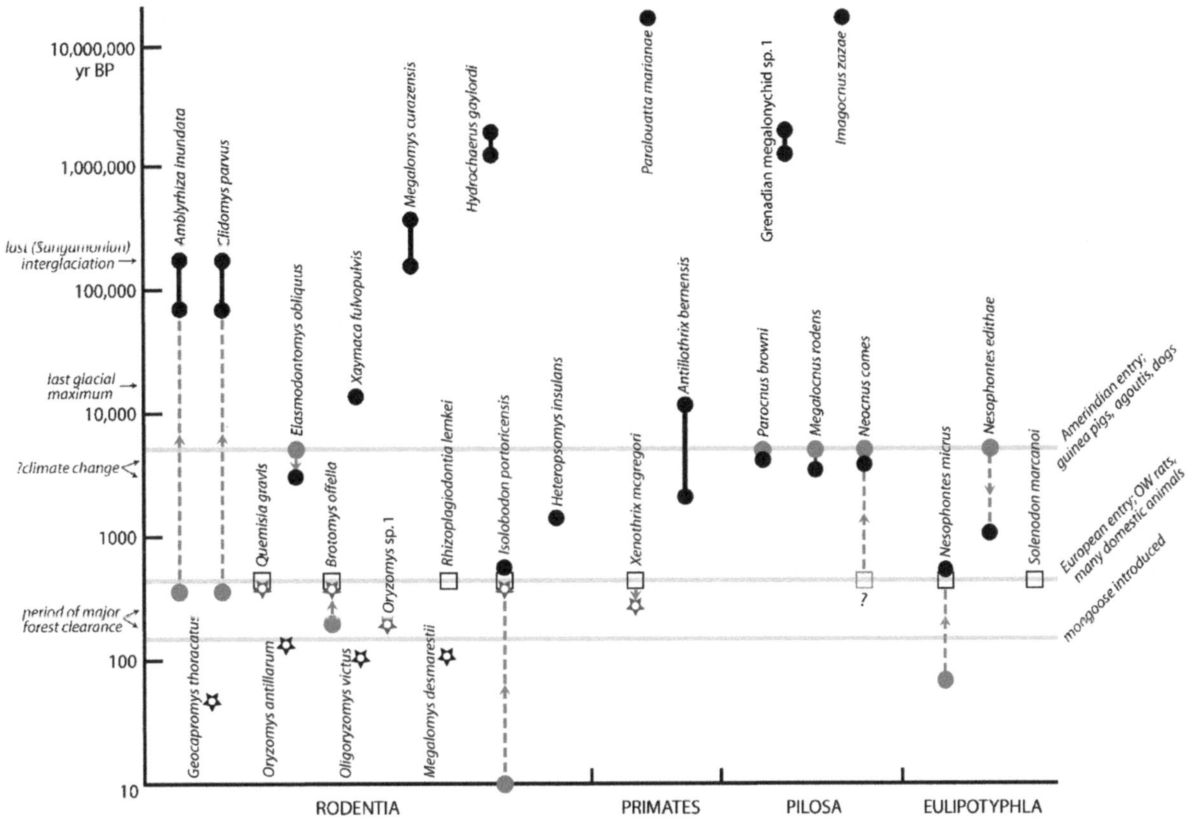

FIGURE 9.1. Representative taxa of West Indian rodents, primates, sloths, and insectivorans: changing views of extinction schedules. Taxa and data were selected to represent different levels of evidence typically used to frame explanations for these extinctions. Note that the vertical axis is logarithmic; ages based on radiocarbon dating have not been calibrated, which makes little or no difference on this scale. Symbols (circles, stars, and squares) in black indicate current LADs ("last" appearance dates) for identified taxa, while gray symbols indicate extinction dates that are either outmoded or based on ambiguous or dubious evidence. Specifically, solid black circles (●) indicate that a given LAD is based on direct radiometric and/or good stratigraphic evidence; if the LAD estimate is a lengthy interval, end member values are represented by linked circles. Gray circles (●) represent earlier views on LAD positions that are no longer supportable; and the dashed lines (---) joining older and newer concepts of disappearance times give some sense of the scale of reassessment in different cases. (Arrow indicates direction of reassessment.) Black stars (☆) indicate that a given current LAD is based on plausible, dated observational evidence for last sighting or collection of the target taxon; gray stars (☆) implies that such evidence is suspect or vague. Finally, squares (□) in black or gray represent varying qualities of associational or indirect evidence (typically, co-occurrence of remains of target and *Rattus*). Thus an earlier view of the extinction of *Amblyrhiza* (e.g., Anderson, 1984; Cole et al., 1994) held that this taxon was still extant in 1600 CE; U-series dates and other considerations (McFarlane et al., 1998a), however, indicate that *Amblyrhiza* probably died out in the later phases of Marine Isotope Stage 5. The transition from the last (Sangamonian) interglaciation to the Eo-Wisconsinan glaciation in North America may have been time-transgressive in different areas, the changeover as a whole lasting between 125 and 75 ka (Lourens et al., 2004). Here the arrows define the warmest episode, MIS 5e, 130–115 ka. The Last Glacial Maximum occurred ca. 20,000 cal bp in North America. Evidence for relative sea level rise (and, doubtfully, a minor amount of associated climate change) in the northern Caribbean between 6,200 and 4,000 BP comes from investigations by Peros et al. (2007) at Laguna de la Leche in western Cuba. The period of major forest clearance, for sugar cane, valuable woods, and related purposes, would have varied from island to island, but the period of greatest activity occurred between the late 17th and early 19th centuries (Wilson, 2001). "*Oryzomys* sp. 1" refers to an unnamed species from Barbados (cf. MacPhee and Flemming, 1999: their table 3), possibly seen by Richard Schomburgk in 1840s (Ray, 1964).

Turning now to Table 9.2, in establishing the list of valid species ("Taxa") I have mostly followed the opinions of modern authorities concerning systematic issues except where they substantially diverge from my own. Synonymizations not mentioned here can be found in relevant recent compendia (e.g., White and MacPhee, 2001; Woods and Kilpatrick, 2005). I have excluded all species incontrovertibly introduced by humans, such as the so-called "mute dogs" of Hispaniola and the highly questionable *Cubacyon transversidens* of Arredondo and Varona (1974). I have also excluded those

TABLE 9.2. Species list, geographic distribution, and last appearance date (if applicable) of endemic non-volant West Indian mammals. (See Table 9.1 and Fig. 9.2 for abbreviations)

Taxa	Range	Body Size[1]	Time Interval[2]	Age Determ.[2]	Notes	Chronometric Dates & Material Dated (If Applicable)[2]
PRIMATES: Pitheciidae						
Paralouatta varonai	Cuba	III	NU	*g*	3a, 4	
Xenothrix mcgregori	Jamaica	III	?ME	*b, j*	3b, 5	2,145 ± 200 BP (unid. bone)
Antillothrix bernensis	Hispaniola	III	AI	*b, k*	3q, 6	3,850 ± 135 BP (charcoal)
SORICOMORPHA: Nesophontidae						
Nesophontes major	Cuba	I	ME	*j*	3c	
Nesophontes micrus	Cuba	I	ME	*j*	3w	‡590 ± 50 BP (skulls)
Nesophontes hypomicrus	Hispaniola	I	ME	*j*	3w	‡790 ± 50 BP (hemimandibles)
Nesophontes paramicrus	Hispaniola	I	ME	*j, ?k*	3w	‡680 ± 50 BP (limb bones)
*Nesophontes zamicrus**	Hispaniola	I	ME	*j*	3k	
Nesophontes edithae	Puerto Rico	II	ME	*a*	3d	‡990 ± 20 BP (hemimandible)
Nesophontes und. sp. 1	Grand Cayman	I	ME	*j*	3e	
Nesophontes und. sp. 2	Cayman Brac	I	ME	*j*	3e	
SORICOMORPHA: Solenodontidae						
Solenodon cubanus	Cuba	II	extant	–	–	
Solenodon paradoxus	Hispaniola	II	extant	–	–	
Solenodon marcanoi	Hispaniola	II	ME	*j*	3f	
Solenodon arredondoi	Cuba	III	LQU	*f*	3g, 9	
PILOSA: Megalonychidae						
Acratocnus odontrigonus	Puerto Rico	IV	LQU	*f*	8	
Acratocnus antillensis	Cuba	IV	LQU	*f*	9	
Acratocnus ye	Hispaniola	IV	HU	*b*	10	10,320 ± 170 to 3,715 ± 175 BP (range, unid. bone)
*Acratocnus simorhynchus**	Hispaniola	IV	PI	*b*	11	21,500 to 18,500 BP (inorganic carbon)
Neocnus gliriformis	Cuba	III	LQU	*f*	9	
Neocnus major	Cuba	III	LQU	*f*	9	
Neocnus comes	Hispaniola	III	AI	*a*	3h	‡4,390 ± 40 BP (humerus, ulna)
Neocnus dousman	Hispaniola	III	HU	*b*	10	10,320 ± 170 to 3,715 ± 175 BP (range, unid. bone)
Neocnus toupiti	Hispaniola	III	HU	*b*	10	10,320 ± 170 to 3,715 ± 175 BP (range, unid. bone)
Megalocnus rodens	Cuba	V	AI	*a*	3i, 12	‡4,190 ± 40 BP (molariform)
Megalocnus zile	Hispaniola	V	PI	*b*	3j, 12	21,170 ± 525 to 17,405 ± 900 BP (range, unid. bone)
Parocnus serus	Hispaniola	IV	HU	*b*	10	10,320 ± 170 to 3,715 ± 175 BP (range, unid. bone)
Parocnus browni	Cuba	IV	AI	*a*	3h	‡4,960 ± 280 BP (humerus)
Paulocnus petrifactus	Curacao	IV	LQU	*g*	13	400–130 ka
Megalonychid und. sp. 1	Grenada	?	NU	*e*	14	3.6 ± 0.4 to 2.7 ± 0.3 Ma (hornblende)
RODENTIA: Heptaxodontidae						
Quemisia gravis	Hispaniola	III	ME	*?h, k*	3k	
Elasmodontomys obliquus	Puerto Rico	IV	AI	*b*	3d, 15	‡3,510 ± 30 to 2,410 ± 30 BP (range, charcoal)
Xaymaca fulvopulvis	Jamaica	II	HU	*b*	31	‡11,260 ± 80 to 10,250 ± 80 BP (range, chitin)
Clidomys osborni	Jamaica	IV	LQU	*c*	3m	130–70 ka
Amblyrhiza inundata	Anguilla/St. Martin	V	LQU	*c*	3m	130–70 ka
RODENTIA: Cricetidae, Sigmodontinae						
*Oryzomys curasoae**	Curaçao	I	ME	*j*	3o, 16	
Oryzomys antillarum	Jamaica	I	ME	*h*	3c	1877
*Oryzomys hypenemus**	Barbuda, Antigua	I	ME	*j*	17	
Oryzomys und. sp. 1	Barbados	I	ME	*h*	3c	pre-1890

(continued)

TABLE 9.2. (continued)

Taxa	Range	Body Size[1]	Time Interval[2]	Age Determ.[2]	Notes	Chronometric Dates & Material Dated (If Applicable)[2]
Oryzomys und. sp. 2	Grenada	I	ME	*k*	3n, 18	
Oryzomys und. sp. 3*	Grenada	I	ME	*k*	3n, 18	
Oligoryzomys victus	St. Vincent	I	ME	*h*	3c	1892
Megalomys desmarestii	Martinique	II	ME	*h*	3c	1902
Megalomys luciae	St. Lucia	II	ME	*h*	3c	pre-1881
Megalomys audreyae	Barbuda	II	LQU	*?*	9	
Megalomys curazensis	Curaçao	II	LQU	*g*	3o	400–130 ka
Megalomys und. sp. 1	Antigua etc.	I	ME	*k*	3n, 19	
Megalomys und. sp. 2	Anguilla etc.	I	ME	*k*	3n, 20	
RODENTIA:						
Echimyidae						
Brotomys voratus	Hispaniola	I	ME	*a, j, k*	3k, 3p	340 ± 60 BP, mandible
*Brotomys contractus**	Hispaniola	I	?ME	*?*	9	
Boromys offella	Cuba	I	ME	*k*	3q	
Boromys torrei	Cuba	I	?ME	*?j*	21	
Heteropsomys antillensis	Puerto Rico	II	?ME	*?*	9	
Heteropsomys insulans	Puerto Rico	II	AI	*a*	3d	‡1,220 ± 30 BP; mandible
Puertoricomys corozalus	Puerto Rico	II	NU	*g*	3r, 4	
RODENTIA:						
Capromyidae						
Capromys gundlachianus	Cuba	III	extant	–	–	
Capromys pilorides	Cuba	III	extant	–	22	
Capromys und. sp. 1	Caymans (all)	III	ME	*b, j*	3e	375 ± 50 BP; mangrove peat containing bone
Capromys latus	Cuba	III	LQU	*?*	9, 23	
Geocapromys brownii	Jamaica	III	extant	–	–	
Geocapromys ingrahami	Bahamas	II	extant	–	–	
Geocapromys thoracatus	Little Swan I	II	ME	*h*	3c	1950s
Geocapromys und. sp 1	Grand Cayman	III	ME	*j*	3e	
Geocapromys und. sp 2	Cayman Brac	III	ME	*?j*	3e	
*Geocapromys columbianus**	Cuba	II	ME	*j*	24	
*Geocapromys pleistocenicus**	Cuba	II	ME	*?j*	25	
Mesocapromys angelcabrerai	Cuba	II	extant	–	–	
Mesocapromys auritus	Cuba	II	extant			
Mesocapromys melanurus	Cuba	II	extant	–	–	
Mesocapromys nanus	Cuba	II	extant	–	26	
Mesocapromys sanfelipensis	Cuba	II	extant	–	–	
*Mesocapromys barbouri**	Cuba	II	LQU	*?*	27	
*Mesocapromys beatrizae**	Cuba	II	LQU	*?*	27	
*Mesocapromys gracilis**	Cuba	II	LQU	*?*	27	
*Mesocapromys kraglievichi**	Cuba	II	LQU	*?*	27	
*Mesocapromys minimus**	Cuba	II	LQU	*?*	27	
Mysateles garridoi	Cuba	III	extant	–	–	
Mysateles meridionalis	Cuba	III	extant	–	–	
Mysateles prehensilis	Cuba	III	extant	–	28	
Hexolobodon phenax	Hispaniola	III	ME	*j*	29	
Hexalobodon sp. 1	Hispaniola	III	?ME	*?*	30	
Isolobodon montanus	Hispaniola	III	ME	*j*	29	
Isolobodon portoricensis	Hispaniola etc.	III	ME	*a, j, k*	3s, 31	‡620 ± 60 BP; hemimandible
Plagiodontia aedium	Hispaniola	III	extant	–	–	
Plagiodontia ipnaeum	Hispaniola	III	ME	*j*	3k	
Plagiodontia araeum	Hispaniola	III	HU	*?*	9	
Rhizoplagiodontia lemkei	Hispaniola	III	ME	*j*	32	
RODENTIA:						
Hydrochoeridae						
Hydrochoerus gaylordi	Grenada	IV	NU	*e*	14	3.6 ± 0.4 to 2.7 ± 0.3 Ma (hornblende)
UNASSIGNED:						
Enigmatic Taxa/Morphs						
"Sheep Pen ?primate femur"	Jamaica	?	LQU	*d*	3t	>100 ka (snail shell)

(continued)

TABLE 9.2. (continued)

Taxa	Range	Body	Time	Age	Notes	Chronometric Dates & Material Dated (If Applicable)[2]
		Size[1]	Interval[2]	Determ.[2]		
"Trou Wòch Sa Wo ?primate femur"	Hispaniola	?	HU	*b*	10	10,320 ± 170 to 3,715 ± 175 BP (range, unid. bone)
"Coco Ree ?primate femur"	Jamaica	?	LQU	*d*	3u	50–30 ka (snail shell)
"Sheep Pen ?caviomorph femur"	Jamaica	V	NU	*g*	4, 12	
Tainotherium valei (Caviomorpha, inc. sed.)	Puerto Rico	III	LQU	*f*	8	

Notes

[1] Body size estimates based where applicable on data collected by Silva and Downing (1995).

[2] Abbreviations defined in Table 9.1. Symbol ‡ before a radiocarbon age estimate signifies that it has been isotopically corrected for $^{13}C/^{12}C$. All estimates and 1-sigma errors rounded to nearest decade.

[3] a, MacPhee and Meldrum (2006); b, MacPhee (1996); c, MacPhee and Flemming (1999); d, Turvey et al. (2007); e, Morgan (1994b); f, Ottenwalder (2001); g, Morgan and Ottenwalder (1993); h, Steadman et al. (2005); Jull et al. (2004); i, MacPhee et al. (2007); j, MacPhee et al. (2000a); k, Miller (1930); l, MacPhee and Flemming (2003); m, McFarlane et al. (1998a); n, Pregill et al. (1994); o, McFarlane and Debrot (2001); p, McFarlane et al. (2000); q, Miller (1929a, b); r, MacPhee and Wyss (1990); s, Flemming and MacPhee (1999); t, Ford (1990); u, Goodfriend and Mitterer (1987); v, Turvey et al. (2006); w, MacPhee et al. (1999c).

[4] Although assumed to be a Quaternary taxon, chronological position is uncertain.

[5] Circumstantial evidence of survival into colonial times.

[6] Age estimate on charcoal published by Rímoli (1977).

[7] Includes *N. longirostris*, *submicrus*, and *superstes* (Condis Fernandez et al., 2005).

[8] Estimate "late Pleistocene or early Holocene" based on faunal association at loc. Cueva de Vaca (Turvey et al., 2006).

[9] No radiometric estimates or cultural or OW murine associations known for this taxon.

[10] Range of ^{14}C age estimates (which are not in stratigraphic order) reported by Woods (1989a) on unid. bones for loc. Trou Wòch Sa Wo, where this species was present.

[11] Range based on dating of associated inorganic radiocarbon as discussed by Rega et al. (2002)

[12] Body size after Burness et al. (2001).

[13] Assumed to be within same age range as *Megalomys curazensis* as dated by McFarlane and Debrot (2001).

[14] 40K/40A dating, accuracy questionable (see MacPhee et al., 2000b); best estimate is L. Pliocene-E. Pleistocene.

[15] Associated date based on charcoal collected at 2–4 cm below surface at loc. Cueva del Perro, where an in situ *Elasmodontomys* cheektooth plate was recovered 2 cm below surface.

[16] Junior synonym of extant *O. gorgasi* (Weksler et al., 2006)?

[17] *Rattus* association according to Ray (1962); species not listed by Musser and Carleton (2005).

[18] Not included in list of Borroto-Páez et al. (in press).

[19] Also Barbuda, Montserrat, Guadaloupe, Marie Galante (Pregill et al., 1994).

[20] Also Montserrat, St. Eustatius, St. Kitts (Pregill et al., 1994).

[21] OW murid association possible but not confirmed (Koopman and Ruibal, 1955).

[22] Includes alleged extinct "species" *Geocapromys megas* (Diáz-Franco, 2001); also *Capromys acevedo, fourniere, intermedius, pappus* (Borroto-Páez et al., in press).

[23] Includes alleged extinct "species" *Capromys antiquus, robustus, arredondoi* (Borroto-Páez et al., in press)

[24] *Rattus* association (Fischer, 1977).

[25] Extinction was before human contact according to Morgan and Woods (1986), but Borroto-Páez et al. (in press) state that some remains are post-Columbian.

[26] Includes alleged extinct "species" *Mesocapromys silvai, delicatus* (Borroto-Páez et al., in press); accepted as extant even though last specimen collected in 1937 according to Woods and Kilpatrick (2005).

[27] Validity uncertain and review needed (Borroto-Páez et al., in press).

[28] Includes *Mysateles jaumei* (Borroto-Páez et al., in press).

[29] *Rattus* association (Woods, 1989a).

[30] Listed as "unnamed *Hexalobodon* of southern Hispaniola" by Woods et al. (2001).

[31] Also Puerto Rico, Virgin islands, Mona, Ile de Tortue, La Gonave; still possibly extant on Tortue according to Woods et al. (2001).

[32] Woods (1989a:72) stated that *R. lemkei* survived "until the time *Rattus* remains are present in the deposits."

whose endemicity is dubious, such as the Grenadian population of *Dasyprocta leporina* or the several "species" of *Procyon lotor* formerly regarded as endemic WI raccoons (Helgen and Wilson, 2003), although I have retained a few others whose status is ambiguous or unsettled (e.g., *Oryzomys curasoae* of McFarlane and Debrot, 2001).

Morgan and Woods (1986; see also Morgan, 2001) estimated that 76 terrestrial species of mammals became extinct during the LQ of the West Indies, while only nine survived (extinction rate, 88%). In this paper, the number of such species, living and extinct, is considered to be somewhere between 73 and 93, depending on how doubtful cases are treated. (Doubts apply to at least 15 taxa or provisionally recognized entities that may turn out to be synonyms of something else, and an additional five whose LADs cannot be more narrowly constrained than "unspecified Neogene".) Fifteen members of this fauna are extant according to the count used here, yielding a recalculated extinction rate of 79% to 84% – for terrestrial mammals, one of the highest in the world (cf. MacPhee and Flemming, 1999). Incidentally, even introduced species have suffered extinctions in the West Indies (cf. Allen, 1942) – a point which indicates just how pervasive, if not omniselective, the process of faunal collapse has been on these islands (see Discussion below).

For fossil taxa, entries under "Body Size" are usually based on considerations of narrow allometry, i.e., the empirical or estimated body-size range of acknowledged members of the same phylogenetic group. For certain fossil taxa statistical approaches have been worked out that are of some help in confining probable ranges (e.g., Biknevicius et al., 1993); these are noted in the appropriate places in the Appendix.

Body size is relevant to any consideration of exploitation and extinction, although it has only been casually examined in most previous treatments of WI extinctions (see Discussion). Figure 9.2 is intended to depict the distribution of body size "classes" for all WI land mammals listed in Table 9.2 that are known to be of late Quaternary age (including living species), for a total of 89 species. Omitted are three enigmatic ?primate-like finds (Table 9.2, Unassigned) and five other taxa whose loss date(s) may be earlier than the LQ (*Paralouatta varonai*, *Puertoricomys corozalus*, *Hydrochoeris gaylordi*, unnamed Grenadian megalonychid, and Sheep Pen ?caviomorph). Although other taxa of dubious status and all unnamed entities could have been omitted as well, I have not done so because differences in interpretation would be slight under all plausible scenarios.

Had no extinctions occurred in the West Indies during the LQ, the distribution of species numbers by body size would have generally corresponded to the histogram on the left-hand side of Fig. 9.2. The right-hand side of the figure shows the present-day distribution after all extinct entities have been removed. It is apparent that large-bodied taxa (classes IV and V) comprise only a small fraction (~16%) of the LQ fauna. Class III (1–10 kg) is the biggest single grouping (33% of total), but half of all LQ taxa fall within the two smallest categories (body sizes ≤ 1 kg). The 15 surviving species are about equally distributed in classes II and III (figures in parentheses, right side of Fig. 9.2). Looked at as a whole, then,

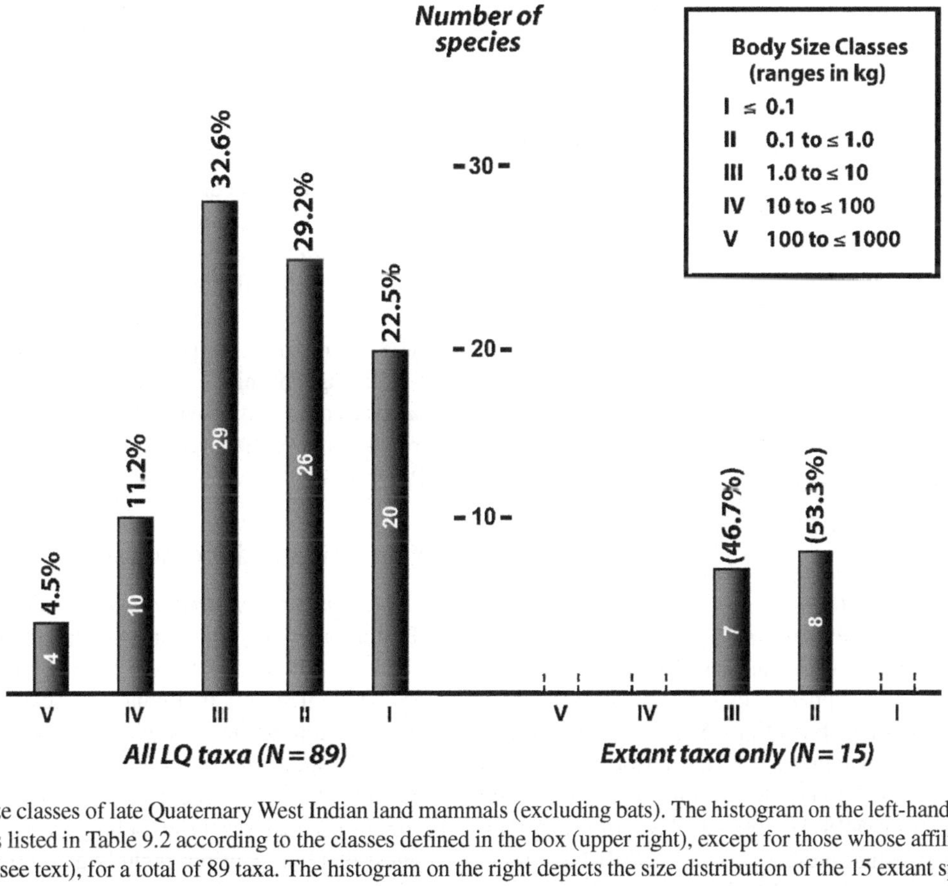

FIGURE 9.2. Body size classes of late Quaternary West Indian land mammals (excluding bats). The histogram on the left-hand side of the figure distributes all species listed in Table 9.2 according to the classes defined in the box (upper right), except for those whose affiliation and age are in substantial doubt (see text), for a total of 89 taxa. The histogram on the right depicts the size distribution of the 15 extant species of WI land mammals. Numbers in white, actual number of species per class; percentages in black, contribution each class makes to its respective total. Large species (classes IV and V) formed a small fraction of the total diversity (~16%); indeed, half of all species were less than 1 kg (classes I and II). As the result of extinction, all members of the largest and smallest classes have disappeared (right-hand side). Even in middle ranks (classes II and III) devastation has been severe: only two insectivores and 13 capromyine rodents have survived to the present.

the LQ land mammals of the West Indies were predominantly meso-to-microfaunal in body size, with but few taxa that could be considered even moderately large. This has implications regarding plausible causes of their disappearance (see Discussion).

Abbreviations and codes under "Time Interval" and "Age Determination" are defined in Table 9.1. Taxa are relegated to one of the "unspecified" periods only if there is no acceptable information that ties them to a particular, more narrowly constrained interval. For example, while it is possible to say on the basis of several radiocarbon dates from the Haitian site of Trou Wòch Sa Wo that the megalonychid *Acratocnus ye* persisted into the Holocene, the spread of the dates with a bearing on this point is very large and does not permit a narrower LAD than HU (i.e., survival within last 10,000 years). By contrast, the direct date of 4,390 ± 40 BP on a specimen of *Neocnus comes* permits us to place this taxon in a more constrained time frame, namely the AI (i.e., Amerindian interval, ~6,000 to 500 BP). Generally speaking, only the particular age determination that provides the currently accepted LAD for a species is indicated, even when there are multiple dates or methods of age determination available. Finally, under the heading "Notes", various sorts of additional information on individual taxa are provided. Some important localities are mentioned, but it is beyond the scope of this chapter to provide detailed information about distributions of species within islands. Although locality records can give the impression that many now-extinct species enjoyed islandwide distributions, because most paleontological sites in most parts of the Caribbean have never been radiometrically dated, it is difficult to determine whether ranges of extinct species were regularly coextensive with island area, or waxed/waned over long periods of time in response to changing environmental conditions (cf. Pregill and Olson, 1981) – another question that awaits future, sophisticated treatment.

Discussion

Contrasts Between Mainland and Insular Extinctions During the New World Holocene

1. *There is no evidence for an end-Pleistocene extinction event in the West Indies ca. 10,000 BP, or for a blitzkrieg at any time.*

Progressive improvement in LAD determinations during the past 20 years makes it less and less likely that anything of major biotic import happened in the West Indies during the Pleistocene/Holocene transition. Nor is there any evidence for significant mammal losses during or immediately after pleniglacial time. Indeed, the overall biotic picture for the West Indies, from MIS 5a time ca. 75 ka to the earliest evidence for Amerindian presence at ca. 6,000 BP, is one of retention rather than collapse. In view of the fact that this range covers most of Wisconsinan time, even in the subjunctive it hardly need be said that climate change may not be quite the extinction agent some commentators think it was.

Things changed for the WI fauna after 6,000 BP, but not in the way classically conceived by the blitzkrieg model of late Quaternary vertebrate extinction. However much Amerindian practices may have affected local environments in the West Indies, with regard to extinction scheduling their effect was either much delayed or even nugatory – an unexpected finding for an island context in which biological first contact has long been assumed to have had a drastic, immediate impact. Paradoxically, on the New World mainlands, the radiocarbon evidence for a very narrow extinction window – one lasting perhaps not more than 400 years and affecting dozens of species simultaneously – continues to build (Fiedel, Chapter 2).

2. *Holocene lineage pruning by complete extinction has been negligible on the mainlands, but severe on the islands.*

MacPhee and Marx (1997) noted that it is implicit in the treatment of generally accepted LADs for LQ mammals in the New World (e.g., Martin, 1984; Faunmap Working Group, 1994) that the extinction rate for continental North American mammals dropped to zero ca. 10 ka, because typically no species-level losses are registered for the pre-ME Holocene. (This statement disregards very recent losses on shelf islands associated with North America, such as those in the Gulf of California). Of course it is unlikely that the rate on the mainland actually dropped that low after 10 ka, and as noted earlier at least one North American taxon previously regarded as a terminal Pleistocene casualty (*Mammuthus primigenius*) managed to survive until the mid-Holocene on islands in the Bering Sea (Guthrie, 2004; Yesner et al., 2005). In fact many losses doubtless occurred within the Holocene, but at the subspecies or population level (cf. the "extinct" sea mink, *Mustela "macrodon"* – probably a subspecies of extant *M. vison*, but see Mead et al. [2000]). Such losses do not qualify as *species* losses because the lineages still persist. The WI case is essentially the polar opposite: *all* major clades suffered lineage pruning, frequently to the point of complete disappearance, and – with the exception of the sloths and perhaps some of the larger rodents – *all* suffered the majority of their losses in the post-1500 CE period.

Exploitation and Extinction

1. *Does competition/predation ever cause extinction?*

Historically, the literature on the theoretical ecology of extinction is divided on the question of "over-competent predators" (i.e., predators that hunt on a scale that forces extirpation/extinction of their prey): do they exist in nature, and are they responsible for some proportion of all extinctions, especially those that occur on islands? These questions are pertinent because recent models (e.g., Schoener et al., 2003) suggest that competition, including competition produced by predation, can result in the permanent or intermittent rarity of affected species but rarely causes outright loss. Instead, it is extrinsic, stochastic factors – accidents of all types – that act to push a rare taxon over the edge. And both theoretical and empirical considerations indicate that such accidents can be a long time in coming or producing their deleterious effect (Rosenzweig, 1995).

Of course, extinctions occurring at more or less the same time may be linked ecologically, in the sense that loss of central players in the local ecological web of relationships may produce a cascade of knock-on effects that result ultimately in many more disappearances or range reductions (cf. "keystone herbivore hypothesis" of Owen-Smith, 1999). This has become a central theme in metanalyses of species richness and ecosystem function. As Cardinale et al. (2006) have pointed out, it has long been recognized that some species are more effective than others in controlling ecological processes within a specific environment, but which species can be lost without causing more widespread collapse turns out to be a significantly difficult problem. Whether such studies are possible and productive in the context of prehistoric island extinctions in the West Indies remains to be seen, but there is active interest in this area (e.g., Turvey et al., 2007).

2. *What is the evidence for human exploitation of endemic WI species during pre-Columbian times?*

Apart from certain rodent taxa, discussed below, the vast majority of extinct WI mammals do not seem to have been subject to predation by humans in any substantial way. Whether this assessment will change with greater attention to the problem remains to be seen. As a kind of parallel to the weak evidentiary record for megafaunal hunting on the mainlands (cf. Fiedel, Chapter 2; Surovell and Waguespack, Chapter 5; Borrero, Chapter 8), the archeological record for the islands may never be good enough to reveal very much about the nature and intensity of exploitation of now-extinct species.

The outstanding WI example of probable human exploitation of a now-extinct species is the capromyid *Isolobodon portoricensis*. (Here and elsewhere in this paper, traditional concepts of Capromyidae and Echimyidae are retained, despite the fact that under strict monophyly their contents would probably be apportioned differently; see Appendix.) Despite the range implication given by its trivial name, this rodent has not been reported from any context in Puerto Rico that demonstrably antedates the Holocene, which suggests its tenure on that island was extremely short (Flemming and MacPhee, 1999). Indeed, as Miller (1918) originally noted, virtually all reported occurrences of *Isolobodon* in Puerto Rico relate to kitchen middens or human occupation horizons; equally late examples of *Isolobodon* have also been found in midden settings in Mona, Vieques, and some of the Virgin Islands. By contrast, in Hispaniola this rodent frequently occurs in cave localities lacking any sign of human occupation. This implies that the latter island was its original home, whence it was purposely introduced by Amerindians into Puerto Rico and nearby islands and perhaps even domesticated by them to some degree (cf. Flemming and MacPhee, 1999).

Even so, domestication was evidently no hedge against extinction in the West Indies, because *Isolobodon* eventually disappeared (or has become so rare that it has avoided detection for several centuries; see Table 9.2). Guinea pigs (*Cavia* sp.), presumably obtained from domesticated South American stocks and introduced during aboriginal times into Hispaniola, Puerto Rico, Antigua, and elsewhere (Allen, 1942; Newsom and Wing, 2004), are not remarked upon by early chroniclers and may have died out before European times. Amerindian introductions of agoutis (*Dasyprocta*) were even more widespread (Miller, 1930), especially in the Lesser Antilles. Although some of these agouti introductions survived into the modern era, most are now thought to be extinct, having perhaps succumbed not only to competition with OW rats but also to predation by mongooses and endemic snakes (cf. pitvipers of Martinique and St. Lucia, which prey specifically on rodents; [Schwartz and Henderson, 1991; Wing, 1989]) – a perfect storm of "accidents". Interestingly, introductions have continued: Honduran *D. punctata*, brought into Grand Cayman about 1900 CE, not only survives there but is regarded as being sufficiently numerous to qualify as an agricultural pest (Morgan, 1994a)!

In the Lesser Antilles, native oryzomyines occasionally formed a substantial portion of the human diet (Wing, 1989), as did capromyids in Cuba (Crespo Díaz and Jiménez Vázquez, 2004). Wing (1989) noted that there is some tendency for sites falling within the early part of late prehistoric time (900–1500 CE) to exhibit a greater relative abundance of terrestrial species, including rodents, than do later sites. In light of this, as already noted it is significant that the largest rodents – *Clidomys* and *Amblyrhiza* – have yet to be found in any context in which coeval human presence has been conclusively demonstrated (Watters, 1989; McFarlane et al., 1998a). The same point applies a fortiori to the endemic sloths: some species are known on the basis of excellent radiocarbon evidence to have survived well into Amerindian times (Steadman et al., 2005; MacPhee et al., 2007). Although Antillean sloth species would not be considered particularly "megafaunal" if arrayed against their truly gigantic mainland relatives, apart from species of *Geochelone* and *Crocodylus* they were either the largest or next to largest species on the islands on which they occurred. In the typical island-loss scenario, they should have been among the first mammals to disappear. That they did not disappear quickly, at least in some cases, seems paradoxical when viewed through traditional expectations and the lens of blitzkrieg. Either the humans that came to these islands during the mid-Holocene were uninterested in hunting mammals, or sloths were so rarely seen that they never became a part of the search image of the local human predators.

Body Size and Extinction

1. *In the WI extinctions, all body size classes were seriously affected, not just the largest.*

Although it is often noted (e.g., Morgan and Woods, 1986) that a major consequence of the LQ extinctions in the West

Indies was collapse of the body size spectrum, unlike the mainlands the loss of larger species was not responsible for most of the drop. Indeed, in the case of the LQ fauna of the West Indies, a better description of what happened would be that both ends collapsed toward the middle: no species are left in the two largest or the smallest classes (I, IV, and V), and only 25–30% of class II (7/29) and III (8/26) species are still living. If the extinctions were simply random with respect to size, "survival in the middle" might be expected because there were more species in the mid-range to begin with. However, stochastic effects aside, losses in the WI were evidently not driven in the same manner as those on the mainlands, where three-quarters of LQ extinctions were in size class IV or above and smaller species were virtually unaffected (cf. Martin and Steadman, 1999).

2. *Why have some capromyids persisted while all oryzomyines have disappeared in the West Indies?*

In general, larger capromyids have survived while all rodent members of class I and most of class II have not. This could reflect the interaction of body size with competition/predation, with the result that endemics similar in size to or smaller than invading murines and mongooses were virtually eliminated. Judging from extant capromyids (average body size, 3.78 kg according to Borroto-Páez et al. [in press]), most extinct hutias would have been larger than even very large (~ 400 g) black and brown rats; *Herpestes auropunctatus*, at 600–700 g, is only a quarter the size of *Capromys pilorides* and other large hutias. By contrast, all oryzomyines, with the doubtful exception of some members of the extinct *Megalomys* group, are or were much smaller, with recorded weights for most living species falling in the range ~100–250 g (Silva and Downing, 1995).

The case for a strong positive correlation between rice rat extinctions and OW rat introductions is compelling – all the more so because oryzomyin collapse in island contexts is not restricted to the Caribbean. During the modern era (ME), island rice rats have suffered catastrophic reduction wherever they have co-occurred with introduced *Rattus*, having disappeared not only from the Caribbean but also Fernando de Noronha in the western Atlantic, islands in the Gulf of California, and most of the Galápagos (MacPhee and Flemming, 1999). In the Galápagos, three of the four surviving endemic oryzomyins are found on islands on which no introduced rodents (or cats) occur (Clark, 1980; Dowler et al., 2000). Interestingly, *Nesoryzomys swarthi* coexists with *Rattus* and *Mus* on Santiago I., although at such a low population size that, following initial description in 1906, its presence was not detected again until 1997 (Dowler et al., 2000). The only other acknowledged exception (which in effect proves the rule) is survival of *Oryzomys palustris* on some of the Florida Keys (G. S. Morgan, 2006, personal communication). Incidentally, if OW rats are mostly to blame for losses of small mammals on islands, then it seems reasonable to believe that other small vertebrates living on the same islands ought to have been affected as well. However, while there is good evidence for reduction in average body size in several WI lizard species, little outright extinction can be documented (Pregill, 1986). Whether this apparent lack of LQ losses among squamates is real or due to our persistent ignorance remains to be established.

Ecology and Extinction

1. *To narrow extinction dates yet further, where should "last survivors" be sought?*

The evidence is increasing that a number of smaller species, previously undated or poorly dated, persisted much closer to the present than previously suspected. Nevertheless, 1500 CE remains a dating Rubicon that few of them can be shown to have actually crossed. A possible explanation is that OW rats underwent ecological release in the early 16th century as they began to spread out within the islands, resulting in enormous increases in their populations in a short time and placing endemic rodents and insectivores under great pressure. In this scenario, many extinctions should have occurred soon after 1500 CE (cf. MacPhee et al., 1999), although endemic populations living in remote areas might have persisted for a longer period.

Environmental degradation rather than competition or predation could be the additional "accident" that tipped the scales for some taxa. Because significant forest clearance for the exploitation of tropical hardwoods, sugar cane agriculture, and other purposes did not occur until the early 18th century (or later) on most islands, it is not out of the question that secure evidence will be found for the survival of some species into truly recent times (cf. Turvey et al., 2007). One possibility would be to look for localities in potential refugial areas, the preservational role of which is mostly unexamined. For example, little paleontological work has been conducted in areas such as the Sierra de Cristal, home of the last of the Cuban solenodons, or Cienaga de Zapata, where the ivory-billed woodpecker may still exist. To be sure, the few expeditions that have tried to search for representatives of "lost" taxa (e.g., Woods et al., 1985; MacPhee et al., 1999) have returned with essentially negative results, but these were brief, intensive studies that did not cover a wide range of biotopes.

2. *When did Antillean sloths become ecologically extinct, and when did they completely disappear?*

Given continuing interest in the fate of large mammals in the LQ, it is appropriate to close with some final remarks on the loss of the Antillean sloths. Whether or not extreme rarity is always coincidental with ecological extinction, the latter is not normally thought of as a long-continued process: species reduced to this condition are usually considered to be only a few years to a few decades away from complete disappearance (cf. Pimm et al., 2006). Although risk factors may vary from species to species, variables such as narrow niche breadth, restricted resource distribution, and poor dispersal ability are frequently cited as important predisposing factors (e.g., Kotiaho et al., 2005). Whether any or all of these risk factors affected Antillean sloths can only be estimated on theoretical grounds in the absence of even the barest empirical evidence that, for example, the species in question suffered range collapse after the arrival of people and for *that* reason were rarely seen or interacted with thereafter. (There are even fewer grounds – indeed, I can think of none – for the logical but unsupportable alternative that severe attenuation occurred *before* Amerindian arrival; but if so, how, and on what evidence?) Solving this conundrum should be an attractive undertaking for those interested in improving understanding of the paleoecology of mammalian extinctions on these islands. It would also help to test the "long fuse" model currently emerging in the study of late Quaternary extinctions in the West Indies – that, for whatever reason, the extinction of the endemic fauna occurred on a longer time scale than previously believed (Steadman et al. 2005; MacPhee et al., 2007; Turvey et al., 2007).

Conclusion: Not "Blitzkrieg," Not Climate Change, Just Collateral Damage

This update and reanalysis of data relating to WI mammal extinctions has shown that, despite progressive improvement in LAD determinations during the past 20 years, there is still no evidence for an end-Pleistocene extinction event in the West Indies. Nor is there any evidence for significant mammal losses during or immediately subsequent to pleniglacial time. Indeed, the overall picture for the entire latest Pleistocene/early Holocene in the West Indies is one of retention rather than collapse of biodiversity. Any influence of climate change was, if anything, positive, as neither the LGM nor events during the Pleistocene/Holocene transition seem to have forced known losses.

Things changed for the mammal fauna after 6 ka, but not in the way classically conceived by the strict blitzkrieg model of LQ vertebrate extinction. However much Amerindian practices may have affected local environments in the West Indies, with regard to extinction their effect may have been largely a non-event – an unexpected finding for any island context, and especially astonishing for this one, where biological first contact was long assumed to have had a severe, immediate impact. Many now-extinct species may have survived long after the beginning of European inroads, their ultimate demise being a sort of by-play within the context of larger environmental impacts. When the end came in such cases will probably remain hard to document empirically, although processually there can be little doubt, as Turvey et al. (2007) have emphasized, that the ultimate driver was the culminating effects of centuries of human-induced distress.

Acknowledgments. It is a pleasure to record my thanks to Pere Bover and Clare Flemming for reading and commenting on a draft of this chapter, and to Gary Morgan and Sam Turvey for their thoughtful reviews. I also thank Gary Haynes for his invitation to submit a chapter for this important volume. This is Contribution 13 to the series "Origin of the Antillean Land Mammal Fauna".

References

Adam PJ (2004) *Monachus tropicalis*. Mamm Species 747:1–9

Alcover JA, Campillo X, Macias M, Sans A (1998) Mammal species of the world: Additional data on insular mammals. Am Mus Novitates 3248:1–29

Allaire L (1997) The Lesser Antilles before Columbus. In: Wilson SM (ed) The indigenous people of the Caribbean. University of Florida Press, Gainesville, FL, pp 20–28

Allen GM (1942) Extinct and vanishing mammals of the western hemisphere with the marine species of all the oceans. Cooper Square, New York

Anderson E (1984) Who's who in the Pleistocene: A mammalian bestiary. In: Martin PS, Klein RG (eds) Quaternary extinctions: A prehistoric revolution. University of Arizona Press, Tucson, pp 40–89

Arredondo C, Arredondo O (2000) Nuevo género y especie de perezoso (Edentata: Megalonychidae) del Pleistoceno de Cuba. Revista Biología 14:66–72

Arredondo O, Rivero M (1997) Nuevo género y especie de Megalonychidae del Cuaternario Cubano. Revista Biología 11:105–112

Arredondo O, Varona LS (1974) Nuevos género y especie de mamífero (Carnivora: Canidae) del Cuaternario de Cuba. Poeyana 131:1–12

Ayes Suares C (1989) Excavaciones arqueológicas en Angostura, Bo. Florida Afuera, Barceloneta, Puerto Rico. Ecos de Plazuela 1(3):9–11

Biknevicius AR, McFarlane DA, MacPhee RDE (1993) Body size in *Amblyrhiza inundata* (Heptaxodontidae, Caviomorpha), an extinct megafaunal rodent from the Anguilla Bank, West Indies: Estimates and implications. Am Mus Novitates 3079:1–25

Borroto-Páez R, Mancina C, Woods CA, Kilpatrick CW (in press) Updated checklist of endemic terrestrial mammals of the West Indies. In: Borroto-Páez R, Woods CA, Sergile FE (eds) Terrestrial mammals of the West Indies: Contributions. Wacahoota Press/University of Vermont, Burlington

Burness GP, Diamond J, Flannery T (2001) Dinosaurs, dragons, and dwarfs: The evolution of maximal body size. Proc Natl Acad Sci USA 98:14518–14523

Burney DA, Burney LP, MacPhee RDE (1994) Holocene charcoal stratigraphy from Laguna Tortuguero, Puerto Rico, and the timing of human arrival on the island. J Arch Sci 21:273–281

Camacho A, Borroto R, Ramos I (1995) Los capromidos de Cuba: Estado actual y perspectives de las investigaciones sobre su sistemática. Marmosiana 1:43–56

Cardinale BJ, Srivastava DS, Duffy JE, Wright JP, Downing AL, Sankaran M, Jouseau C (2006) Effects of biodiversity on the functioning of trophic groups and ecosystems. Nature 443:989–982

Carleton MD, Olson SL (1999) Amerigo Vespucci and the rat of Fernando de Noronha: A new genus and species of rodent

(Muridae: Sigmodontinae) from a volcanic island off Brazil's continental shelf. Am Mus Novitates 3256:1–59

Caughley G, Gunn A (1996) Conservation biology in theory and practice. Blackwell Science, Cambridge, MA

Clark DB (1980) Population ecology of an endemic neotropical island rodent: *Oryzomys bauri* of Sante Fe Island, Galapagos, Ecuador. J Ani Ecol 49:185–198

Cole FR, Reeder DM, Wilson DE (1994) A synopsis of distribution patterns and the conservation of mammal species. J Mammal 75:266–276

Condis Fernández MM, Jiménez Vázquez O, Arredondo O (2005) Revisión taxonómica del género *Nesophontes* (Insectivora: Nesophontidae) en Cuba: Análisis de los caracteres diagnóstico. Monografies de la Societat d'Història Natural de las Balears 12:95–100

Crespo Díaz R, Jiménez Vázquez O (2004) Arqueología precolombina del municipio Boyeros. Revista de Gabinete de Arqueología 3:67–74

Day D (1989) Vanished species. Gallery Books, New York

Díaz-Franco S (2001) Situación taxonómica de *Geocapromys megas* (Rodentia: Capromyidae). Caribbean J Sci 37:72–80

Dowler RC, Carroll DS, Edwards CW (2000) Rediscovery of rodents (Genus *Nesoryzomys*) considered extinct in the Galapagós Islands. Oryx 34:109–117

Dunn PD (1994) Oceanic islands. Blackwell, Oxford

Emmons LH (2005) A revision of the genera of arboreal Echimyidae (Rodentia: Echimyidae, Echimyinae), with descriptions of two new genera. Univ California Publ Zool 133:247–310

Faunmap Working Group (1994) Faunmap: A database documenting Late Quaternary distributions of mammal species in the United States. Illinois State Museum Scientific Papers 35(1), 2 vol.

Fischer K (1977) Quartäre mikromammalia Cubas, vorwiegend aus der Höhle San José de la Lamas, Santa Fé, Provinz Habana. Zeitschrift für geologische Wissenschaft 5:213–255

Flemming C, MacPhee RDE (1999) Redetermination of holotype of *Isolobodon portoricensis* (Rodentia, Capromyidae), with notes on recent mammalian extinctions in Puerto Rico. Am Mus Novitates 3278:1–11

Ford S (1990) Platyrrhine evolution in the West Indies. J Hum Evol 19:237–254

Ford SM, Morgan GS (1986) A new ceboid femur from the late Pleistocene of Jamaica. J Verteb Paleontol 6:281–289

Ford SM, Morgan GS (1988) Earliest primate fossil from the West Indies. American J Phys Anthropol 75:209

Galewski T, Mauffrey J-F, Leite YLR, Patton JL, Douzery EJP (2005) Ecomorphological diversification among South American spiny rats (Rodentia; Echimyidae): A phylogenetic and chronological approach. Mol Phylogenet Evol 34:601–615

Goodfriend GA, MittererRM (1987) Age of the ceboid femur from Coco Ree, Jamaica. J Vert Paleontol 7:344–345

Guthrie R (2004) Radiocarbon evidence of mid-Holocene mammoths stranded on an Alaskan Bering Sea island. Nature 429:746–749

Harrington M (1921) Cuba before Columbus. Museum of the American Indian/Heye Foundation, New York

Helgen K, Wilson DE (2003) Taxonomic status and conservation relevance of the raccoons (*Procyon* spp.) of the West Indies. J Zool Soc Lond 259: 69–76

Higuera-Gundy A, Brenner M, Hodell DA, Curtis JH, Leyden BW, Binford MW (1999) A 10,300 ^{14}C record of climate and vegetation change from Haiti. Quat Res 52:159–170

Holdaway RN (1999) Differential vulnerability in the New Zealand vertebrate fauna. In: MacPhee RDE (ed) Extinctions in near time: Causes, context, and consequences. Kluwer/Plenum, New York, pp 189–238

Hooijer D (1959) Fossil rodents from Curaçao and Bonaire. Studies on the Fauna of Curaçao and Other Caribbean Islands 9(35):1–27

Horn SP, Orvis KH, Kennedy LM, Clark GM (2000) Prehistoric fires in the highlands of the Dominican Republic: Evidence from charcoal in soils and sediments. Caribbean J Sci 36:10–18

Hughen KA, and 26 others (2004) Marine04 marine radiocarbon age calibration, 1–26 cal kyr BP. Radiocarbon 46:1059–1086

Jaimez Salgado E, Gutiérrez Calvache D, MacPhee RDE, Gould GC (1992) The monkey caves of Cuba. Cave Science, Trans Brit Cave Res Assoc 19:25–30

Jull AJT, Iturralde-Vinent MA, O'Malley JM, MacPhee RDE, McDonald G, Martin PS, Moody J, Rincón A (2004) Radiocarbon dating of extinct fauna in the Americas recovered from tar pits. Nucl Instrume Meth Phys Res B 223/224:668–671

Koopman KF, Ruibal R (1955) Cave-fossil vertebrates from Camaguey, Cuba. Breviora 46:1–8

Kotiaho JS, Kaitala V, Komonen A, Päivinen J (2005) Predicting the risk of extinction from shared ecological characteristics. Proc Natl Acad Sci USA 102:1963–1967

Kozlowski JK (1974) Preceramic cultures of the Caribbean. Uniwertytetu Jagiellonskiego, Kraków

Kraglievich L (1926) Los grandes roedores terciarios de la Argentina y sus relaciones con ciertos géneros Pleistocenos de las Antillas. Anales del Museo Nacional de Historia Natural "Bernardino Rivadavia" 34:121–135

Kuzmin YV, Orlova LA (2004) Radiocarbon chronology and environment of woolly mammoth (*Mammuthus primigenius* Blum.) in northern Asia: Results and perspectives. Earth-Sci Rev 68:133–169

Leite YL, Patton JL (2002) Evolution of South American spiny rats (Rodentia, Echimyidae): The star-phylogeny hypothesis revisited. Mol Phylogenet Evol 25:455–464

Lourens L, Hilgren F, Shackleton NJ, Laskar J, Wilson D (2004) The Neogene period. In: Gradstein FM, Ogg JG, Smith AG (eds) A geologic time scale 2004. Cambridge University Press, Cambridge

MacLeod N (2003) Extinctions, causes and evolutionary significance. In: Rothschild L, Lister AM (eds) Evolution on planet earth: The impact of the physical environment. Academic Press, London, pp 1–11

MacPhee RDE (1984) Quaternary mammal localities and heptaxodontid rodents of Jamaica. Am Mus Novitates 2803:1–34

MacPhee RDE (1996) The Greater Antillean monkeys. Revista de Ciència 18:13–32

MacPhee RDE, Burney DA (1991) Dating of modified femora of extinct dwarf *Hippopotamus* from southern Madagascar: Implications for constraining human colonization and vertebrate extinction events. J Arch Sci 18:695–706

MacPhee RDE, Fleagle JG (1991). Postcranial remains of *Xenothrix mcgregori* (Primates, Xenotrichidae) and other late Quaternary mammals from Long Mile Cave, Jamaica. Bull Am Mus Nat Hist 206:287–321

MacPhee RDE, Flemming C (1999) *Requiem aeternum*: The last five hundred years of mammalian species extinctions. In: MacPhee RDE (ed) Extinctions in near time: Causes, contexts, and consequences. Kluwer/Plenum, New York, pp 333–372

MacPhee RDE, Flemming C (2003) A possible heptaxodontine and other caviidan rodents from the late Quaternary of Jamaica. Am Mus Novitates 3422:1–42

MacPhee RDE, Flemming C, Lunde DP (1999) "Last occurrence" of the Antillean insectivore *Nesophontes*: New radiometric dates and their interpretation. Am Mus Novitates 3264:1–19

MacPhee RDE, Ford DC, McFarlane DA (1989) Pre-Wisconsinan mammals from Jamaica and models of late Quaternary extinctions in the Greater Antilles. Quat Res 31:94–106

MacPhee RDE, Marx PA (1997) The 40,000-year plague: Humans, hyperdisease, and first-contact extinctions. In: Goodman S, Patterson B (eds) Natural change and human impact in Madagascar. Smithsonian Institution Press, Washington, DC, pp 169–217

MacPhee RDE, Meldrum J (2006) Postcranial remains of the extinct monkeys of the Greater Antilles, with evidence for semiterrestriality in *Paralouatta*. Am Mus Novitates 3516:1–65

MacPhee RDE, White JL, Woods CA (2000a) New species of Megalonychidae (Xenarthra, Phyllophaga) from the Quaternary of Hispaniola. Am Mus Novitates 3303:1–32

MacPhee RDE, Singer R, Diamond M (2000b) Late Cenozoic land mammals from Grenada, Lesser Antilles Island-Arc. Am Mus Novitates 3302:1–20

MacPhee RDE, Tikhonov AN, Mol D, de Marliave C, van der Plicht H, Greenwood AD, Flemming C, Agenbroad L (2002) Radiocarbon chronologies and extinction dynamics of the late Quaternary mammalian megafauna of the Taimyr Peninsula, Russian Federation. J Arch Sci 29:1017–1042

MacPhee RDE, Iturralde-Vinent M, Gaffney E (2003) Domo de Zaza, an Early Miocene vertebrate locality in south-central Cuba, with notes on the tectonic evolution of Puerto Rico and Mona Passage. Am Mus Novitates 3394:1–42

MacPhee RDE, Iturralde-Vinent MA, Jiménez-Vázquez O (2007) Prehistoric sloth extinctions in Cuba: Implications of a new "last" appearance date. Caribbean J Sci 41:94–98

MacPhee RDE, Woods CA (1982) A new fossil cebine from Hispaniola. Am J Phys Anthropol 58:419–436

MacPhee RDE, Wyss AR (1990) Oligo-Miocene vertebrates from Puerto Rico, with a catalog of localities. Am Mus Novitates 2965:1–45

Mancina CA, Garcia-Rivera L (2005) New genus and species of fossil bat (Chiroptera: Phyllostomidae) from Cuba. Caribbean J Sci 41:22–27

Martin PS (1984) Prehistoric overkill: the global model. In: Martin PS, Klein RG (eds) Quaternary extinctions: A prehistoric revolution. Tucson, University of Arizona Press, pp 354–403

Martin PS, Steadman DW (1999) Prehistoric extinctions on islands and continents. In: MacPhee RDE (ed) Extinctions in near time: Causes, contexts, and consequences. New York, Kluwer/Plenum, pp 17–56

Matthew WD, Paula Couto C de (1959) The Cuban edentates. Bull Am Mus Nat Hist 117:5–56

Mayr E (1967) The challenge of island faunas. Australian Nat Hist 15:359–374

McFarlane DA (1999a) A note on dimorphism in *Nesophontes edithae* (Mammalia: Insectivora), an extinct island-shrew from Puerto Rico. Caribbean J Sci 35:142–143

McFarlane DA (1999b) Late Quaternary fossil mammals and last occurrence dates from caves at Barahona, Puerto Rico. Caribbean J Sci 35:238–248

McFarlane DA, Debrot A (2001) An extinct oryzomyine rodent from the Quaternary of Curaçao, Netherlands Antilles. Caribbean J Sci 37:182–184

McFarlane DA, Lundberg J (2004) *Reliquae diluvianae alter*: Last Interglacial flood deposits in the caves of the West Indies. In: Sasowsky ID, Mylroie J (eds) Studies of cave sediments: Physical and chemical records of paleoclimate. New York, Kluwer/Plenum, pp 313–322

McFarlane DA, MacPhee RDE, Ford DC (1998a) Body size variability and a Sangamonian extinction model for *Amblyrhiza*, a West Indian megafaunal rodent. Quat Res 50:80–89

McFarlane DA, Lundberg J, Flemming C, MacPhee RDE, Lauritzen S-E (1998b) A second pre-Wisconsinan locality for the extinct Jamaican rodent *Clidomys* (Rodentia: Heptaxodontidae). Caribbean J Sci 34:315–317

McFarlane DA, Vale A, Christenson K, Lundberg J, Atilles G, Lauritzen S-E (2000) New specimens of Late Quaternary extinct mammals from caves in Sanchez Ramirez Province, Dominican Republic. Caribbean J Sci 36:163–166

McKenna MC, Bell SK (1997) Classification of mammals above the species level. Columbia University Press, New York

Mead JI, Spiess AE, Sobolik KD (2000) Skeleton of extinct North American sea mink (*Mustela macrodon*). Quat Res 53:247–262

Meltzer D, Mead JI (1985) Dating the Pleistocene extinctions: Theoretical issues, analytical bias, and substantive results. In: Mead JI, Meltzer D (eds) Environments and extinction: Man in Late Glacial North America. Center for the Study of Early Man, Orono, pp 145–172

Miller GS (1916) Bones of mammals from Indian sites in Cuba and Santo Domingo. Smithsonian Misc Coll 66(12):1–10

Miller GS (1918) Mammals and reptiles collected by Theodoor de Booy in the Virgin Islands. Proc US Natl Mus 54:507–511

Miller GS (1929a) A second collection of mammals from caves near St. Michel, Haiti. Smithsonian Misc Coll 81(9):1–30

Miller GS (1929b) Mammals eaten by Indians, owls, and Spaniards in the coast region of the Dominican Republic. Smithsonian Misc Coll 82(5):1–16

Miller GS (1930) Three small collections of mammals from Hispaniola. Smithsonian Misc Coll 82(15):1–9

Moore C (1991) Cabaret: Lithic workshop sites in Haiti. Reports of the Archaeological-Anthropological Institute of the Netherlands Antilles 9, pp 922–1004

Morgan GS (1985) Taxonomic status and relationships of the Swan Island hutia, *Geocapromys thoracatus* (Mammalia, Rodentia, Capromyidae), and the zoogeography of the Swan Islands vertebrate fauna. Proc Biol Soc Washington 98:29–46

Morgan GS (1994a) Mammals of the Cayman Islands. In: Brunt MA, Davies JE (eds) The Cayman Islands: Natural history and biogeography. Kluwer Academic, Boston, pp 435–463

Morgan GS (1994b) Late Quaternary fossil vertebrates from the Cayman Islands. In: Brunt MA, Davies JE (eds) The Cayman Islands: Natural history and biogeography. Kluwer, Boston, pp 465–508

Morgan GS (2001) Patterns of extinction in West Indian bats. In: Woods CA, Sergile FE (eds) Biogeography of the West Indies: Patterns and perspectives. CRC Press, Boca Raton, FL, pp 369–407

Morgan GS, Ottenwalder JA (1993) A new extinct species of *Solenodon* (Mammalia: Insectivora: Solenodontidae) from the Late Quaternary of Cuba. Annals Carnegie Mus 62:151–164

Morgan GS, Wilkins L (2003) The extinct rodent *Clidomys* (Mammalia: Rodentia: Heptaxodontidae) from a Pleistocene cave deposit in Jamaica. Caribbean J Sci 39:34–41

Morgan GS, Woods CA (1986) Extinction and the zoogeography of West Indian land mammals. Biol J Linn Soc 28:167–203

Musser GG, Carleton MD (2005) Superfamily Muroidea. In: Wilson DE, Reeder DM (eds) Mammal species of the world: A taxonomic and geographic reference. Johns Hopkins, Baltimore, pp 894–1531

Newsom LA, Wing ES (2004) On land and sea: Native American uses of biological resources in the West Indies. University of Alabama Press, Tuscaloosa AL

Ottenwalder JA (2001) Systematics and biogeography of the West Indian genus *Solenodon*. In: Woods CA, Sergile FE (eds) Biogeography of the West Indies: Patterns and perspectives. CRC Press, Boca Raton, FL, pp 253–330

Oviedo y Valdés GF de (1535) Historia general y natural de las Indias. Juam Cromberger, Seville [trans. and ed. by Stroudemire SA (1959) University of North Carolina Studies in Romance Languages and Literature 32, pp 1–140]

Owen-Smith N (1999) The interaction of humans, megaherbivores, and habitats in the late Pleistocene extinction event. In: MacPhee RDE (ed) Extinctions in near time: Causes, contexts, and consequences. Kluwer/Plenum, New York, pp 57–70

Pascual R, Vucetich MG, Scillato-Yané GJ (1990) Extinct and recent South American and Caribbean edentates and rodents, outstanding examples of isolation. Azzaroli A (ed) Biogeographical aspects of insularity, Atti Convegni Lincei 87:627–640

Patterson B, Wood AE (1982) Rodents from the Deseadan Oligocene of Bolivia and the relationships of the Caviomorpha. Bull Mus Compar Zool 149:371–543

Perez VR, Godfrey LR, Nowak-Kemp M, Burney DA, Ratsimbazafy J, Vasey N (2005) Evidence of early butchery of giant lemurs in Madagascar. J Human Evol 49:722–742

Peros MC, Reinhardt EG, Davis AM (2007) A 6000-year record of ecological and hydrological changes from Laguna de la Leche, north coastal Cuba. Quat Res 67:69–82

Pimm S, Raven P, Peterson A, Sekercioglu CH, Ehrlich PR (2006) Human impacts on the rates of recent, present, and future bird extinctions. Proc Natl Acad Sci USA 103:10941–10946

Pino M, Castellanos N (1985) Acerca de la asociación de perezosos Cubanos extinguidos con evidencias culturales de aborígenes Cubanos. Reporte de investigación del Instituto Ciencias Sociales 4

Pregill GK (1981) Late Pleistocene herpetofaunas from Puerto Rico. University of Kansas Mus Nat Hist Misc Publ 71:1–72

Pregill GK (1986) Body size of insular lizards: A pattern of Holocene dwarfism. Evolution 40:997–1008

Pregill GK, Olson SL (1981) Zoogeography of West Indian vertebrates in relation to Pleistocene climatic cycles. Ann Rev Ecol Syst 12:75–98

Pregill GK, Steadman DW, Olson SL, Grady FV (1988) Late Holocene fossil vertebrates from Burma Quarry, Antigua, Lesser Antilles. Smithsonian Contr Zool 463:1–27

Pregill GK, Steadman DW, Watters DR (1994) Late Quaternary vertebrate faunas of the Lesser Antilles: Historical components of Caribbean biogeography. Bull Carnegie Mus Natl Hist 30:1–51

Purvis A, Gittleman JL, Cowlishaw G, Mace GM (2000) Predicting extinction risk in declining species. Proc Roy Soc B 267:1947–1952

Ray CE (1962) Oryzomyine rodents of the Antillean subregion. Ph.D. Dissertation, Harvard University, Cambridge, MA

Ray CE (1964) A small assemblage of vertebrate fossils from Spring Bay, Barbados. J Barbados Mus Hist Soc 31:11–21

Rega E, McFarlane DA, Lundberg J, Christenson K (2002) A new megalonychid sloth from the Late Wisconsinan of the Dominican Republic. Caribbean J Sci 38:11–19

Rímoli R (1977) Una nueva especie de mono (Cebidae: Saimirinae: *Saimiri*). Cuadernos de CENDIA, Universidad Autonomia de Santo Domingo 242:5–14

Rodríguez M (1999) Excavations at Maruca, a Preceramic site in southern Puerto Rico. In: Winter JH (ed) Proceedings of the Seventeenth Congress of the International Association for Caribbean Archaeology. Molloy College, Rockville Centre, New York, pp 166–180

Rouse I (1989) Peopling and repeopling of the West Indies. In: Woods CA (ed) Biogeography of the West Indies: Past, present, and future. Sandhill Crane Press, Gainesville, FL, pp 119–136

Rouse I (1992) The Tainos: Rise and decline of the people who greeted Columbus. Yale University Press, New Haven, CT

Rouse I, Allaire L (1978) Caribbean. In: Taylor RE, Meighan CW (eds) Chronologies in New World archaeology. Academic Press, New York, pp 431–481

Rosenzweig ML (1995) Species diversity in space and time. Cambridge University Press, Cambridge

Sánchez-Villagra MR, Aquilera O, Horovitz I (2003) The anatomy of the world's largest extinct rodent. Science 301:1708–1710

Schoener TW, Clobert J, Legendre S, Spiller DA (2003) Life-history models of extinction: A test with island spiders. Am Naturalist 162:558–573

Schwartz A, Henderson RW (1991) West Indian reptiles and amphibians: Descriptions, distributions, and natural history. University of Florida Press, Gainesville, FL

Silva M, Downing JA (1995) CRC handbook of mammalian body masses. CRC Press, New York

Steadman DW, Ray CE (1982) The relationship of *Megaoryzomys curioi*, an extinct cricetine rodent (Muroidea: Muridae) from the Galapagos Islands, Ecuador. Smithsonian Contr Paleobiol 51:1–23

Steadman DW, Martin PS, MacPhee RDE, Jull AJT, McDonald G, Woods CA, Iturralde-Vinent MA, Hodgkins G (2005) Asynchronous extinction of Late Quaternary sloths on continents and islands. Proc Natl Acad Sci USA 102:11763–11768

Stuart AJ, Sulerzhitsky LD, Orlova LA, Kuzmin YV, Lister AM (2002) The latest woolly mammoths (*Mammuthus primigenius* Blumenbach) in Europe and Asia: A review of the current evidence. Quat Sci Rev 21:1559–1569

Stuart AJ, Kosintsev PA, Higham TFG, Lister AM (2004) Pleistocene to Holocene extinction dynamics in giant deer and woolly mammoth. Nature 431:684–689

Suárez W (2005) Taxonomic status of the Cuban vampire bat (Chiroptera: Phyllostomidae: Desmodontinae: *Desmodus*). Caribbean J Sci 41:761–767

Suárez W, Díaz-Franco S (2003) A new fossil bat (Chiroptera: Phyllostomidae) from a Quaternary cave deposit in Cuba. Caribbean J Sci 39:371–377

Suárez RR, Fernandez-Leyva O, Vento-Canosa E (1984) La convivencia de la fauna de desentados extinguidos con el aborigen de Cuba. Kobie, Bilbao Revista de Ciencias 14:561–566

Sulerzhitsky LD, Romanenko FA (1999) The "twilight" of the mammoth fauna in the Asiatic Arctic. Ambio 28:251–255

Turvey ST, Grady FV, Rye P (2006) A new genus and species of "giant hutia" (*Tainotherium valei*) from the Quaternary of Puerto Rico: An extinct arboreal quadruped? J Zool 270:585–594

Turvey ST, Oliver JR, Narganes Storde YM, Rye P (2007) Late Holocene extinction of Puerto Rican native land mammals. Biol Lett 3:193–196

Varona LS (1974) Catálogo de los mamíferos vivientes y extinguidos de las Antillas. Academia de Ciencias de Cuba, La Habana

Vartanyan SL, Garutt VE, Sher AV (1993) Holocene dwarf mammoths from Wrangel Island in the Siberian Arctic. Nature 362:337–340

Vartanyan SL, Arslanov KA, Tertychnaya TV, Chernov SB (1995) Radiocarbon dating evidence for mammoths on Wrangel Island, Arctic Ocean, until 2000 BC. Radiocarbon 37:1–6

Watters DR (1989) Archaeological implications for Lesser Antilles biogeography: The small island perspective. In: Woods CA (ed) Biogeography of the West Indies: Past, present, and future. Sandhill Crane Press, Gainesville, FL, pp 153–166

Weksler M, Percequillo AR, Voss RS (2006) Ten new genera of oryzomyine rodents (Cricetidae: Sigmodontinae). Am Mus Novitates 3537:1–29

White JL (1993) Functional and phylogenetic implications of the postcranial skeleton of fossil sloths for the evolution of arboreality in tree sloths. Ph.D. Dissertation, Stony Brook University, NY

White JL, MacPhee RDE (2001) The sloths of the West Indies: A systematic and phylogenetic review. In: Woods CA, Sergile FE (eds) Biogeography of the West Indies: Patterns and perspectives. CRC Press, Boca Raton, FL, pp 201–236

Whittaker RJ (1998) Island biogeography: Ecology, evolution, and conservation. Oxford University Press, New York

Wilson DE, Reeder DM (eds) (2005) Mammal species of the world: A taxonomic and geographic reference. Johns Hopkins University Press, Baltimore, MD

Wilson SM (1997) Introduction to the study of the indigenous people of the Caribbean. In: Wilson SM (ed) The indigenous people of the Caribbean. University Press of Florida, Gainesville, pp 1–8

Wilson SM (2001) The prehistory and early history of the Caribbean. In: Woods CA, Sergile FE (eds) Biogeography of the West Indies: Patterns and perspectives. CRC Press, Boca Raton, FL, pp 519–528

Wing ES (1989) Human exploitation of animal resources in the Caribbean. In: Woods CA (ed) Biogeography of the West Indies: Past, present, and future. Sandhill Crane Press, Gainesville, FL, pp 137–152

Wing ES (2001) Native American use of animals in the Caribbean. In: Woods CA, Sergile FE (eds) Biogeography of the West Indies: Patterns and perspectives. CRC Press, Boca Raton, FL, pp 481–518

Woods CA (1982) The history and classification of South American hystricognath rodents: Reflections on the far away and long ago. In: Mares MA, Genoways HH (eds) Mammalian biology in South America. University of Pittsburgh, Pittsburgh, PA, pp 377–392

Woods CA (1989a). A new capromyid rodent from Haiti: The origin, evolution, and extinction of West Indian rodents and their bearing on the origin of the New World hystricognaths. In: Black CC, Dawson MR (eds) Papers on fossil rodents in honor of Albert Elmer Wood. Natural History Museum of Los Angeles County, Los Angeles, pp 58–89

Woods CA (1989b) The biogeography of West Indian rodents. In: Woods CA (ed) Biogeography of the West Indies: Past, present, and future. Sandhill Crane Press, Gainesville, FL, pp 741–798

Woods CA (1993) Suborder hystricognathi. In: Wilson DE, Reeder DM (eds) Mammal species of the world: A taxonomic and geographic reference. Smithsonian Institution Press, Washington DC, pp 771–806

Woods CA, Kilpatrick CW (2005) Infraorder Hystricognathi Brandt, 1855. In: Wilson DE, Reeder DM (eds) Mammal species of the world: A taxonomic and geographic reference. Smithsonian Institution Press, Washington DC, pp 1538–1600

Woods CA, Ottenwalder JA, Oliver WLR (1985) Lost mammals of the Greater Antilles: The summarised findings of a ten weeks field survey in the Dominican Republic, Haiti and Puerto Rico. Dodo, J Wildlife Preserv Trust 22:23–42

Woods CA, Borroto-Páez R, Kilpatrick CW (2001) Insular patterns and radiations of West Indian rodents. In: Woods CA, Sergile FE (eds) Biogeography of the West Indies: Patterns and perspectives. CRC Press, Boca Raton, FL, pp 335–353

Worthy T, Holdaway RN (2002) The lost world of the moa: Prehistoric life of New Zealand. Indiana University Press, Bloomington

Yesner DR, Veltre DW, Crossen KJ, Graham RW (2005) 5,700-year-old mammoth remains from Qagnax Cave, Pribilof Islands, Alaska. Short Papers and Abstracts, The World of Elephants 2nd International Congress, 22–25 September, 2005, Hot Springs, SD, pp 200–204

Appendix: Notes on Extinct Taxa and Human Arrival in the West Indies

Introduction

In this appendix, the major clades of endemic WI land mammals are considered sequentially with reference to taxonomic matters, estimated body size, LAD, and the likelihood and timing of human interference/predation. The names of major taxa (orders) are those utilized in the most recent edition of Wilson and Reeder's (2005) checklist. LADs based on radiometric dates are listed where available, but for most species an empirically based disappearance time can be stated only in vague terms. "Pleistocene" and "Holocene" have their usual temporal extents. Although there is a movement to remove "Quaternary" as an interval name even in informal usage (Lourens et al., 2004), I retain it here as the most recent part of the chronostratigraphically defined Neogene Period.

Body size estimates are precisely that: good conjectures exist for only a handful of extant taxa, and to date very little effort has been expended on inferring the sizes of extinct species (but see Biknevicius et al., 1993; Turvey et al., 2007). To avoid false precision, size classes (see Fig. 2) rather than "real" figures for body sizes are provided.

There are several definitions of the "West Indies" (e.g., Morgan, 2001). In this paper I define it as the geographical grouping that consists of all of the landmasses in the Caribbean Sea that would have remained as islands (i.e., unconnected to any mainland) at the height of the last glaciation ca. 18,000 BP. This definition includes all of the Greater and Lesser Antilles and their satellites as conventionally defined, plus Bahamas and all of the small islands in the western Caribbean that supported endemic land mammals (e.g., Swan I.). Although Barbados, Tobago, and Trinidad are geographically in the western Atlantic, biogeographically they are mostly related to the (other) WI islands, and it would be pedantic not to include them here. It would be equally pedantic to exclude the shelf islands that dot the Caribbean-facing coasts of Central and South America, including among others Isla Margarita and Curaçao.

West Indian representatives of Pitheciidae, Nesophontidae, Solenodontidae, Megalonychidae, Heptaxodontidae, Echimyidae, and Hydrochoeridae presently have reasonably stable taxonomies, thanks to revisionary work by various authors in recent decades, but Oryzomyini and Capromyidae remain greatly in need of attention. Oryzomyin cricetids comprised the only widely distributed LQ land mammal group in the Lesser Antilles, but fossil oryzomyin collections from most of these islands, where they exist at all, were either worked on decades ago or have never been investigated. As to capromyids, so many dubiously distinct extinct species have been named from Cuban localities that it is difficult to say at this moment whether hutias on this island suffered a major extinction by losing as much as half of their LQ diversity, or actually passed through this interval nearly unscathed.

In view of the fact that a high proportion of the WI fauna is already extinct, it is unsurprising that the rate of loss has diminished considerably in the past century. The only terrestrial extinction at the species level in the West Indies during the last ~50 years for which there is secure evidence is that of *Geocapromys thoracatus*, which disappeared from Swan I. in the 1950s. However, several Cuban capromyids (including nominal subspecies) have not been collected or seen for many years, and they too may be extinct (e.g., *Mesocapromys nanus* on Isle of Pines; Woods et al., 2001).

Soricomorpha

The WI fauna includes two soricomorph families: the Solenodontidae, housing the extant but endangered genus *Solenodon* (almiquí), and the Nesophontidae (nesophonts, island-shrews), also restricted to a single genus (*Nesophontes*) which is generally thought to be completely extinct. *Solenodon* was limited to the main island and satellites of Cuba and Hispaniola, but *Nesophontes* had a wider and apparently natural distribution that included both of the foregoing islands as well as Puerto Rico, Vieques, St. Thomas, St. John, and the Cayman Islands.

While there has been no recent change in the number of species of *Solenodon* considered valid (cf. Ottenwalder, 2001), it is probable that the content of genus *Nesophontes* will be reduced in the near future. Condis Fernandez et al. (2005) have already submerged several nominal Cuban species within *N. micrus*, and it is likely that the Hispaniolan diversity is also oversplit (H. Whidden, 2000, personal communication). The Cayman Islands nesophonts, from Grand Cayman and Cayman Brac, appear to be good species as described (but not yet named) by Morgan (1994b).

Most of the shrewlike nesophonts were probably less than 25 g in body size; even the largest, *N. edithae*, was probably less than 200 g (cf. McFarlane, 1999a). Species of *Solenodon* were considerably larger: the largest, the "giant" solenodon *S. arredondoi*, is approximately one-third larger for linear measurements than *S. cubanus* (Ottenwalder, 2001).

Ottenwalder (2001) noted that *S. cubanus* and S. *paradoxus* remains have been recovered from numerous Amerindian sites in Cuba and Hispaniola, including localities specifically interpreted as middens. However, no such associations have yet been reported for the extinct species *S. arredondoi* and *S. marcanoi*. Whether nesophonts were hunted is moot (but see claim by Miller, 1930); their small size, coupled with the presence of much larger mammals at least theoretically available for hunting, could not have made them especially attractive targets. The mandible of *N. edithae* dated to ca. 1,000 BP by Turvey et al. (2007) was found in fill associated with a funerary site, but this was not considered by the authors to be evidence of predation. The fact that most or all of the WI insectivores were still extant at least as recently as the beginning of European occupation (cf. MacPhee et al., 1999) suggests that predation by Amerindians (or their dogs) was in fact negligible.

Dates not listed in Table 9.2 for *Nesophontes* species are all earlier than 1500 CE (MacPhee et al., 1999), and LADs are generally based on the fact that most (?all) species except *N. edithae* have been found in association with *Rattus*/*Mus* in owl pellets. Although there are reasons for believing that some species persisted into very recent times (e.g., Woods et al., 1985), very lengthy survival has yet to be demonstrated by specimen evidence. Given the very few dates available for nesophonts it is not possible to assess whether they suffered a lengthy period of range collapse prior to extinction. By contrast, in the case of *Solenodon* in both Cuba and Hispaniola, subfossil evidence provides clear indications that surviving species have been relegated to a small fraction of their previous ranges (Ottenwalder, 2001).

Pilosa

All LQ sloths recovered from Caribbean landmasses belong to a single family, Megalonychidae (Phyllophaga: Pilosa), also widely represented in North, Central, and South America during the Neogene. The internal phylogenetic relationships of the Antillean sloths have long been controversial (see review in White and MacPhee, 2001), but authorities are unanimous that the Antillean taxa are closely related to the sole surviving megalonychid, the two-toed sloth *Choloepus*.

Although recent synonymizations have reduced the plethora of named Antillean sloth taxa to more manageable (and believable) proportions, discovery continues apace. Four new species from the LQ of Haiti have recently been described in detail on the basis of both cranial and postcranial characters (see MacPhee et al., 2000a; White and MacPhee, 2001). *Acratocnus simorhynchus* (Rega et al., 2002) from the Dominican Republic has been provisionally included in the list of extinct LQ megalonychids, although it may eventually prove to be a synonym of *Acratocnus ye*. By contrast, the nominal species *Galerocnus jaimezi* (Arredondo and Rivero, 1997) and *Paramiocnus riveroi* (Arredondo and Arredondo, 2000), denoted as Species B and C by White and MacPhee (2001:225), have been omitted because they are based on extremely scanty remains of uncertain age.

Three of the 15 sloth species listed in Table 9.2 have been directly dated to the Amerindian period; four others, recovered from association-dated horizons at Trou Wòch Sa Wo, Haiti (Woods, 1989a) can be broadly dated to the Holocene. LADs for the eight remaining species are earlier (PI) or unresolved (LQU, NU). This includes *Acratocnus odontrigonus*, the only LQ mammal from Puerto Rico for which positive evidence of survival into the AI is still lacking (cf. Turvey et al., 2007). In my opinion, unequivocal associations of OW murines and sloths have never been identified. However, Woods (1989a) mentioned that rats and megalonychids co-occur at some natural trap sites on the Plain Formon, Département du Sud, Haiti (e.g., Trouing Jérémie #5). Fossiliferous sediments at this site are very shallow and interspersed with large amounts of breakdown, making it difficult to assess what any putative "associations" might mean. Also, owl pellets are very rare or absent, indicating that any OW murines represented at these sites had to have fallen in, something which could have occurred at any time in the past 500 yr. Even at Trou Wòch Sa Wo there is an appreciable gap (>20 cm) between layers containing rats and mice and those containing the majority of megalonychid fossils (cf. MacPhee et al., 2000a: their table 5).

Date of loss of the Curaçao sloth, *Paulocnus petrifactus*, will remain uncertain until more conclusive information is available for this species. Extensive remains of this sloth have come from now-destroyed cave fills on Tafelberg Santa Barbara, a feature located on an erosional terrace that roughly dates on geomorphological grounds to the late Middle Pleistocene or Late Pleistocene (400,000–130,000 BP) according to McFarlane and Debrot (2001). Claimed recovery of a single immature sloth vertebra from Sint Jan I, an archeological site dated to 500–1,000 BP, has not been confirmed (Steadman et al., 2005). If *Paulocnus* broadly corresponded in age to *Megalomys curazensis* from similar contexts on the same island, then it might not have persisted later than the last interglaciation, in which case PI would be the most appropriate LAD for this species. The unnamed megalonychid from a lahar deposit on Grenada described by MacPhee et al. (2000b) is significantly older (Late Pliocene or Early Pleistocene).

It is misleading to refer corporately to the Antillean megalonychids as "ground" sloths, as was often done in the past (e.g., Matthew and de Paula Couto, 1959). None of the WI species was even remotely as large as the largest mainland members of this family (e.g., *Megalonyx*), and White (1993) has determined that most were arboreal to a greater or lesser degree. Only species of *Megalocnus* and *Parocnus* are likely to have spent a considerable amount of time on the ground, and even these taxa display morphological indicators consistent with some amount of climbing or other arboreal activity. Interestingly, *Neocnus*, the most arboreal genus, included at least one species (*N. toupiti*) that was probably even smaller in body size (< 5 kg) than living tree sloths (White, 1993).

Harrington (1921) believed that he had recovered sloth bones in association with human remains or artifacts at various localities in Cuba, and concluded that Amerindians hunted (and may therefore have contributed to the extinction of) *Megalocnus* shortly before European discovery of the island. Although other investigators have made similar claims (e.g., Miller, 1929b, Suárez et al., 1984; Pino and Castellanos, 1985), none has yet stood up to critical evaluation (MacPhee et al., 2007). The rarity or, indeed, the apparent absence of modified sloth bone in Amerindian occupation sites strongly indicates that megalonychids could not have been hunted very frequently, possibly because they were becoming extremely rare after the beginning of AI time as I argue in the text. Also, in contrast to the Hispaniolan situation (which is best described as ambiguous), there are no reports of sloth bones being commingled with those of OW murines at any Cuban or Puerto Rican sites.

Primates

Three markedly different but related platyrrhines lived in the Greater Antilles during the LQ. Among platyrrhines generally, *Antillothrix bernensis* (Hispaniola) and *Xenothrix mcgregori* (Jamaica) would be considered middle-sized species, but *Paralouatta varonai* (Cuba) was the size of a large howler monkey and thus would have been one of the largest species in the entire New World platyrrhine radiation (MacPhee and Meldrum, 2006). *Xenothrix* has been reconstructed as a slow-moving arborealist (MacPhee and Fleagle, 1991; MacPhee and Meldrum, 2006). Limb bones of *Antillothrix* are much rarer, although an undescribed distal humerus from Trou Wòch Sa Wo, found with some teeth referable to *A. bernensis*, resembles that of typical platyrrhine arboreal quadrupeds for such features as the low degree of medial epicondylar retroflexion (unpub. obser., R. MacPhee, 2007). Surprisingly, several morphological features of *Paralouatta* indicate that it may have been semiterrestrial, a locomotor pattern otherwise unprecedented among New World monkeys (MacPhee and Meldrum, 2006).

Whereas *Rattus* associations and radiometric determinations establish that the Hispaniolan and Jamaican monkeys survived into the late Holocene (perhaps into the late 1600s or early 1700s in the case of *Xenothrix*; MacPhee, 1996), the situation with regard to *Paralouatta varonai* is unclear. All of the material referred to this species comes from fissure fillings exposed in two caves in prov. Pinar del Río. The fills are of uncertain age (or ages), although they include a variety of species that have been found at other late Quaternary contexts (Jaimez Salgado et al., 1992). Nevertheless, analysis is complicated by the fact that the type of *P. marianae*, an astragalus from the Early Miocene locality of Domo de Zaza, is almost identical to that of *P. varonai* (MacPhee et al., 2003). Until dateable material comes to hand it cannot be ruled out that *Paralouatta* lived in an epoch earlier than the LQ, and for that reason its extinction date is relegated to NU in Table 9.2.

There is some evidence that monkeys are depicted in various petroglyphs, pot lugs, and other artifacts that have been occasionally found in Cuba and Hispaniola (MacPhee and Woods, 1982). So are frogs, bats, and lizards, which suggests that Amerindians often took their inspiration from the local vertebrate fauna. This is another line of evidence that at least some monkey species persisted well into the AI period.

Several femora regarded by Ford (1990 and earlier papers) as possibly primate are too enigmatic for clear assignment to this group (see below, Unassigned).

Rodentia

In the West Indies, losses among LQ rodents occurred across the entire spectrum of body sizes, from truly megafaunal *Amblyrhiza inundata* (a species as large as some of the largest Antillean sloths) to mouse-sized *Boromys offella* and *Oryzomys antillarum*. Unfortunately, in-depth study of the adaptations or paleoecology of extinct WI rodents is a woefully underdeveloped area, although beginnings have been made (e.g., Biknevicius et al. [1993] on body size and population structure in *Amblyrhiza*; Turvey et al. [2006] on possible locomotor adaptations of *Tainotherium*).

With the exception of the muroid Oryzomyini, all endemic Antillean rodents were members of Caviomorpha. Capromyidae is the only family that retains living species in the West Indies; extant members include semifossorial, terrestrial, and arboreal forms (Woods et al., 2001; Borroto-Páez et al., in press).

Oryzomyini

This relatively speciose tribe of cricetids, well represented in the extant faunas of North, Central, and South America (for recent literature, see Weksler et al. [2006]), also inhabited a surprisingly large number of islands in the temperate and tropical parts of the New World (Steadman and Ray, 1982; Woods, 1989b; Carleton and Olson, 1999). Nearly all insular rice rats have become extinct, including the entire WI diversity.

The rice rats of the West Indies were overwhelmingly Lesser Antillean in distribution: they lived on many, if not most, of the islands in the chain (Pregill et al. 1994). However, the only wild-caught specimens, mostly collected around the turn of the previous century, are from Jamaica, St. Vincent, St. Lucia, and Martinique. All other occurrences (minimally including Montserrat, Anguilla, St. Eustatius, St. Kitts, Antigua, Barbuda, Guadeloupe, Marie Galante, Grenada, Curaçao, and Barbados) are based on skeletal remains. The only confirmed Greater Antillean species is the Jamaican rice rat *Oryzomys antillarum* (sometimes listed as a subspecies of *O. couesi*), which became extinct in the last quarter of the 19th century (MacPhee and Flemming, 1999).

Musser and Carleton (2005:1144) have aptly characterized the genus *Oryzomys* in its present form as a mere "polyphyletic shell". Progress in understanding *Oryzomys* phylogeny fundamentally depends on sorting out how the various lineages currently housed in this wastebasket taxon are actually related to one another (Weksler et al., 2006). A comprehensive study of WI rice rats has never been completed. In addition to the unsettled taxonomy of taxa recovered on different islands, the degree to which some species distributions may have been influenced by human transport is thoroughly unclear. For example, Montserrat may have (improbably) supported as many as four oryzomyin species according to preliminary observations by Pregill et al. (1994), but how these taxa relate to each other or to oryzomyin populations on other islands is uninvestigated. Finally, most Lesser Antillean oryzomyin populations were probably post-1500 CE casualties, but adequate dating treatments are sorely needed; only the best-documented cases are listed in the table, which surely underrepresents the group's diversity.

Most WI oryzomyins were comparatively small, probably <100 g. Although there appear to be no published body weights for the "giant" rice rats (e.g., *Megalomys*), they are assumed to have been ca. 1 kg in body size, similar to small capromyids (e.g., *Mesocapromys*).

Echimyidae

The WI spiny rats comprise a comparatively small radiation (Heteropsomyinae) distinct from South/Central American echimyids; their sister-group within the latter assemblage is obscure (for recent molecular work on extant diversity, see Leite and Patton [2002]; Galewski et al. [2005]; Emmons [2005]; and comments under Capromyidae below). Ignoring undated *Puertoricomys corozalus* (see below), it is of interest that a species "pair" occurs on each of the three Greater Antillean islands that possessed echimyids, and that both species often occur in the same sites. Are these species pairs or are they simply morphs (sexes) of one species?

Systematic treatment of the extinct Antillean species has been somewhat variable in the literature (cf. Varona, 1974; Woods, 1989b; Woods and Kilpatrick, 2005). Table 9.2 lists seven species, which is the conventional number, although *Brotomys contractus* is represented only by its holotype, an edentulous palate, and may be conspecific with *Brotomys voratus*. Species of Cuban *Boromys* differ mostly in size rather than morphology, although I retain their separate status here. Varona (1974) submerged *Brotomys* and *Boromys* in *Heteropsomys*, but retained all four species. (*Heteropsomys antillensis* and *H. insulans* are also retained as separate species here, following Woods and Kilpatrick [2005], but they too may be sex morphs of one species.)

Earlier workers such as Miller (1930) and Allen (1942) concluded that WI echimyids must have become extinct "within the last century," a statement that has often been repeated (e.g., Day, 1989:239; Woods, 1993) as though there were conclusive evidence for it. The latest radiometric date for a member of this group (in this case, Hispaniolan *Brotomys voratus*) is ca. 350 BP (McFarlane et al., 2000). *Brotomys* may have been the "mohuy" (Woods, 1989b), one of several endemic rodents of uncertain affiliation hunted by Hispaniolan Amerindians according to de Oviedo y Valdes (1535; see Miller, 1929b).

Puerto Rican *Heteropsomys* is much larger than Cuban and Hispaniolan *Boromys* and *Brotomys*, possibly because in Puerto Rico heteropsomyines were able to fill niches occupied by capromyids elsewhere (Woods, 1989b). Although Turvey et al. (2007) recovered a date of ca. 1200 BP on a specimen of *Heteropsomys insulans*, it is surprising that (in contrast to *Brotomys*) neither of the nominal species in the former genus has been found in association with exotic murine markers.

The remaining taxon, *Puertoricomys corozalus*, from a site no longer identifiable and represented only by its holotype, may in fact be Pleistocene or even Tertiary in age (MacPhee and Wyss, 1990).

Heptaxodontidae

The WI heptaxodontids are usually treated as though they formed a monophyletic group, either alone or in combination with certain of the large to truly gigantic caviomorphs recovered at Mio-Pliocene fossil localities in South America (cf. Kraglievich, 1926; McKenna and Bell, 1997; Sánchez-Villagra et al., 2003). Woods (1982, 1989b) developed the alternative argument that the group (or at least some taxa assigned to it, such as *Quemisia*) may have been derived from within Capromyidae, which would render the latter paraphyletic. This point underlines the problem that Heptaxodontidae *sensu lato* has never been revised adequately and its contents and sister-group relationships are, unsurprisingly, poorly explored (Patterson and Wood, 1982; MacPhee and Flemming, 2003). For the purposes of this paper I will treat the Antillean group as organized in Table 9.2, although it should be noted that some authorities believe that the various WI taxa originated from different mainland clades (e.g., Pascual et al., 1990).

The Jamaican LQ paleontological record recently lost one heptaxodontid (*Clidomys parvus*) to synonymy (Morgan and Wilkins, 2003) but gained another in the form of *Xaymaca fulvopulvis*, based on a single specimen radiometrically dated to the terminal Pleistocene (MacPhee and Flemming, 2003). If *Xaymaca* was truly a heptaxodontid, it was a tiny one, not appreciably larger than the echimyids *Brotomys* and *Boromys* which it morphologically resembles in a few (presumably convergent) respects.

Rattus associations and documentary evidence, if correctly attributed, indicate that *Quemisia* was extant at the start of the modern era (Miller, 1929a,b). Associational dates on wood charcoal in a thin stratigraphic layer at the site of Cueva del Perro in central Puerto Rico imply that *Elasmodontomys* remains recovered there are between 2,500–3,500 BP, according to Turvey et al. (2007). There are no examples of *Elasmodontomys* remains from middens.

There are no radiometric records younger than late Sangamonian/early Eo-Wisconsinan time for either *Clidomys* or *Amblyrhiza*, a fact of some interest in light of the unsupported statement (e.g., Anderson, 1984; Cole et al., 1994) that these large-bodied rodents persisted into the late Holocene. Recent efforts to date *Clidomys* and *Amblyrhiza* by U-series disequilibrium dating of associated speleothem (MacPhee et al., 1989; McFarlane et al., 1998a, b) produced a variety of estimates, partly because of pervasive problems with leaching. Nevertheless, the majority of non-infinite dates lie within the range 70-130 ka, which may stand as a reasonable interim LAD in view of the lack of evidence for later occurrence of these rodents. Although sealevel rise during the Sangamonian may help to explain the disappearance of *Amblyrhiza* on St. Martin/Anguilla (Biknevicius et al., 1993; McFarlane and Lundberg, 2004), how a few meters' rise would have contributed to the loss of *Clidomys* living on a much larger, higher island like Jamaica is difficult to imagine. In any case, for once climate change may have the edge as the default explanation for the cause of these extinctions, in part because nothing of any consequence is known about them.

Most heptaxodontids were large by the standard of recent caviomorphs; some were huge by any standard. Using predictive equations based on humeral and femoral anteroposterior dimensions and cross-sectional areas, Biknevicus et al. (1993) calculated body mass estimates ranging between 50–200 kg for individuals of *Amblyrhiza inundata*. So wide a range implies the presence of a complicating factor, such as multiple coeval species, temporal variation in mixed-age samples, or marked sexual size dimorphism (see below). Although given sample limitations none of these possibilities can be conclusively ruled out, the fact remains that *Amblyrhiza*, from the tiny islands of Anguilla and St. Martin, is the largest island rodent ever found, and thus represents a problem in macroecological interpretation of a very interesting kind. How did it maintain its numbers, and over what span of time following initial arrival?

Using the ordinary least square equation (OLS) of Biknevicius et al. (1993) based on proximal femoral AP diameter data, McFarlane (1999b) estimated a body mass for *Elasmodontomys* of 13.7 kg. In Table 9.2, it may be seen that the same equation yields an estimate of 3.8 kg for the only known distal femur of *Quemisia* (USNM 253176, 7.9 mm

AP diameter), while another OLS equation (based on distal humerus AP diameter) provides a body mass of 28.5 kg for a specimen of *Clidomys* (AMNH 108581, minimum AP diameter 15.0 mm). At least for the first two taxa, the estimates seem low: they suggest that *Elasmodontomys* corresponded in body size to a large pacarana (*Dinomys*), while *Quemisia* was only the size of a large agouti (*Dasyprocta*). The result for *Quemisia* may be influenced by the extreme AP flattening of the femur, a peculiarity of this taxon (Miller, 1929a). The skull and long bones of *Elasmodontomys* are much more robust than those of *Dinomys*, suggesting that the equations of Biknevicus et al. (1993) may not perform well when predicting the size of unknown species at the lower end of the "large" caviomorph body mass range (cf. similar finding by Turvey et al. [2006] for *Tainotherium valei*). Skeletal elements of *Clidomys* are notably variable in size, with a suggestion of bimodality in some instances (MacPhee, 1984). Although sexual dimorphism is not a prominent feature of living species of caviomorphs (an exception is the capromyid *Mesocapromys angelcabrerai*), this may be the most parsimonious explanation for the scale of variation seen in *Clidomys* (see also Morgan and Wilkins, 2003).

Capromyidae

Sequence information (e.g., Leite and Patton, 2002; Galewski et al., 2005) has confirmed that hutias, spiny rats, and nutrias form a distinct clade within Caviomorpha, but lack of agreement in fine analyses of the molecular data makes it difficult to determine how these three groups are related *inter se* or how they should be organized systematically (although the view that *Capromys/Geocapromys* and *Myocastor* are sister taxa, as suggested by Leite and Patton (2002), is unlikely to be correct according to Galewski et al. [2005]). In light of the prevailing uncertainty, I retain Capromyidae and its constituent subfamilies as higher-taxon nomina for the purposes of this paper, but recognize at the same time that their probable fate is to be absorbed into a vastly expanded Echimyidae.

With the exception of ubiquitous *Capromys pilorides* in Cuba, most of the other extant capromyids are classed as vulnerable, endangered, or critically endangered by the IUCN. Hutias are (or were recently) found on Jamaica, Cuba, Hispaniola, and several smaller islands in Caribbean Sea. A fairly diverse group of WI endemics, capromyids are currently divided into four subfamilies:

Capromyinae: Of the (unrevised) Quaternary diversity of capromyids listed by Woods et al. (2001), amounting to 43 species, only 13 survive, all of which are members of Capromyinae. Accepted at face value, these numbers imply that, with an apparent extinction factor of 70%, hutias must have suffered a catastrophic collapse during the LQU. However, this analysis is misleading, as it is driven by the fact that subfamily Capromyinae in particular is replete with poorly diagnosed, undated nominal species that have never been formally re-evaluated as a group (cf. Camacho et al., 1995). Fortunately, revisionary work is being undertaken presently by Borroto-Páez and colleagues, whose current synonymy concepts for Cuban capromyids are reflected in Table 9.2.

The unnamed taxon from Grand Cayman, Cayman Brac, and Little Cayman (Morgan, 1994b), referenced here as *Capromys* sp. 1, is thought to have become extinct from "predation, habitat alteration and introduction of exotic mammals by humans during the last 500 years" (Morgan, 1994b:503). Its extinction is believed to have occurred prior to the 1800s because the existence of an animal as large as the Cayman hutia would have been noted, and there are no reports of such according to Morgan (1994b). There is a date of 375 ± 50 BP on mangrove peat associated with a *Capromys* femur from Crocodile Canal, Grand Cayman. At Furtherland Farms and other sites, *Capromys* remains are reported from the same (undated) levels as *Rattus*. If humans did not settle the Caymans prior to European discovery (see Morgan, 1994b), then all of the land mammal extinctions on these islands may have occurred in the ME. Said to be most similar to Cuban *C. pilorides* and possibly derived from it, the Cayman hutia has been recovered from levels dated to end-Pleistocene, suggesting that its presence is natural and antedates human arrival (Morgan, 1994b).

Geocapromys sp. 1 and sp. 2 are from Grand Cayman and Cayman Brac respectively. There is an apparent *Rattus* association for *Geocapromys* sp. 1; Morgan (1994b) makes no explicit reference to such an association for sp. 2, although both species are thought by him to have survived until a few hundred years ago.

Geocapromys thoracatus from Little Swan Island, Honduras, was last collected in 1937, although individuals were seen alive in the 1950s. Surveys in 1960 and 1974 failed to reveal either living animals or scat (Morgan, 1985). Regarded as a subspecies of Jamaican *Geocapromys brownii* by some authors, *G. thoracatus* is considered a valid species by most commentators (e.g., Woods and Kilpatrick, 2005).

Hexolobodontinae: The only named species in this subfamily, *Hexolobodon phenax*, is associated with *Rattus* at Trouing Jérémie #5 according to Woods (1989a: table 3). Another, as yet unnamed, species from southern Hispaniola was noted by Woods et al. (2001) and is listed here as *Hexalobodon* sp. 1. Also according to Woods (1993:803), *H. phenax* includes *H. poolei*.

Isolobodontinae: Both *Isolobon portoricensis* and *I. montanus* are known from Hispaniola and its offshore islands (including La Gonave), which constitute their original range. Varona (1974) regarded *Isolobodon* as a subgenus of *Plagiodontia*, but he has not been followed in this by other commentators. *Isolobodon* is possibly the "hutia" or "cori" which de Oviedo y Valdes (1535) described as having been hunted by Hispaniolan Indians (see Miller, 1929b). There is no evidence that either species survived much beyond 1,500 CE. The possibility that *I. portoricensis* still survives in

Puerto Rico or Ile de Tortue is speculative; searches in 1980s failed to produce any sign of it (Woods et al., 1985).

Plagiodontinae: Numerous species of *Plagiodontia* have been named in the past, but only three are currently recognized (including extant *P. aedium*). *Plagiodontia ipnaeum*, closely related to *P. aedium*, has been found in association with *Rattus* and *Mus* at Trouing Jérémie #5 according to Woods (1989a: table 3, under the synonym *P. velozi*). *Rhizoplagiodontia lemkei*, known from only one locality, is also said by Woods (1989a) to have been found in association with *Rattus*. No similar association occurs with *P. araeum* at its type locality. Woods (1989a) argued that at least some smaller Hispaniolan endemics survived as late as the last few centuries on the Massif de la Selle in Haiti, where *P. araeum* has been recovered in an undated context.

Hydrochoeridae

The named species *Hydrochoeris gaylordi* (MacPhee et al., 2000b), from the same Grenadian lahar deposit as the unnamed megalonychid discussed earlier, is represented only by a partial maxillary dentition which slightly differs from that of the living species, *H. hydrochaeris*. Hooijer (1959) briefly noted the existence of some juvenile capybara teeth recovered from a phosphate deposit of unknown age in eastern Curaçao. He thought the teeth could be attributed to the living species, and wondered whether they represented an animal introduced by humans. Hooijer's find is omitted from the table because it apparently concerns a non-endemic.

Unassigned: Enigmatic Taxa and Morphs

This section might have been entitled with equal merit "Enigmatic Femora", as all of the provisional taxa discussed under this heading are represented exclusively by thigh bones. This situation is presumably not reflective of a taphonomic peculiarity of certain WI islands, but instead implies that other elements pertaining to these taxa have probably been recovered but go unrecognized in collections (cf. discovery of jaw of *Xaymaca fulvopulvis*; MacPhee and Flemming, 2003).

Incomplete but distinctive femora from the localities of Trou Wòch Sa Wo in Hispaniola and Sheep Pen and Coco Ree in Jamaica have been favorably compared to those of various platyrrhines by Ford (1990; see also Ford and Morgan, 1986, 1988). The Trou Wòch Sa Wo femur is from an immature animal and lacks the distal epiphysis, although it is otherwise intact and quite fresh-looking. By contrast, the Sheep Pen and Coco Ree elements are heavily mineralized, and there is some evidence that they are of significant age (Ford, 1990; Ford and Morgan, 1986, 1988). Whereas Coco Ree is a typical cave site, the Sheep Pen locality is a hillside consisting of exposed, indurated fissure fillings (MacPhee, 1984).

How many taxa these three elements actually represent is unclear. Ford (1990:247) stated that the Coco Ree and Trou Wòch Sa Wo specimens are so similar that they might "represent the same species". This would be unusual inasmuch as the only examples of the same endemics occurring on two or more widely separated islands in the Greater Antilles seem to be the result of human transport (e.g., *Isolobodon portoricensis*). As to relationships, MacPhee and Fleagle (1991) and MacPhee and Flemming (2003) wondered whether the femora might represent some taxon other than Primates (?perhaps Caviomorpha). Whatever the real facts may eventually prove to be, the mere existence of these unallocated elements is a further indication that basic systematic discoveries in the LQ fauna of these islands still remain to be made.

The other (?caviomorph) Sheep Pen specimen was found in a museum collection dating from the 1960s and can only be tied to this locality circumstantially (MacPhee and Flemming, 2003). No systematic allocation for this specimen can be suggested at this time. In size and to a certain degree in shape this heavily mineralized fossil resembles femora of the giant caviomorph *Amblyrhiza*, but this comparison is of limited significance given that *Amblyrhiza* lived on islands situated at the opposite end of the Caribbean Sea. It is far too large to belong to *Clidomys osborni*, although it may represent a related species. An age assignment of "unspecified Neogene" is simply a default; the femur may well be late Pleistocene, but this would need to be established on grounds that are more meaningful than its alleged Fundort.

Tainotherium valei is included in this section because Turvey et al. (2006) were unable to assign the holotype and only known specimen to a secure familial position, although there is every reason to believe that a caviomorph is represented. Heptaxodontid affinity is certainly possible, but more material will be needed to establish this allocation. There are no relevant dates from Cueva de la Vaca, the type locality in north-central Puerto Rico, and it is therefore unknown whether *Tainotherium* disappeared early or late (Turvey et al., 2006).

Other Taxa

Bat extinctions will not be reviewed in detail in this paper; a magisterial review of chiropteran losses in the West Indies was recently published by Morgan (2001), to which a few additions may now be made as the result of the recent work of Suárez and Diáz-Franco (2003), Suárez (2005), and Mancina and Garcia-Rivera (2005). Bats present special problems in biogeographical and extinction studies, since they tend to be more widely distributed than non-volant taxa, and may disappear in one part of their range but persist elsewhere in places where the extinction driver is ineffective. Also, in an ecological sense bats are sharply set apart from terrestrial mammals, suggesting that anything happening to the latter fauna would not necessarily affect the former, and conversely.

At present, there is evidence for nine complete extinctions of bats in the West Indies during the LQ; to this number may be added the local extirpation of 20 other species, the majority of which are or were obligate cave dwellers. It has often been suggested that the comparatively low extinction rate of WI bats compared to terrestrial taxa is due to their high vagility, fecundity, and re-colonization abilities. Still, it is rather more likely that the resilience of the bat fauna as a whole to extinction is a function of their ecological isolation from the impacts supplied by introduced rats, mongooses, cats, and goats. In the case of specialized cave-dwelling taxa, Morgan (2001) related losses to changes in the size, distribution, and local ecology of caves in the West Indies, perhaps as a consequence of post-Pleistocene climate change. For the rest the reason for loss or extirpation is unclear, although it has been suggested that, for example, the Cuban vampire bat *Desmodus puntajudensis* (formerly considered a subspecies of mainland *D. rotundus*) may have died out in conjunction with the collapse of the megalonychids. Most other bat losses may be AI (or later) as well, but in the absence of extensive dating this is impossible to confirm. The Puerto Rican flower bat, *Phyllonycteris major*, certainly survived into the late Holocene: its current LAD is 3,330 ± 50 BP, based on material from Burma Quarry, Antigua (Pregill et al., 1988). However, whether this date is representative of a time of heightened loss of chiropterans is simply unknown.

For completeness it may be briefly noted that LQ losses within the Caribbean basin include one sea mammal, the Caribbean monk seal (*Monachus tropicalis*), last seen in the 1950s (Adam, 2004). This seal had the dubious distinction of being the first WI endemic to directly experience the impact of European arrival: sailors on Columbus' second voyage in 1494 killed several basking on rocks near Hispaniola. Seals of the genus *Monachus* seem particularly vulnerable to habitat disturbance; both surviving species (*M. shauinslandi* and *M. monachus*) are considered endangered by the IUCN (http://www.iucnredlist.org).

Human Arrival

Of all the exotic species that entered the islands during the recent past, none was more portentous for the ultimate fate of the endemic fauna than *Homo sapiens* – with, of course, the possible exception of its facilitated co-invaders, *Rattus, Mus,* and, later on, *Herpestes.*

In view of the limited scale of prehistoric archeological investigations in the West Indies (Watters, 1989), it is an open question whether we should consider first indications of humans on a specific island to be evidence of mere passing landfall or actual colonization. The implications of these terms are notably different: occasional visitation of islands by sea-faring Amerindians would presumably have had a lesser impact on terrestrial faunas than true colonization, with its implication of continuing settlement and the inevitable human penchant for consuming available resources. In addition there is the problem of the "repeopling," in Rouse's (1989) sense, of parts of the West Indies by different cultural groups. The archeological evidence shows quite clearly that several distinct Amerindian migrations occurred prior to European times, involving different regional/cultural origins on the continents and associated levels of technological complexity. Some archeologically defined groups seem to have made greater use of terrestrial resources and the interiors of islands than did others, although the examples cited by Rouse (1989) are all rather late (BCE/CE transition or later). Whether this point has any significance for the extinction story cannot be determined at this stage of our knowledge, although it is certainly compelling to think that we might one day be able to associate specific declines/extinctions of taxa with specific cultural groups (cf. Turvey et al., 2007). In this paper I sidestep this problem as beyond current resolution and regard any pre-European archeological evidence of human presence as "Amerindian".

This is not the place to review in detail the evidence for and timing of human migrations into the West Indies (see Rouse and Allaire, 1978; Rouse, 1989, 1992; Wilson, 1997, 2001). According to scanty empirical evidence, briefly detailed below, humans first reached the larger islands in the West Indies ca. 6,000 BP, perhaps from Yucatán (Rouse, 1992; Wilson, 2001). Much earlier entry has also been claimed, especially in the case of Cuba (see discussion by Rouse, 1992), but in the absence of unimpeachable radiometric evidence for earlier incursions this claim is hard to support.

Cuba: The oldest dated archeological site yet discovered on this island is a rock shelter in the Levisa River basin (prov. Holguín), with a basal radiocarbon age of 5,140 ± 170 BP (5,590–6,280 cal bp)(Kozlowski, 1974). Earlier dates for the settlement of Cuba exist, but these are nonradiometric and have not been confirmed by rigorous testing.

Hispaniola: The oldest radiocarbon estimate for this island, 5,580 ± 80 BP (6,210–6,550 cal bp), comes from the Haitian coastal site of Vignier III (Moore, 1991). As Allaire (1997:21) points out, however, this date is derived from surface-collected seashells, "perhaps not the most reliable context." Corrected for marine reservoir effects using the global ocean correction factor of Hughen et al. (2004), the date can be calibrated to 5,780–6,180 cal bp – still old, but younger than the uncorrected version by nearly half a millennium.

Puerto Rico: Burney et al. (1994) claimed a date of ca. 5,300 cal bp for human arrival in Puerto Rico on the basis of the timing of a marked change in the charcoal-influx signal in a core recovered from Laguna Tortuguero on the island's northern coast. On the other hand, a similar spike in charcoal abundance ca. 6,000 BP in a core from Lac Miragôane (Haiti) has been ascribed instead to climate change (Higuera-Gundy et al., 1999). Also, Horn et al. (2000), having found abundant macroscopic charcoal in samples of late Pleistocene through late Holocene age collected in the highlands of the Dominican Republic, observed that fire must be considered a natural disturbance factor. In any case, the earliest accepted archeological site on the island, Maruca, is considerably younger than the date published by Burney et al. (1994): Rodríguez (1999) cited an interval of 3,895–4,840 BP for occupation of this site based on ranges of eight radiocarbon dates (data not presented). An even earlier date, 5,960 BP (> 6,700 cal bp), exists for the site of Angostura (Ayes Suares, 1989); but as this is even earlier than the estimate of Burney et al. (1994), it requires additional corroboration.

Jamaica: The oldest age estimate for humans in Jamaica is very young, ca. 1,500 BP (Rouse, 1992), which may signify no more than that the search for early evidence of human settlement has been less productive on this island than elsewhere in the Greater Antilles.

Lesser Antilles: Although it is logical to imagine that humans dispersing from northern South America would have traveled up the Lesser Antillean chain, the earliest dates hover around 4,000 BP, for sites in Antigua (Allaire, 1997).

While remains of terrestrial endemics are not rare in Amerindian sites, the best-documented ones are mostly rather late and virtually always small-bodied (e.g., Wing, 2001). In a strong parallel to the continuing paradox that is the absence of evidence for extensive megafaunal hunting on the American mainlands, in the West Indies clear evidence of human manipulation of large species (with the interesting and pertinent exception of the manatee, *Trichechus manatus*) is exceptionally rare and perhaps even non-existent (see main text).

10. Afterword, and Thoughts About the Future Literature

Gary Haynes*
Anthropology Department,
University of Nevada, Reno
Reno, Nevada 89557-0096
gahaynes@unr.edu

This Collection

This book has collected the thoughts of expert archeologists, paleontologists, and paleoecologists, in addition to a specialist in ancient DNA studies, but in the end it is not possible for me as editor to identify a clear majority opinion (and hence solve the puzzle of extinction). Borrero (Chapter 8) sees little useful evidence in South America that human foragers had an impact on megafaunal populations; yet, referring to much of the same and other sets of fossil evidence in South America, Cione and associates (Chapter 7) lay the ultimate blame on the human factor, which was added to the ecological stresses imposed on megafauna by climate changes. Fisher (Chapter 4) proposes that central USA mastodont populations were in an apparently healthy phase just before they became extinct, which supports the idea that human hunting could have been the abrupt and fatal factor that caused extinctions. Surovell and Waguespack (Chapter 5) argue that human preference for killing the largest animals in North America is rational and supported by ethnographic, theoretical, and archeological data. My own paper (Chapter 3) suggests that megafaunal populations in North America were fragmented and much more vulnerable to human hunting during the Late Glacial. Fiedel (Chapter 2) presents a case based on the chronologies of extinction that the first indications of human hunting more closely track the disappearance dates of many species than do climate-changes, in both North and South America. MacPhee, on the other hand, while agreeing with other authors (who examined data from the late Pleistocene) that climate changes alone cannot cleanly account for extinctions in the West Indies, presents arguments that first human contacts in the Holocene did not wipe out species – instead the process was extended over many centuries, implying that they resulted from the synergy of habitat alterations, the introduction of exotic or competing species, and sustained human hunting pressures.

Readers may scratch their heads in wonder that so much scientific evidence can lead to so many variations and disagreements. Perhaps future directions in research and the uncovering of new data will remove some of the nagging ambiguity.

* Address for correspondence: gahaynes@unr.edu

G. Haynes (ed.), *American Megafaunal Extinctions at the End of the Pleistocene,* 195–197.

The Future Literature

The future literature will be similar in many ways but of course it will also be different, because literature tends to be cumulative. It will provide more details about megafaunal diets, reveal more reliable dating of particular species, present more archeological discoveries of megafaunal bones, and so forth. But if it is to increase in promise and power it must also be different in even more important ways. Each species that became extinct or changed its range dramatically when it disappeared from the Americas must be studied intensively and equally, from the rarest to the most abundant. A set of standards will be established (I hope) for evaluating extinction risks of the different taxa at different times in their late Pleistocene existence. The key question that will be addressed will be this: "What is the benchmark of healthy megafaunal populations in their regional ecosystems, against which declines can be measured?" Of course, identifying the possible answers to such a simple question is complicated by the cycles of change affecting climates and paleoenvironments in the Late Glacial. Each different phase of the cycles would be different but "healthy" in its own way, if we assume that stability was regularly reached from phase to phase. The health of different

TABLE 10.1. Factors affecting extinction risks for individual taxa.

Relative endemicity	Extent of occurrence	Taxonomic distinctiveness	Hunted by humans
1. Very localized	1. Marginal range everywhere (widespread in hemisphere but not abundant locally)	1. One species in an entire order or family	1. Frequently throughout continent
2. Found in one or more regions but in less than ½ of continent	2. Limited good range regionally, hence abundant locally, but only in up to ½ of continent	2. One species in one genus	2. Sometimes throughout continent
3. Found in ½ of continent or more	3. Good range and abundance in up to ½ continent	3. Multiple species in one genus	3. Sometimes, but regionally only (in less than ½ continent)
4. Found hemisphere-wide	4. Good range and abundance everywhere	4. Multiple species in multiple genera	4. Never

sets of communities would have to be relatively measured somehow. One possibility is to approach the question from the modern situation, in which ecosystems are healthiest when most heterogenous, that is when characterized by mosaics of patches in different states of stability and succession. Heterogeneity is required for biotic diversity of large ecosystems, and certainly improves species' chances for long-term survival. Each successional phase during the late Pleistocene should have had its own kind of heterogeneity.

I think the most important focus in the future studies of extinct ecosystems should be on defining the nature of habitat fragmentation in quantified ways. For example, I think we can begin discovering very specific details about large- and small-scale refugia defined first on the basis of fossil occurrences, taphonomic histories, and studies of hydrology, depositional macro-environments, genetic exchanges, and isotope geochemistry. These sorts of collaborative studies can identify individual fragmented subpopulations, as Hoppe (2004) and others have demonstrated, and locate them within the ecological constraints of regional conditions, namely precipitation, seasonality, plant communities, nutrient availability, megafaunal demography, and so forth.

North America was an enormously spotty expanse of megafauna sources and sinks, in which animal densities and distributions were extremely variable. A few studies support the idea of small and large refugia existing in the continent during the Late Glacial; for example, in Agenbroad (2005) North American mammoth distribution maps of occurrences at 5,000-year intervals indicate that fairly gross-scale refugia must have existed, and that there was a wider distribution during the Late Glacial than before. A wider distribution does not necessarily indicate greater abundance overall; there may have been relatively low densities of dispersed subpopulations. In Europe, mtDNA studies of cave bears, brown bears, cave hyenas, and Neandertals before the last glaciation (for example, Hofreiter et al., 2004) also suggest cycles of species retreat into refugia during glacials followed by incomplete re-dispersal during interglacials.

Another direction for research would be to learn how to evaluate 'extinction risk' for an entire ecosystem, as a whole, as well as for each individual species (as per Grayson and Wroe recommendations, discussed in Chapter 1). This cannot be done until all the extinct species have been adequately dated, taphonomically and demographically studied site by site and region by region, and isotopically sampled for evidence about diet.

Of course, risk must be evaluated at three levels: the refugial, the regional, and the continental. Current efforts towards biotic conservation in modern ecosystems can inform future studies of extinct American megafauna. For example, Kruger National Park has recently examined the possible indicators of ecosystems in trouble (Kruger National Park Scientific Services, n.d.) and developed a set of methods for rating risks to wildlife in that South African park. Similar rating systems can be applied to extinct American megafauna. The factors that could be 'graded' are summarized in Table 10.1 (note that low numbers = less likely to survive [high extinction risk], and high numbers = more likely to survive).

Concluding Thoughts

Rating schemes are ideas that may or may not be put into practice; but before they are dismissed as impractical or quixotic, we should keep looking for the necessary data, which very well might be preserved in some places, and might be discovered from time to time. Of course, we also need to reach agreement about the meaning of the data before we can adequately understand late Pleistocene paleoecology and gain the ability to unambiguously reconstruct the dramatic changes that removed so many genera of large terrestrial mammals. Only step by step and lucid reconstructive snapshots of the great dying-out will lead us to a reliable allocation of blame.

Acknowledgments I am grateful – beyond adequate words – to the manuscript reviewers whose comments allowed the authors of this book's chapters to revise, clarify, and improve their contributions. I most warmly acknowledge Stuart Fiedel for helpful and near tireless assistance, Larry Agenbroad for rapid advice and comments, and Paul Martin for encouragement and advice. I would like to thank Ross MacPhee for persuading me to undertake the editing of this book, although it has not been without difficulties and delays. I thank the authors of this book's chapters, who should be proud of their work, and I give my most heartfelt thanks to Janis Klimowicz for editorial assistance.

References

Agenbroad LD (2005) North American proboscideans: Mammoths: The state of knowledge, 2003. Quat Intl 126–128:73–92

Hofreiter ML, Serre D, Rhland N, Rabeder G, Nagel D, Conard N, Münzel S, Pääbo S (2004) Lack of phylogeography in European mammals before the last glaciation. Proc Natl Acad Sci USA 101(35):2963–12968

Hoppe KA (2004) Late Pleistocene mammoth herd structure, migration patterns, and Clovis hunting strategies inferred from isotopic analyses of multiple death assemblages. Paleobiology 30(1): 129–145

Kruger National Park Scientific Services and Scientific Reference Group (n.d.) The Development of Thresholds of Potential Concern (TPC's) in the Kruger National Park for Biodiversity Outcomes for which Elephant May Act as Drivers. Manuscript

Index

Printed by Printforce, the Netherlands